COMMERCIAL AIRCRAFT PROJECTS

Commercial Aircraft Projects
Managing the Development of Highly Complex Products

HANS-HENRICH ALTFELD

Routledge
Taylor & Francis Group

LONDON AND NEW YORK

First published in paperback 2024

First published 2010 by Ashgate publishing

Published 2016 by Routledge
4 Park Square, Milton Park, Abingdon, Oxon OX14 4RN

and by Routledge
605 Third Avenue, New York, NY 10158

Routledge is an imprint of the Taylor & Francis Group, an informa business

British Library Cataloguing in Publication Data
Altfeld, Hans-Henrich.
    Commercial aircraft projects : managing the development of highly complex products.
    1. Aeronautics, Commercial–Management. 2. Transport planes–Design and construction–
Management.
    I. Title
    629.1'33340423-dc22

Library of Congress Cataloging-in-Publication Data
Altfeld, Hans-Henrich.
    Commercial aircraft projects : managing the development of highly complex products / by
Hans-Henrich Altfeld.
        p. cm.
    Includes bibliographical references and index.
    ISBN 978-0-7546-7753-6 (hardback)
  1. Transport planes–Design and construction. 2. Production management. 3. Aircraft industry–
Management. 4. Aeronautics, Commercial. I. Title.

    TL671.28.A48 2010
    629.133068'4–dc22

                                                                            2010009660

ISBN: 978-0-7546-7753-6 (hbk)
ISBN: 978-1-03-283828-1 (pbk)
ISBN: 978-1-315-57283-3 (ebk)

DOI: 10.4324/9781315572833

Contents

List of Figures

Foreword

In 1999, Dr. Hans-Henrich Altfeld joined my team to prepare the industrial launch and development of the Airbus A380, the biggest commercial airliner in the world. It was clear from the onset that this development project would represent significant technical, technological, environmental and infrastructure challenges. But the teams participating in the project would also enter new territories in the management of such a huge and complex commercial aircraft project, especially when considering the international and inter-cultural character of Airbus.

However, it quickly became obvious that the application of classical project management techniques to the development of the A380 would not be sufficient to ensure the success of this tremendous project. For one, the sheer complexity of the project would make classical project management very elaborate to an extent which seemed unaffordable for a commercial undertaking. In addition, it had for some time been recognised worldwide that developments in project management were lagging behind the evolution of modern IT- and Concurrent Engineering-based product design and development processes.

With the programme launch imminent, Hans-Henrich proposed new schemes for highly integrated processes for the development phase of the A380. Astonishing results were achieved whenever these new schemes were rigorously implemented and thoroughly applied. This was especially the case in the area of the A380 wing design in the UK, where Hans-Henrich played a leading role.

But Hans-Henrich also recognised that these new schemes would not close the unacceptable and still growing gap between the capabilities of project management on the one hand and product development processes on the other. Many large-scale and complex projects, commercial and military, and including the A380 development project, continued to live with this lack of full compatibility, and, as a result, suffered delays and cost overruns.

Along with many others, Hans-Henrich was not satisfied with the situation. But he took personal initiative by looking into possibilities of how to further reduce the compatibility gap while keeping focused on commercial affordability. After several years of work during his scarce private time, I am very pleased indeed to see that his ideas of how to project manage commercial aircraft developments are now available as a book. As he stresses in many parts of the book, the secret in closing the capability gap lies in integrative management aspects. This comprises integrating different disciplines, which are required to develop complex projects, integrated project architectures and structures, as well as rigor and discipline in adhering to established processes. How exactly this works within the constraints of commercial affordability is the theme of this book.

It is definitely worth noting that many of his findings do not only apply to commercial aircraft developments, although this is the area Hans-Henrich knows best, but also to many other technological fields where complex product developments are taking place. I therefore wish this book a large reading public and a wide circulation – within the aerospace community and beyond.

Jürgen Thomas
Toulouse, France

Preface

All commercially funded product development projects of very high complexity share a common dilemma. Being complex, it is not sufficient to apply standard project management techniques to manage them and to keep them under control. Instead, they need a much wider management approach which is perfectly adapted to their complex nature. This, however, may generate additional cost. The dilemma arises because in commercially-driven product developments there is the natural tendency to limit the management-related costs associated with such projects.

The development of a new commercial transport aircraft is no exception to this. It certainly is a very challenging and complex undertaking. In fact, in many ways it features additional complexities not found in other projects, even in most other complex projects. The development of a new commercial transport aircraft can therefore be regarded as an extreme example for a commercially-driven, complex product development. This is why I found it especially useful to analyse what is needed in terms of capabilities and practices to manage commercial aircraft developments.

The theme of this book is that any management approach for complex, commercially-driven product developments should seek to integrate the strengths of state-of-the-art management disciplines while limiting their application to some basic essentials. It is therefore integrative and essential at the same time. In other words: whatever the detailed management approach, it should be limited to the minimum amount of capabilities and practices which are necessary to bring together people, processes, methods and tools into a harmonious interoperation to maximise their efficiency and effectiveness.

Wherever in this book a management approach for complex, commercially-driven product developments is discussed, it is intentional to concentrate on the *management* challenges related to such developments. Other crucial contributors to the success of a complex development project, such as for example the mastering of technical challenges, are therefore not addressed in any depth.

There is an immense amount of literature available on individual management techniques for developing products, such as Project Management, Systems Engineering and Supply Chain Management, to name but a few. However, they tend to deal with the principles of *either* Project Management *or* Systems Engineering *or* Supply Chain Management. The interdependencies between these disciplines are often overlooked while in real projects they do play an important role. For example, essentially all project management activities occur within some kind of systems engineering context, and vice versa. This book tries to move further in the field of interconnectedness between individual management disciplines by explicitly considering the matter of integrative management. How exactly integrative, essential management works will be described in this book, using commercial aircraft developments as an extreme example.

Over the last 20 years or so it has been the evolution of Information Technologies (IT), which has dramatically changed the way aircraft and many other complex products are designed, developed, built and project managed. How should modern commercial aircraft development be managed and organised to fully exploit the advantages offered by today's IT?

What would the required project architectures look like to integrate teams, schedule plans, three-dimensional-design models and so on? This book is about providing some answers to these questions.

In Chapter 1, some arguments will be provided in support of the hypothesis that commercial aircraft development projects are indeed different from most other projects. After demonstrating how IT has become a major driver for aircraft development in Chapter 2, the generic development process for commercial aircraft will be introduced in Chapter 3. The reader will be familiarised with some of the key terminologies used in managing an aircraft development and will gain a comprehension of its major phases. This is followed by Chapter 4, which provides an overview of the different management disciplines required in complex product developments. However, as there are a large number of books available for each of those disciplines, this book will not enter into any details wherever possible because this would be a repetitive effort. But, of course, those essential aspects, which are required for integrative management, will be described. They include first of all:

- the building of efficient and effective multi-functional design-build teams (Chapter 5);
- holistic, interdisciplinary and intercultural leadership (Chapter 5);
- the need for co-locating team members (Chapter 6);
- the right organisational balance (Chapter 6);
- the sequencing of project priorities (Chapter 7);
- the essential aspects of Requirements Engineering and Verification & Validation (Chapter 8); as well as of
- Risk Management (Chapter 9).

All of the aforementioned essential, integrative elements are important not only for a company's own internal product development activities, but also for the management of its supply chain. As there is again lots of literature available on Supply Chain Management, this book concentrates on the above mentioned list of essential, integrative elements when describing how to integrate suppliers into the processes of developing complex products.

However, all of the aforementioned aspects relevant to the management of complex, commercially-driven product development projects are still not sufficient to master the challenges associated with such projects. Where this book goes beyond management essentials found elsewhere in literature is therefore the aspect of architectures' integration, necessary to interlink product, process and resources data. Only if based on an integrated project architecture, the interoperation of the management essentials described in Chapters 5 to 9 will yield maximum efficiency and effectiveness. A proposition for such an architecture and how it can be used for integrative management of complex product developments, such as commercial aircraft, is outlined step-by-step in Chapters 10-15. It includes:

- architectures required to allocate costs to projects (Chapter 10);
- the establishment of an integrated project architecture (Chapter 11);
- ways to keep changes to this architecture under control, as well as the data attached to it (Chapter 12);
- integrated project set-up based on this architecture (Chapter 13 and Chapter 14); as well as
- project execution, monitoring and control, also based on this architecture (Chapter 15).

Establishing an integrated project architecture and using it for project set-up, monitoring, control, for all design activities, deliverable integration and realisation, for quality control and achievement of maturity and so on, is not an easy job at all. I found it also challenging to explain it. Throughout this book I have tried hard to support the reader by using an understandable language and many figures supporting complex subjects. The reader should nevertheless not be surprised that, in particular, Chapters 10–15 need dedicated patience and concentration to read. Leaving out these chapters would not have been an option as an integrated project architecture is after all definitely required for an integrated management approach. I am personally very convinced that those who make their way through these chapters will be rewarded with a new vision about how the product- and process-related architectures, traditionally used, for example, by systems engineers and project managers, should in future merge to come to a single, integrated project architecture.

Along the course of this book, which is summarised in Chapter 16, it should become clear that commercial aircraft companies no longer only compete on technological grounds when developing new aircraft. Instead, the mastering of integrated management practices has become a decisive factor for superior competitiveness too. In the fiercely competitive world of plane-making, it has become integrated management that companies will give their planes an additional competitive edge. Both Airbus and Boeing have invested heavily in training, staff development and tools to ensure integration. However, a lot remains to be done in this area. The recent delays in some bigger commercial aircraft development projects, generating substantial problems for the entire aviation sector, is more than proof for this.

This book has primarily been written for everyone seeking to know more about suitable practices for the management of complex and commercially-driven product developments. As commercial aircraft developments are taken as extreme examples for the latter, the book should also be highly beneficial to project leaders, team leaders and all other team members of multi-functional teams which lie at the heart of modern commercial aircraft developments, covering airframes, systems and engines. It should also deliver benefits to everyone who intends to gain a general overview on the commercial aircraft development business. This includes individuals working in the classical industry Functions, such as Engineering, Manufacturing and Procurement, as well as at academia and other research establishments. Finally, the book should be of interest to everyone involved in developing IT tools in support of complex product developments, such as computer-aided design, project management, enterprise resource planning or life-cycle management tools.

Hans-Henrich Altfeld
Hamburg, Germany

Acknowledgements

I wish to thankfully acknowledge the most appreciated contributions – such as providing ideas, suggestions, texts, figures, proof reading, secretarial support and many more – of the following individuals: Jane Adams, Susanne Altstaedt, Stephen Ash, Fabienne At, Claudine Aubert, Kevin Baker, Pierre De-Chazelles, Martine Delpech, Walter Dolezal, Burkhard Domke, Dirk Effenkammer, Philip Ellwood, Debbie Ewens, Henrik Fransson, Raphael Giesecke, Gunnar Gross, Davena Hyatt-Jones, Friedrich Kerchnawe, Sabine Klauke, Colin Ingamells, Serge Leichter, Gerry McArdle, Karl-Heinz Muehlnickel, Ralf Myska, Philip Lawrence, Alain Ramier, Mark C. Robinson, Geoffrey Shuman, Bernd Siepen, Matthias Spengler, Suja Sreedharan, Thomas Thiele, Mark M. Turner as well as André Wegener.

A very special 'thank you' goes to my wife Annette, who not only read the manuscript and provided me with numerous suggestions for improvements but also was a source of invaluable support during the eight years it took me to complete this book.

Despite all the help and support I received, I remain soley responsible for the content of the book. Except for quotations from other sources, all messages contained therein represent my personal view and not the view of anyone else or of any company mentioned in the text.

Hans-Henrich Altfeld

List of Acronyms

ACWP	Actual Cost of Work Performed
AMM	Aircraft Maintenance Manual
AOG	Aircraft On Ground
AS	Assembly Stage
ASO	Assembly Stage Operation
ATA	Air Transport Association
A-thread	Aggregated thread
ATO	Authorisation to Offer
BCWS	Budgeted Cost of Work Scheduled
BoM	Bill of Material
BPA	Build Process Architecture
CA	Constituent Assembly
CAD	Computer Aided Design
CAE	Computer Aided Engineering
CAM	Computer Aided Manufacturing
CBS	Cost Breakdown Structure
CCB	Change Control Board
CDF	Contractual Definition Freeze / Customer Definition Freeze
CE	Concurrent Engineering
CFD	Computational Fluid Dynamics
CI	Configuration Item
CI-PS	Configuration Item – PAT-Solution
CI-PS-CI	Configuration Item – PAT-Solution – Configuration Item
CM	Configuration Management
CoS	Condition of Supply
CP	Critical Path
CPA	Critical Path Analysis
CPD	Concurrent Product Definition
CPI	Cost Performance Index
CRD	Change Repercussion Document
CS	Certification Specification
CTCD	Complete Technical Change Description
CV	Cost Variance
DBD	Data Basis for Design
DBT	Design-Build Team
DfM	Data for Manufacture
DMU	Digital Mock-Up
DtC	Design to Cost
DtX	Design to X

EASA   European Aviation Safety Agency
ECN   Engineering Change Note
EIRD   Equipment Installation Requirements Dossier
EIS   Entry Into Service
ERD   Equipment Requirements Document
EV   Earned Value
EVM   Earned Value Management
FAA   Federal Aviation Administration
FAI   First Article Inspection
FAL   Final Assembly Line
FAR   Federal Aviation Regulations
FBS   Functionality Breakdown Structure
FCAC   Forecast Cost at Completion
FDAC   Forecast Duration at Completion
FEA   Finite Element Analysis
FEM   Finite Element Model
FMC   First Metal Cut
FTWC   Forecast Time of Work Complete
GATT   General Agreement on Tariffs and Trade
GDP   Gross Domestic Product
GLARE   'GLAss-REinforced' Fibre Metal Laminate
GPA   Generic Process Architecture
GR   Goods Receive
GRM   Geometric Reference Model
gWBS   generic Work Breakdown Structure
IBS   Infrastructure Breakdown Structure
ICM   Institute of Configuration Management
ICY   Interchangeability Drawing
IPC   Illustrated Parts Catalogue
IPT   Integrated Product Team
IRR   Internal Rate of Return
IT   Information Technology
ITP   Instruction to Proceed
JIT   Just in Time
KBE   Knowledge Based Engineering
KPI   Key Performance Indicator
LCC   Life-Cycle Cost
LDL   Loads-Design-Loop
MBWA   Management by Walking About
ME   Manufacturing Engineering
MFD   Must Finish Date
MS   Manufacturing Stage
MSD   Must Start Date
MSN   Manufacturing Serial Number
MSO   Manufacturing Stage Operation
MTOW   Maximum Take-off Weight
MWE   Maximum Weight Empty
NASA   National Aeronautics and Space Administration

NC	Numerically Controlled
NPV	Net Present Value
NRC	Non-Recurring Cost
OBS	Organisation Breakdown Structure
OCC	Opportunity Cost of Capital
PAT	Product & Assembly Tree
PBS	Product Breakdown Structure
PCWC	Planned Cost of Work Complete
PCWP	Planned Cost of Work Performed
PCWS	Planned Cost of Work Scheduled
PD	Principle Diagram
PDAC	Planned Duration at Completion
PDM	Product Data Management
PERT	Program Evaluation and Review Technique
PLM	Product Life-cycle Management
PLM	Programme Life-cycle Management
PMO	Project Management Office
PO	Purchase Order
PoE	Point of Embodiment
PPP	Phased Project Planning
PR	Public Relations
PRA	Particular Risk Analysis
PS	PAT-solution
PTCD	Primary Technical Change Description
PTWC	Planned Time of Work Complete
R&T	Research & Technology
RASCI	Responsible, Accountable, Supported, Consulted, Informed
RC	Recurring Cost
RoI	Return on Investment
RSP	Risk Sharing Partner
RSS	Risk Sharing Supplier
RTD(t)	Remain-To-Do at time t
SAM	Space Allocation Model
SBS	Systems Breakdown Structure
SDRB	Stress Design Reference Base
SIRD	System Installation Requirements Dossier
SoW	Statement of Work
SPI	Schedule Performance Index
SRD	Systems Requirements Document
SRM	Structural Repairs Manual
S-thread	Summary thread
SV	Schedule Variance
sWBS	specific Work Breakdown Structure
TCAC	Target Cost at Completion
TLARD	Top Level Aircraft Requirements Document
TLMRD	Top Level Maturity Requirements Document
TLMRSRD	Top Level Maintenance, Reliability, Supportability Requirements Document
TLQRD	Top Level Quality Requirements Document

TLSRD	Top Level Systems Requirements Document
TLStRD	Top Level Structural Requirements Document
TMC	Target Must Cost
TPA	Team for Product Architecting
TQM	Total Quality Management
V&V	Verification and Validation
WBS	Work Breakdown Structure
WD	Wiring Diagram
WP	work package
WP%(t)	Percentage of Work Complete until time t
WP(t)	Work Performed at time t
WPD	Work Package Description
WQN	Work Query Note
WS(t)	Work Scheduled until time t
WTO	World Trade Organization

PART I

Introducing Basics in Project Managing Aircraft Development

1

Why Commercial Aircraft Development is so Special

In order to justify a dedicated management approach for it, it is firstly necessary to explain why the development of a commercial airliner is distinctively different – if not unique – compared to most other product developments and what is actually meant by 'commercial aircraft development'?

DEFINING COMMERCIAL AIRCRAFT DEVELOPMENT PROJECTS

Commercial aircraft development is about developing new aircraft designed for the purpose of people[1] or cargo transport operating in a primarily commercial environment whereby the development is commercially managed as well.

The question whether such development can generally be regarded as a *project* or rather a *programme* leads to more than just an argument about semantics: the precise perimeters must be defined, as this has many technical and managerial consequences.[2] For the purpose of this book the total scope of designing, developing, supporting, maintaining and so on of an aircraft is defined as a programme rather than a project. A programme then covers a huge variety of components and stretches across the whole life-cycle, covering many years as well as different phases such as Research, Development, Production, among other. Projects generated for the purpose of developing an aircraft stretch across the actual Development phase only and apply to major component assemblies like the wing or the hydraulics system as well as the integration of the entire aircraft. Thus, developing a commercial aircraft is a large project consisting of smaller projects or multi-projects, which each can still represent billions of dollars of business volume.

Developing such projects commercially means that they are financed commercially. This definition excludes, for example, military aircraft, which usually are entirely government-

---

1   In practice, the lower level of number of passengers to be transported commercially is not strictly defined. At the lower limit, a two-seater aircraft may have one paying passenger. But this usually would fall under the category of what is called General Aviation. However, the main focus of this book is on commercial aircraft above 30 seats or so as well as their cargo equivalents.
2   'Understanding of these key words is the first hurdle to be crossed in achieving good programme or project management. For example, the British word *programme* frequently refers to a plan, and the Spanish *proyecto*, the French *projet*, and the Italian *progette* are frequently used to identify technical designs or plans ....' From: Archibald, R.D. (1992), *Managing High-Technology Programs and Projects*, 2nd Edition, (New York: John Wiley & Sons), p. 25.

funded.[3, 4] This is because governments want their indigenous industry to produce aircraft with superior technologies, the development of which industry cannot afford to finance commercially. This is especially the case for technologies that are *significantly* different compared to those embodied on existing products.

> 'By almost any measure, the [B-2] bomber's development was one of the largest, most technically complex, expensive and demanding programs in aerospace history. … Whatever resources were deemed necessary to meet national security goals, they were made available, despite the cost. "We kept a top-10 list of [B-2 concerns] on the briefing-room wall," [B-2 chief project engineer Albert F.] Myers recalls. "We were seven years into the program before "cost" made that list."[5]

But are commercial transport aircraft developed by Airbus, Boeing and other aircraft manufacturers really developed commercially or isn't commercial aircraft development generally funded by governments, too?

The answer to this question lies at the heart of a fierce debate between Europe and the US, which lasted for about 20 years and – after some quiet years – recently erupted again.[6] The US Government accused Europe of unfair subsidies for its commercial aircraft developments. In the sixties, where the skies were almost exclusively dominated by US commercial aircraft products, European Governments were willing to invest into an industry, which looked promising from a high-technology perspective with all its derived strategic benefits but where the entry into the market had to be enforced. This was done through the granting of reimbursable loans, repayable over a period of time with interest as well with royalties based on the number of aircraft sold. In the beginning, no one in the US took the European initiative to develop new commercial aircraft seriously, and therefore no one was worried about this funding approach. But with the breakthrough of the Airbus A320 and subsequent gaining of more and more market share, the US administration became alerted about this perceived direct support mechanism.

In reply, Europe claimed that the US was using unfair subsidy practices by applying technologies, methods, processes and tools to its commercial airliners, which were developed for and funded by the US Department of Defense, and under contracts from NASA. This was and is called indirect support. Examples for this are the Boeing 707, which was primarily developed as a tanker aircraft for the US Air Force, as well as the Boeing 747, which was largely based on Boeing's proposition for a super military transport aircraft.[7]

> 'An impressive list of technologies, design tools, management techniques, manufacturing processes and myriad lessons spawned by the B-2 [bomber] program are contributing to many commercial and military programs. … even Boeing's 777 airliner have close ties to the B-2.'[8]

---

3   Although for military transport aircraft considerations put forward in this book may also be valid.
4   The definition used here does, however, include military aircraft in general and military derivatives of commercial aircraft in particular if developed commercially.
5   Scott, W.B. (2006), 'Stealthy Genesis', *Aviation Week & Space Technology* 27 March 2006, p. 57.
6   For a summary of the history of this debate see: McGuire, S. (1997), *Airbus Industrie – Conflict and Cooperation in US-EC Trade relations* (London: MacMillan Press).
7   In the latter case, the US Government decided against Boeing, giving preference to the design of Lookheed which subsequently built the C-5 Galaxy. However, Boeing used its design to build the 747. Both design propositions were fully funded by the Department of Defense. For details see: Aris, S. (2002), *Close to the Sun – How Airbus Challenged America's Domination of the Skies* (London: Aurum Press).
8   Scott, W.B. (2006), 'Stealthy Genesis', *Aviation Week & Space Technology* 27 March 2006, pp. 56–7.

The discussions between Europe and the US resulted in various agreements, the most important one being the General Agreement on Tariffs and Trade (GATT) agreement on trade in large civil aircraft, which limited the application of both, direct and indirect support mechanisms.[9] It was signed and took effect on July 17, 1992 and became widely known as the 1992 agreement. Under the terms of this Agreement, only one-third of the total development cost could since then be funded through reimbursable loans at rates at least equivalent to the cost of government borrowing, plus 1 per cent. Indirect support in excess of 3 per cent of the annual commercial turnover of the civil aircraft industry for large civil aircraft[10] was also prohibited. As a result, development of modern commercial aircraft must be regarded as a predominantly commercial venture rather than a governmentally funded one, even if both Europe and the US continue to claim each other's violation of the 1992 agreement, which incidentally was unilaterally abrogated by the US in 2004, leading to trade cases being put by both sides before the World Trade Organisation (WTO).[11]

ABOUT THE REQUEST TO MINIMISE COST OF MANAGEMENT

Where governments are themselves customers of products, they have introduced – and are prepared to fund – Management disciplines like Project Management and Systems Engineering to ensure a high probability of achieving development objectives in terms of schedule, cost, quality and performance. In fact, the development of most of the Management disciplines mentioned in the Preface has been funded by the US Department of Defense as well as by NASA during the 1950s and 1960s. Since then, they have been *contractually* requested as Management disciplines to be applied for military aircraft or large space project developments by both, US and non-US governmental customers. This is because government administrations need to be able to tightly control budgets allocated to them through the annual parliamentary budgetary process.

As a result of governmental interests in meeting objectives, in particular cost objectives, the application of state-of-the-art Management disciplines has reached a level of sophistication with the involved contractor companies, which is unmatched by other industries – apart maybe from the nuclear industry. However, such disciplines do not come for free as they are usually quite resource intensive. About 10–15 per cent of total development cost is acceptable to governments to be absorbed by them.

For a commercial development of any product, this 10–15 per cent would come directly off the annual profit of the company. As in commercial development there is no contractual obligation of whatever kind to apply Project Management, Systems Engineering or any of the other Management disciplines, there is a natural tendency to limit the costs associated with them. This can also be observed with commercial aircraft developers. However, commercial aircraft are very complex products and their development is equally

---

9   'Agreement between the European Economic Community and the Government of the United States of America concerning the application of the GATT Agreement on Trade in Civil Aircraft on trade in large civil aircraft', *Official Journal of the European Communities* No L 301/32, 17 October 1992.

10 Or of 4 per cent of the annual commercial turnover of any one company producing large civil aircraft, whatever applies.

11 'Since … 1992, the two major issues concerning the … [GATT] agreement … included: (1) the US contention that the Airbus A380 program was receiving over 33 per cent development funding, and (2) the EU Commission's contention that Boeing benefited from NASA research and development programs and export subsidies … amounting to between 5.2 per cent and 7.4 per cent of the FY 1998 commercial turnover.' From: Pritchard, D.J. and MacPherson, A. (2003), 'The Trade and Employment Implications of a New Aircraft Launch: the Case of the Boeing 7E7', *Canada – United Sates Trade Center Occasional Paper*, No. 28, December 2003.

complex and challenging. This would require the highest level of Project Management sophistication. Yet it cannot be afforded. There is here a fundamental dilemma typical for all commercially funded projects of very high complexity.

Fortunately, there is a way out of this dilemma: by concentrating on the essential elements of such disciplines – to keep their principal strengths – and combining them in an intelligent and pragmatic way, cost reductions can be achieved. This is why the management of commercial aircraft must be performed on the basis of affordable essentials taken from state-of-the-art Management disciplines combined with a fully integrated approach.

Developing a commercial aircraft really is a mega-project due to the many complexities involved. In fact, there are hardly any other product developments with this level of commercial orientation *and* complexity. What are these complexities all about? In his book *Taming Giant Projects*, Grün defines complexity 'by the number of activities and milestones, and the number of participants who need to be coordinated to achieve the project goals'.[12] In the case of commercial aircraft developments, complexity encompasses some important additional aspects as will be outlined below.

VOLUMETRIC COMPLEXITY

Aircraft for people transport are physically much larger than many other products, for example cars or computers. Significant numbers of people are therefore required to develop an aircraft. In today's business environment thousands of people work together during the Development phase to deliver the aircraft to the first customer within time, budget and of the right quality. At Boeing, about 6,500 employees were tasked with the 777 development.[13] At the peak of the development, around 6,000 people worked at Airbus to make its mega-liner A380 a reality. Another 34,000 are estimated to have been directly involved in the project at suppliers.[14]

In addition, although not directly linked, size is a good indicator for the number of individual parts. 'A car may consist of some 7,000 parts, whereas an airplane can consist of up to 6 million parts'.[15] Boeing engineers working on the 777 described it as more than '4 million parts all moving in close formation'.[16]

> 'I saw Boeing's new jet as 75,000 drawings, 4.5 million parts, 136 miles of electrical wiring, 5 landing gear legs, 4 hydraulic systems, and 10 million labour hours.'[17]

Even if fasteners are not included in the counting of parts, there are many hundreds of thousands of individually-shaped pieces of different types of material on a modern commercial aircraft. Both the number of people and number of components involved represent volumetric complexity. Managing so many people coherently and so many parts logistically represents a major challenge on its own.

---

12 Grün, O. (2004), *Taming Giant Projects* (Berlin: Springer Verlag), p. 4.
13 http://www.boeing.com/commercial/777family/pf/pf_milestones.html, accessed 6 August 2004.
14 Asendorpf, D. (2004), 'Das große Flattern', *Die Zeit* No. 1, 30 December 2004, p. 35.
15 Evans, G. and Foerstemann, M. (2004),, 'Learning from the Automotive Industry', *Aerospace International*, April 2004, p. 28.
16 Anderson Jr, J.D. (2002), *The Airplane - A History of Its Technology* (Reston: AIAA), p. 353
17 Sutter, J. (2006), *747: Creating the World's First Jumbo Jet and Other Adventures from a Life in Aviation* (New York: Smithsonian Books, HarperCollins Publishers), p. 176.

SYSTEMS' COMPLEXITY

A system is generally defined as a unit, which encompasses different interoperable elements with their own individual functionalities. A modern commercial aircraft consists of many different systems, such as the structure (or airframe as it is often called), flight control unit, hydraulics system, electrics and pneumatics systems, avionics systems, cabin systems, and so on.[18] Not only are these systems already individually very complex, they also have to coexist together in a relatively narrow space (compared to most other products).

In addition, commercial aircraft systems have to cope with very stringent conditions concerning high and low temperature limits, vibrations, moisture, liquid contamination and so on.[19] Their installation conditions must be safe throughout the aircraft life. In fact, safety requirements request that in the case of failure of one system component or system component bracket, it should be ensured that the integrity of the aircraft behaviour is not affected.

All of this leads to a very sophisticated installation design, where the installation designer has to deal with specialists' requests from all systems. Complexity increases significantly as a result of this. This is particularly true for some crowded areas, such as the aircraft's nose section, the center fuselage or the overhead crown area, where components of many different systems are co-located, each one having its specific requirements. Installation design has become a real nightmare here.

> 'The systems in commercial jet aircraft have undergone a substantial evolution from [for example] the Boeing 707 to the 777. In the 707, the systems were relatively simple in function, and thus in complexity, and had little interaction with each other. … This allowed each system group to design part of the airplane with only a small degree of systems coordination.
>
> By the time the Boeing 757 and 767 (circa 1980) were being developed, system functionality had grown dramatically in complexity and interdependency. No longer could these airplane systems be developed independently. System designers were required to achieve a higher degree of coordination and communication with designers of other systems. … This increased functionality of individual systems equipment also has significantly increased the interdependency of the various systems, hence overall complexity.'[20]

But a modern commercial aircraft does not only accommodate individual complex systems but is certainly a complex system on its own as everything has to interoperate smoothly without fault. Even more complex, a commercial aircraft is only one element out of the larger aircraft system, which also comprises elements such as the training equipment, the support equipment as well as different types of facilities. Hence, when performing top-level syntheses, it is necessary to synthesise the higher-level system, not the lower-level elements of it.

---

18 For example, there are 120 different systems on an Airbus A380.

19 This is one of the reasons why all system specialists want to have their 'personal' brackets supporting their systems components. However, too much of a bracket variety is very harmful to all kinds of aircraft design and build aspects and results in too high a design, procurement and production effort. There is therefore a strong push from the cycle-time and cost-side of design and build to standardise brackets as much as possible. But despite this, the numbers of different brackets on a commercial aircraft remain very high. As an example, the A380 has around 25,000 different types of brackets which leads to more than 250,000 brackets installed for basic technical and cabin systems.

20 Petersen, T.J. and Sutcliffe, P.L. (Boeing Commercial Airplane Group, Seattle, Washington) (1992), 'Systems Engineering as applied to the Boeing 777', *Aerospace Design Conference* (Irvine, California: AIAA 92-1010), 3–6 February 1992, p. 2 (reprinted by permission of the American Institute of Aeronautics and Astronautics, Inc.).

An aircraft system is also part of an even larger air transport system, which includes airports, air traffic controllers, airworthiness authorities, navigation and communication satellites, maintenance facilities and others, see Figure 1.1. All of these stakeholders play an equally important role. A passenger can only be flown at highest possible standards of safety from A to B if all systems on all levels work perfectly together. For example, without air traffic control, modern commercial air transport cannot be fully safely operated in airspace with very high traffic density.

Thus, there are different levels of systems which an aircraft developer has to consider. Aircraft development therefore also has a strong aspect of overall systems' complexity in addition to the complexity of individual systems accommodated within the aircraft.

DESIGN COMPLEXITY

Highest Possible Safety

The most important aspect of any commercial aircraft development project is to ensure operational safety for the future aircraft. 'Safety first' is the rule of commercial air transport. This is because commercial aircraft accidents potentially represent sad tragedies for a high number of people *at the same time*. As a consequence, they could generate excessive insurance cases for the operating airlines. In addition, any airline's reputation as a safe service provider could be at stake with potentially detrimental repercussions for its whole business. Safety is therefore a top priority for any airline.

While there is a lot an airline can do on its own to ensure operational safety – such as regular maintenance and aircraft checks – it will primarily rely on the aircraft's inherently safe design. Safety is therefore, of course, also of top interest to the aircraft manufacturer. Its reputation as a developer and producer of sound and safe aircraft will be at stake if an aircraft's accident is proven to be due to design errors or production flaws. In the 1950s, for example, there were a number of fatal accidents involving the world's first jet passenger aircraft, the Comet, which were due to error-prone design. They not only damaged the reputation of the De Havilland company but eventually resulted in the total termination of complete aircraft production by this company.[21]

'*Safety* is a broad and often abstract term. You don't design safety *into* an airplane; instead you design an airplane to have excellent airworthiness characteristics so that a pilot can handle it in good conditions and bad, including those unforeseeable occasions when damage is sustained, or the airplane must operate in severe atmospheric conditions, or when highly adverse conditions prevail during takeoff or landing. This in turn dictates a robust structure and engines that are reliable and powerful enough so that major events won't put the airplane into jeopardy. The airplane must be able to absorb punishment, and after it does, its pilot must be able to bring it in for a safe landing.'[22]

'One tenet of the Boeing safe-design philosophy is the idea of *no single failure modes*. In plain English, this means that airplanes shall be designed in such a way that no single system failure or structural failure can ever result in catastrophic consequences. Another tenet is the idea of *no uninspectable*

21 For more details, see for example: Aris, S. (2002), *Close to the Sun – How Airbus Challenged America's Domination of the Skies* (London: Aurum Press), p. 22.
22 Sutter, J. (2006), *747: Creating the World's First Jumbo Jet and Other Adventures from a Life in Aviation* (New York: Smithsonian Books, HarperCollins Publishers), p. 128.

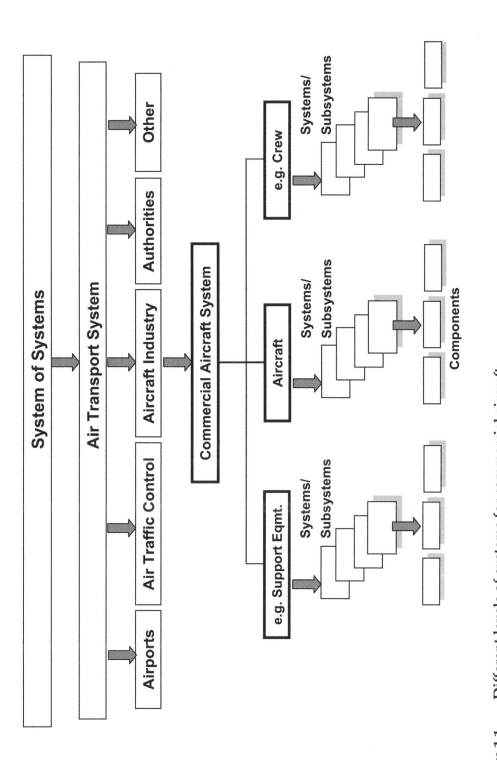

Figure 1.1   Different levels of systems for commercial aircraft

Source: Jackson, S. (1997), Systems Engineering for Commercial Aircraft (Aldershot: Ashgate Publishing), with permission.

limited-life parts. In other words, if a part isn't where an airline can get to it in order to check it, then it must be designed to last the entire life of the airplane.

One approach we use to achieve safety is *redundancy.* This is the term we engineers use to refer to safety enhancements achieved through the use of *backup systems or structure.* When a primary system or structure fails, prudent design ensures that there are backup systems and load paths available to take over, preserving the safety and integrity of the flight.'[23]

But lack of safety may not only stem from immature technologies or error prone designs. Lack of development process adherence may play an equally important role.

'Studies have shown that the common factor in major catastrophes … was not technological but rather organizational. … Factors identified were: time pressures, failure to observe warnings of deterioration and signals of malfunction, lack of an incentive system to handle properly the tradeoffs between productivity and safety, failure to learn from mistakes and motivate reporting of problems, and lack of communication and processing of uncertainties. It is the responsibility of the Program Management to assure that these factors are minimized.'[24]

Because safety is so important, safety standards have been established by airworthiness authorities worldwide. The latter monitor the design, development, production, operation, maintenance and repair of any commercial aircraft to these standards. As there are two major authorities in the world, the US Federal Aviation Administration (FAA) and the European Aviation Safety Agency (EASA) in Europe, it may well be that aircraft manufacturers have to comply with both of their certification requirements, which can differ from each other. While being absolutely mandatory, these safety standards for design clearly lead to increased product complexity.

'For example, [EASA] requirements [on top of the domestic FAA requirements] regarding escape hatches on some of the derivative Boeing 737 programs required extensive redesigns that added several months to the design cycle process.'[25]

In addition, the documentation effort needed to satisfy the requirements by the authorities for processes' and parts' data documentation is tremendous, with the data to be archived for many decades. This, too, increases the complexities on the side of aircraft manufacturers, resulting in high administration costs. The only other industries with similar high standards of safety assurance are the nuclear and chemical industries.

Lowest Possible Weight

The most distinct technological difference to most other products is the necessity of all aerospace products to minimise weight. Any flying machine has to lift into the air its own weight and that of a useful payload (including crew) before it can even begin to do the job it was designed for. If the weight is not right compared to the lift generated, the aircraft will not take off from the ground.

---

23 Sutter, J. (2006), *747: Creating the World's First Jumbo Jet and Other Adventures from a Life in Aviation* (New York: Smithsonian Books, HarperCollins Publishers), p. 129.

24 Jackson, S. (1997), *Systems Engineering for Commercial Aircraft* (Aldershot: Ashgate Publishing), pp. 126–7.

25 Spitz, W., Golaszewski, R., Berardino, F. and Johnson, J. (2001), *Development Cycle Time Simulation for Civil Aircraft*, NASA/CR-2001-210658, pp. 3–6.

In addition, if at the end of the aircraft's Development phase the target take-off weight has been missed by, say, only 2 per cent(!), the excess weight of an aircraft of 100t target take-off weight would be 2t. How many paying passengers would an airline have to take out of the cabin, including seats and luggage, to compensate for this excess weight? About 15! Fifteen paying passengers less on long haul flights equates to at least $1 million per year in loss of revenue for the airline. This is unacceptable to airline customers. In other words, any additional kg on the aircraft has a penalty cost of $500 or more, which needs to be compensated by other cost advantages.

While never compromising safety, the achievement of weight targets is therefore fundamental for the development of a commercial aircraft. It is in fact so fundamental, that it dominates to a large extent the prioritisation of Management decisions during development. As an example, the weight of a new commercial aircraft has to be known reliably very early during the design process, that is about seven to eight years before it enters service with an airline. Saving weight has been a characteristic of aircraft design since the first successful powered flight by the Wright brothers in 1903 and it distinguishes aerospace development fundamentally from the development of non-aerospace products.

Constrained Interfaces Between Structure and Systems

Aerospace development in general and aircraft development in particular requests a perfect blending of structural and systems' functionalities. The structure not only has to ensure the integrity of the product across a large operational envelope, it also needs to have the right shape to generate the necessary lift at minimum drag. Thus, beyond the aspect of being a platform for equipment (like a computer chassis carrying circuit boards) or beyond any aesthetical design aspects, the airframe has a fundamental functionality on its own. This significantly limits designers in their freedom to establish a structure, which serves as a platform to accommodate systems. Space allocation management is therefore a crucial skill developed within aerospace. In addition, systems' functionalities, for example for flight control, are heavily linked with the structural functionalities via the flight control units such as flaps, slats, spoilers, ailerons, elevators and rudders. The relative importance of an aircraft's structure compared to its systems is, thus, higher than, for example, for cars, computers and many others.

CUSTOMISATION COMPLEXITY

Commercial aircraft are some of the most customised products in the world, apart from cars. Cabin customisation alone accounts for about 10 per cent of the value of a commercial aircraft.[26] Modern cabins are extremely complex systems, which offer a lot more for the passenger and the crew than just seats, galleys and restrooms in a pressurised 'tube'. The cabin layout of an airliner is a very important factor for an airline company to differentiate itself from its competitors, see an example in Figure 1.2. The need for differentiation drives the demand for customisation of commercial aircraft.

---

26 Fuchs, R., Airbus Senior Vice President for Cabin and Cargo, in: Wall, R. and Sparaco, P. (2005), 'Cabin Control', *Aviation Week & Space Technology* 5 April 2005, p. 46.

Figure 1.2   Cabin layout for the Airbus A380 (example)
Source: Airbus, with permission.

However, not only do different airlines define different layouts, different batches of aircraft ordered by the same airline usually differ in cabin layout too. The first aircraft of such a batch is traditionally called 'Head of Version'. This term was introduced at a time where all aircraft belonging to the same batch had the same layout, that is, the layout of the Head of Version. This is often no longer the case as airline customers frequently change their minds. Instead, almost every aircraft cabin is unique. This has major consequences for any aircraft manufacturer's design and production processes. As Corky Townsend, Chief Project Engineer for the Boeing 747 programme, puts it: 'There is significant amount of disruption associated with configuring an airplane for each customer...'[27]

To manage the customisation process from the point in time where a customer airline has selected its options (called Contractual Definition Freeze or Customer Definition Freeze (CDF)) until the promised delivery date with all these options correctly implemented on the aircraft is therefore a huge Engineering and Management challenge. An aircraft producing company, which manages this process well and at shortest possible leadtimes has a definite competitive advantage. Thus, not only the classical objectives time, cost and quality are of importance but equally the ability to manage customisation variability.

PROCESSES COMPLEXITY

The development of commercial aircraft requires myriads of processes, by far not all related to customisation. These processes need to be robust as well as known and applied by all the people involved in the development. For the aircraft producing company to understand how efficient processes are, Key Performance Indicators (KPIs) are constantly

---

27 Townsend, C. (2007), '747-8 Cockpit Key Source', *Aviation Week & Space Technology* 12 February 2007, p. 38.

measured. For example, it is rather common among many aircraft companies to measure the number of changes Engineering has to produce after the initial drawing release. Boeing, for example, used such metrics to measure processes performance for its 777 programme. It was called 'design errors reaching the factory', and consisted of change, error and rework items. The latter would show up on rejection tags as physical interference problems and failures to fit (requiring enlarged holes or shimming) among others.[28]

In 1993, a report on a survey covering 60 large development programmes demonstrated that aircraft developments by then typically needed four to ten rework cycles before reaching the desired maturity levels.[29] By comparison, other industries, such as large construction works or software developments, only needed up to three rework cycles. This, of course, is a direct result of all the other complexities mentioned in this chapter.

Since 1993 there have been new aircraft programmes, such as the Boeing 777, which reached rework cycles of below two. However, this still represents a significant financial risk if programme Business Cases are laid out with no allowance for rework at all! But the Boeing 777 example shows, at least, that significant improvements are possible. In a nutshell, Boeing achieved them by introducing multi-functional teams using Concurrent Engineering techniques with full 3D capabilities based on common databases.

LOW FREQUENCY COMPLEXITY

The frequency with which new aircraft are being developed is today much lower than in the past. Commercial aircraft are nowadays low frequency products. Airbus, as well as Boeing, only manage to develop new commercial airliners every five to eight years. Low frequencies in new product generation are a direct consequence of the risen complexity and costs of aircraft.

For today's aircraft engineers this means that they might only see three or four completely new aircraft being developed during their business careers. Many of the best of them may not even be staying that long in the same company. It is becoming increasingly difficult to come down one's individual learning curve of experience. A typical personal learning curve today looks more like a zig-zag line with an overall downward trend than a smooth downward curve. This affects primarily engineers but also applies to most other skills involved in aircraft development, such as Manufacturing or Project Management.

In addition, Information Technology (IT)-based processes and tools, which support, for example, Design, Project Management and Systems Engineering, do change so rapidly themselves that with each new aircraft development, two or more generations of tools have gone by.[30] As a result, for example, Project Management, as a Management discipline applied by individuals which have come sufficiently down on their personal learning curve, realistically does not even have a chance to come to full fruition any more – even if sufficient funding would be available. The same applies to other disciplines such as Systems Engineering. Their strengths cannot be fully exploited as the skillset required for

---

28 From: Hernandez, C.M. (1995), *Challenges and Benefits to the Implementation of Integrated Product Teams on Large Military Procurements* (Boston: Massachusetts Institute of Technology, MSc thesis), p. 36.
29 Cooper, K.G. (1993), 'The Rework Cycle: How It Really Works… And Reworks…', PMNetwork Vol. 7 No. 2, February 1993, pp. 25–28.
30 This problem could be somewhat relaxed if sufficient people within the same company could be shifted back and forth between military and commercial aircraft development projects so as to ensure continuous learning in projects.

in-depth use is not sufficiently abundant within companies developing products at such low frequencies.

> 'A firm's learning economies depreciate at an annual rate as high as 40 percent due to attrition in the workforce, changes in work assignments, and simple losses in proficiency in seldom repeated tasks. It is, therefore, important for the firm to maintain a constant rate of [development and] production... .'[31]

As a result of insufficient numbers of development projects, therefore, the commercial aircraft-developing company has to invest heavily in training because many individuals will not have sufficiently matured to do the job efficiently and effectively. It is relatively easy nowadays to recruit people with basic knowledge in, say, Project Management, but additional training is needed to understand the fundamentals of the wider aspects of aircraft development, teamwork, conflict resolution, project economics and so on and to get introduced to the latest available methods and tools applied within the company. Raising the level of skills through training is, of course, expensive and time consuming. It is estimated that around 100 hours of additional training beyond the functional skills are required for each member of a multi-functional development team.[32] In addition, once the aircraft development project has kicked-off there is not much time available anymore for training. As a result of aircraft development cycles becoming shorter and shorter to ensure better market responsiveness, there are far fewer opportunities for training sessions to take place.

All of this makes product development at low frequencies more difficult and complex. In fact, a recent study by Deloitte mentions the lack of sufficiently experienced people resulting from low frequency products as one of the root causes why aerospace projects often fail and become more and more risky.[33] However, because there will be limitations when it comes to training it makes sense to concentrate on essentials and to ensure that Engineering and Management methods, processes and tools are harmonised and based on an integrated architecture everywhere.

FINANCIAL COMPLEXITY

New commercial aircraft developments are extremely expensive. The historic trend – if calculated on the basis of constant development costs per seat (in order to compare aircraft of different sizes and different years of Entry into Service (EIS)) – seems to indicate a continuous increase in development costs. In the early 1990s the 777 was already 'the largest privately-funded endeavour in the world with the possible exception of the tunnel under the English Channel'.[34] In 1994, one source estimated the final development costs for the 777 to be around U$5.5 billion.[35] However, in 2001, the President and Chief

---

31 Spitz, W., Golaszewski, R., Berardino, F. and Johnson, J. (2001), *Development Cycle Time Simulation for Civil Aircraft*, NASA/CR-2001-210658, p. 4–2.

32 From: Hernandez, C.M. (1995), *Challenges and Benefits to the Implementation of Integrated Product Teams on Large Military Procurements* (Boston: Massachusetts Institute of Technology, MSc thesis), p. 65.

33 Anon. (2008), Can we afford our own future? Why A&D programs are late and over-budget – and what can be done to fix the problem (Deloitte Consulting), p. 8.

34 Condit, P.M. (1994), 'Focusing on the Customer: How Boeing Does it', *Research Technology Management* January–February 1994, p. 33.

35 Anon. (1994), *Aviation Week & Space Technology* April 1994.

Executive Officer of Boeing Commercial Aircraft Group, Alan Mulally, was quoted saying: 'the 777 came in at a much steeper price, … double in fact … The 777 cost US$6 billion more than we first estimated.'[36] Ten years later, the development costs of the passenger version of the A380 was announced to be US$10.7 billion.[37]

Huge sums of money are involved with commercial aircraft developments and recoupment leadtimes are long, even assuming constant revenue streams. But the air transport market is not stable and often shows rapid fluctuations. Thus, new commercial aircraft always represent significant financial risks to the companies which develop them and measures designed to minimise these risks have to be implemented. But these measures further increase the complexity of the development project.

One can easily imagine a variety of different financial risks, which can easily put the aircraft-developing company in jeopardy, such as:

- higher than expected development expenses;
- lower than expected revenues (for example, when the aircraft EIS happens to take place during an unexpected market downturn or the new aircraft becomes less acceptable to airline customers because of changed market conditions); as well as
- financing problems.

'Development costs for the Boeing 747 … have been estimated at $1.2 billion spanning roughly a 4 year period between December 1965 to January 1970. At the time the development of the aircraft commenced in late 1965, total shareholder's equity was only about $372 million. The ratio of development costs to equity was approximately 3.23; that is, the development cost of the B-747 alone was more than three times the value of stockholders' investments. In short, Boeing was required literally to "bet" the company on the success of the B-747.'[38]

'Boeing's problems with the 747 sounds like a litany of the damned … [and almost] threatened the company's survival. … Boeing not only had to pay penalty fees for late deliveries, but, far worse, didn't receive the large last instalments until the deliveries were made. Deprived of an adequate … cash flow, Boeing found itself seriously short of funds yet obliged to finance a huge inventory of partly built 747s.'[39] [40]

McDonnell-Douglas incurred similar risks in developing the DC-10. Development costs for this aircraft have been estimated at $1.1 billion. The value of shareholders equity was only about $364 million in 1967, the year in which development commenced. The ratio of development costs to equity was about 3.02. McDonnell-Douglas, then, was also required to risk the fate of the firm in developing the DC-10.

Also the traditional airplane maker Fokker ran into liquidity problems in the late 1990s while developing the F70, a smaller derivative of the F100. In this case, the company did not survive.

---

36 Anon. (2001), 'People: Alan Mulally', *Business Week* 2 July 2001, p. 7.
37 Forgeard, N. (2005), *Airbus Annual Press Conference* (Paris), 12 January 2005.
38 Spitz, W., Golaszewski, R., Berardino, F. and Johnson, J. (2001), *Development Cycle Time Simulation for Civil Aircraft*, NASA/CR-2001-210658, p. 2–7.
39 Newhouse, J. (1982), *The Sporty Game* (New York: Alfred A. Knopf), pp. 166 & 168–9.
40 Joe Sutter, the 'father' of the Boeing 747, asked by the editor of Aerospace International, Richard Gardner, whether Boeing was betting the company on the 747 in 1966, answered: 'Yes', in: Gardner, R. (2004), 'Joe Sutter', *Aerospace International* Vol. 31 No. 12 December 2004, p. 22.

'An example for changed airline preferences giving trouble to an aircraft programme was the Boeing 717. Born as the MD-95 in 1995 and renamed after the Boeing McDonnell Douglas merger in 1997, it was McDonnell Douglas' newest model and thought of as the successor to the MD-87 and a replacement for the DC-9-30 as the 717 was their size. But by the time it entered the market in 1999, larger single-aisle aircraft were much more popular and the 717 production was terminated in 2006 after only 155 aircraft had been delivered.'[41]

In most cases significant shares of the total development costs are financed through loans borrowed from banks. If those loans are of a long-term nature, there are substantial financial risks to these stakeholders too.[42] They result, for example, from lower than expected sales of the newly developed aircraft. During the very long pay-back periods encountered in the commercial aircraft industry sales trends are inherently more difficult to forecast. Banks try to reduce their risks by claiming higher interest rates on long-term loans compared to short-term ones. Thus, for those commercial aircraft development projects where loans cannot be paid back on the basis of current operations but need to rely on future sales of the newly developed aircraft, financing becomes relatively more complex – and therefore more expensive – compared to product developments with shorter life-cycle times.[43]

A particular additional problem to non-US commercial aircraft makers is the worldwide practice to trade aircraft in US$. There are two reasons for this: First, until about 20 years ago, commercial air transport was dominated by US manufacturers, who sold their aircraft in US$. This generated an 'aviation currency' not only used for original aircraft trading, but also for second-hand transactions, maintenance, repair and overhaul works, payments of airport fees and so on. Second, throughout the world kerosene has always been traded in US$. Fuel represents a significant share of the direct operating costs of an aircraft. An airline, which has an interest in minimising its exposure to exchange rates, therefore tries to trade all its other operating costs and its revenues (such as tickets) in US$ as well. Therefore, when purchasing or leasing a new aircraft, many airlines are reluctant to accept any other trading currency than the US$. Non-US manufacturers, which have their cost base in a different currency, are therefore exposed to substantial exchange rate risks.

PROJECT MANAGEMENT COMPLEXITY

Project Management, as defined by the various bodies of knowledge, (for example, the US-based Project Management Institute) appears to be an appropriate approach for what is referred to as classical projects.[44] There may perhaps be a high level of unknowns at the start of a classical project, which can get resolved early, but there are only few new unknowns arising during execution.

---

41 Anon. (2006), 'Boeing will close the assembly line...', *Aviation Week & Space Technology* 13 February 2006, p. 20.
42 If loans are of a short-term nature, they are usually paid back via the revenues made on other aircraft programmes. The risk is then limited to the usual risk of market demand fluctuations.
43 Note, that this aspect also represents a significant market entry barrier for new entrants.
44 Anon. (2004), *PMI. A Guide to the Project Management Body of Knowledge*, 3rd Edition, (Newtown Square, PA: Project Management Institute).

In contrast to this, the development of a commercial aircraft is a much more dynamic project with a high level of unknowns at the start and a high rate of new unknowns throughout the project. One source of new unknowns is the Concurrent Engineering approach (to be introduced in Chapter 3) widely in use today to allow for a parallel development of airframe, systems definitions, systems installations, jigs & tools and so on. But concurrency implies the necessity to balance the different design maturities of airframe, systems definitions, systems installations and others at any given point in time, as some mature earlier and others later. Finding the right balance is impossible without constantly making some assumptions, which represent new unknowns that later could turn out to be wrongly chosen.

New unknowns could also arise from the introduction of new materials, such as the material GLARE[45] used for the first time on the Airbus A380, or carbon composite materials for aircraft fuselages as firstly introduced with the Boeing 787. They could also arise from a new type of project organisation (as, for example, introduced for the development of the Boeing 777) or a streamlined supply chain architecture (such as the one for the Boeing 787). There are many root causes for unknowns.

The effort to resolve the unknowns, which exist at the beginning of a dynamic project, is severely challenged by the introduction of the additional unknowns along the way. This is because what is learned can become obsolete in less time than it takes to learn. Therefore, the speed of learning during a dynamic project is crucial for its success. It requires highest levels of communication in an already complex development environment, which makes the management of aircraft development projects particularly complex.

As new unknowns are not yet known at start of the project, the management controls put in place at that point to run the project must be flexible enough to deal with 'unknown unknowns'. They also must be flexible enough to handle the inconvenient bursts, in which form new unknowns may appear, and often just after the exercise to establish an agreed Project Plan update has been completed.

However, flexibility is already required even before the management controls are fully established. This is because the launch of a new commercial aircraft development is essentially driven by market needs as well as by what the competition is doing. There is generally a bigger 'hurry' to develop the aircraft and bring it into the market compared to, for example, military aircraft. To this end, projects related to the development of military products usually can afford a sound preparatory phase, in which both, technical *and* managerial problems are addressed. This preparation is also fully paid by governments, as described above. The technical and managerial preparations must be at the required level before governments are willing to launch the next and much more intense development steps.

In commercial aircraft development, too, there is a date at which the technical preparations must be at the required level to be sure that there are no showstoppers. The shareholders of the aircraft manufacturer will be waiting up to this date before being willing to launch the next development steps, otherwise the project risks would be too high. However, the inherent resistance against every expenditure which is perceived to be unnecessarily early, including on the managerial side of developing projects, leads to a situation where the managerial preparations in commercial aircraft development are often far from being complete at that date. For example, the architectures for product, work, costs and project organisation may not be ready by then, team leaders may not be

45 GLARE – GLAss-REinforced Fibre Metal Laminate.

available yet, teams not ready, roles and responsibilities neither clear nor allocated, and so on.

Thus, after this date, when the development enters into its most intense phase with thousands of engineers designing the individual aircraft components, the managerial aspects still have to be worked out for some time. Besides the challenges arising from managing thousands of engineers and their technical work in the first place, there is also the challenge to finish up the managerial preparations while already working on the components' design. This leads to additional friction and results in additional management complexity. The latter is essentially a consequence of the commercial nature of the aircraft development projects discussed here.

However, there is also the aspect of long development leadtimes frequently encountered in commercial aircraft developments. With such long leadtimes, there is an impact on the accountabilities of the many project leaders involved in such developments. This is because only very few of the project leaders will be with the project from its beginning to its end. For example, those who signed-off the Business Case or parts thereof are most probably no longer around when the project is at full steam. Whether the original cost targets put into the Business Case were estimated to be too low or whether the cost overruns encountered later are due to project leaders not managing their part of the project well is a constant debate often found in long duration projects. Compared to short duration projects, where project leaders are likely to remain with the project all the time, the question of project leaders' accountability is more complex to deal with and requires different answers.

MONOPSONY COMPLEXITY

In economics, a monopsony is a market form in which only one buyer faces many sellers. It is an example of imperfect competition, similar to a monopoly, in which only one seller faces many buyers. As the only purchaser of goods or a service, the 'monopsonist' may dictate terms to its suppliers in the same manner that a monopolist controls the market for its buyers. The high development costs of commercial aircraft, coupled with low production numbers and custom-built specifications, shapes the competitive business environment, in which commercial aircraft manufacturers operate, and makes it very similar to a monopsony market situation.

W. Spitz et al. describe the situation as follows:

> 'Stability in the marketplace depends upon the ability of [aircraft producing] firms to differentiate their products and, more specifically, to build different size aircraft with different capabilities that will be attractive to specific niches in the marketplace. When firms build aircraft of the same size with similar capabilities, they often find that the market is too small to yield satisfactory returns on their investments. Competition becomes so vigorous for limited sales opportunities that airlines acquire monopsony power – the ability to dictate the terms of the sale to the seller. This situation can have debilitating effects on competitors and it can reinforce the already existing tendency for one firm to emerge as the dominant competitor during any given era.'[46]

---

46 Spitz, W., Golaszewski, R., Berardino, F. and Johnson, J. (2001), *Development Cycle Time Simulation for Civil Aircraft*, NASA/CR-2001-210658, pp. 2–5.

In this monopsony situation even the Number One aircraft manufacturer has less power over its ability to control prices or aircraft deliveries when comparing it with other industry sectors with high development costs (such as automotive, semi-conductors).

For example, annual commercial aircraft deliveries tend to peak before the summer break because this is when airlines need aircraft the most. However, any peak in the delivery stream represents a challenge to stable production flows, which are required to achieve lowest possible production costs. Thus, monopsony power acts against the inherent interest of the aircraft producing company.

It is even worse to be the Number Two. Staying competitive against a competitor, which has the tendency to become the Number One is, thus, vital for survival. The tendency of airlines to acquire monopsony power over aircraft manufacturers has therefore significant consequences for all units and disciplines of a company producing commercial aircraft: it forces them to constantly and significantly improve processes and quality, and to reduce costs and leadtimes – often to a higher degree than in other industries.

This means that even during the Development phase of a commercial aircraft project, existing technical and managerial business processes can often not be kept as they are but may have to be further optimised. This is even more necessary in view of the very long cycle times encountered during commercial aircraft developments. To manage product development and processes improvement in parallel, without allowing the aircraft development being unfavourably affected by this, is a complex managerial task indeed and requires significant efforts.

MULTI-CULTURAL COMPLEXITY

In the late 1960s no single European country would have been any more able to develop a commercial aircraft of the size of the A300. It was a question of survival to Europe's aeronautical industry against a US-dominated market to pool the resources, eventually resulting in the creation of a truly multi-cultural (that is, British, French, German and Spanish) company called Airbus. Boeing, Airbus' strongest competitor, has always been a national company but it reflects the multi-ethnical background of the US society and is therefore also not free of multi-cultural aspects. The same applies to other aircraft companies in the US and Canada. So, it is fair to say that the bigger commercial aircraft companies are multi-cultural companies.

In addition, while aircraft development started many decades ago as a national venture with structural platform and systems providers essentially located in the same country (notably the US, UK, France, Germany, Japan, and Russia), it has evolved since then into a truly global business with contributing companies from all over the world.[47] Many Second and Third World countries are today capable of producing aerostructures, often at lower prices than in the traditional aeronautics nations.[48] An example for this would be

---

47 A good overview on this subject is presented in: Pritchard, D.J. (2001), 'The Globalisation of Commercial Aircraft Production: Implications for US-based Manufacturing Activity', *International Journal of Aerospace Management* Vol.1, N° 3, October 2001, p. 213. Also see: Anselmo, J.C. (2005), 'Distant (Off)Shores', *Aviation Week & Space Technology* 14 November 2005, pp. 58–59.
48 A good overview on aerostructures manufacture in the Asian Newly Industrialised Economies can be found in:

wing assemblies for the Boeing 717 produced in South Korea[49] or Boeing 787 composite rudders produced in China.[50]

Although aircraft systems and equipment development today still is largely a domain of these traditional aeronautics nations,[51] the trade of aeronautics products between them has significantly increased – and this more on the commercial side than on the military where sensitivities associated with the secret or confidential aspect of key technologies often prevail against free trade. As a result, around 50 per cent of each Airbus A380 is US-made[52] generating around 60,000 jobs in the US.[53] In the late 1990s, Boeing's then Commercial Airplane President Ron Woodard claimed that the Boeing 737 series generated 33,000 jobs in France alone.[54] Reasons for this increase in global aerostructure and aeroequipment trade include:

- the constant pressure to reduce the high cost of commercial aircraft development – or even to achieve its affordability in the first place – by making use of Second and Third World level production costs;
- the necessity to search for risk-sharing partners to share the huge risks associated with aircraft development;
- the availability of specific technologies in other countries;[55]
- the customer airlines' desire to be able to chose between engine suppliers. As engines represent about 20–30 per cent of the value of a complete aircraft, which means that engines have a significant impact on international aeronautics trade; as well as
- the necessity to outsource work packages for reasons of market access or to satisfy offset agreements.

'In the context of the commercial aircraft sector, the first industrial offsets started in the 1960s, when Douglas subcontracted the fuselage assemblies for the DC-9 and DC-10 to Alenia in Italy. These transactions resulted in substantial sales of Douglas aircraft to the flag carrier of Italy. One of Boeing's early offsets was with Japan in 1974, when Mitsubishi was given contracts to produce inboard flaps for the 747, resulting in major sales of 747s to Japan. In virtually every case that has been documented, the goal of an offset agreement is to secure a sale that would not take place in the absence of compensatory provisions.'[56]

Eriksson, S. (1995), *Global Shift in the Aircraft Industry: A Study of Airframe Manufacturing with Special Reference to the Asian NIEs* (Gothenburg: University of Gothenburg, Department of Geography Series B No. 86).

49 MacPherson, A. and Pritchard, D.J. (2003), 'The International Decentralisation of US Commercial Aircraft Production: Implications for US Employment and Trade', *Futures* Vol. 35, No. 3, April 2003, pp. 221–238.

50 Mecham, M. (2004), 'Chartering Sales', *Aviation Week & Space Technology* 12 July 2004, p. 38.

51 With the addition of Bresil and Canada which today produce aircraft and equipment too.

52 Sparaco, P. (2004), 'Time Bomb: Weakening U.S. Dollar Casts Long Shadow over Health of Europe's Aerospace Industry', *Aviation Week & Space Technology* 29 March 2004, p. 24.

53 Anon. (2001), *The A380 – Driving Sustainable Growth* (Toulouse: Airbus), 28; based on a model created by: Altfeld, H.H. (2001), 'A Model for Quantifying Strategic Benefits of Aerospace Programmes', *International Journal of Aerospace Management* Vol. 1, No. 3, October 2001, p. 201.

54 Sparaco, P. (2004), 'Transatlantic 7E7', *Aviation Week & Space Technology* 19 July 2004, p. 78.

55 For example, consumer electronics devices that are also being used in commercial aircraft are today largely built outside the major aerospace countries, for example in Asia. The volumes and learning curves are so hardly in favour of these other countries that the few remaining electronic devices companies in the aircraft producing countries are struggling to keep up technologically and commercially.

56 Mecham, M. (2003), 'Overseas Shipments', *Aviation Week & Space Technology* 24 November 2003, p. 36.

'When Airbus and Boeing recently sold US$10 billion worth of new aircraft to India, they had to deliver an offset commitment of US$3 billion.'[57]

'In 1960, imports of aircraft and parts amounted to only 5 per cent of aircraft exports by value, compared to 45 per cent today. Foreign content for the 7E7 may run as high as 70 per cent. The foreign content of a Boeing 727 was only 2 per cent in the 1960s, compared to nearly 30 per cent for the 777 in the 1990s.'[58]

'Japanese suppliers hold a 21 per cent workshare on the 777 program.'[59]

'In the case of the 777, there is no domestic production for the vertical and horizontal stabilizers, the center wing box, or the aft and forward fuselage sections. The only significant part of the airframe that is domestically manufactured is the wing assembly. And with the proposed launch of the 7E7, Boeing will for the first time be surrendering the wing and composite structure technology to the Japanese by having them design, develop manufacturing processes and undertake final assembly of the wing.'[60]

As a result of both the multi-cultural workforce of the larger commercial aircraft companies as well as because of the increasingly international worksharing and trade, multi-cultural aspects and problems resulting from working in a multi-cultural environment have become an important area of potential risk. Flawless multi-cultural communication is therefore vital for business success. It needs to be fostered:

- between individuals and teams with different cultural backgrounds;
- across language barriers and different corporate cultures;
- despite there being different methods, processes and tools in place.

Thus, commercial aircraft development of today needs to cope with the challenges of multi-cultural complexity.

PUBLIC RELATIONS COMPLEXITY

Commercial aircraft represent huge technological achievements and therefore are quite often regarded as symbols of national pride. They contain a message about the technological level of a country or region and the level of the qualification of its people. They even carry by themselves this message around the globe. For an aircraft-developing country, commercial aircraft are excellent publicity reaching millions of people. Because commercial aircraft represent nations or – like in the case of Airbus – a collaborative effort of nations, they are often used as symbols to represent these nations in a graphical way.

Like other products representing means of transport, such as cars, motorbikes, trains and ships, aircraft provoke genuine emotions, even among people who have nothing to do

57 Anon. (2006), *Managing a Competitive Imperative – Achieving High Performance with Effective Offset Deals* (Accenture).
58 Pritchard, D.J. (2004), 'Are Federal Tax Laws and State Subsidies for Boeing 7E7 Selling America Short?', *Aviation Week & Space Technology* 12 April 2004, p. 74.
59 Mecham, M. (2003), 'Overseas Shipments', *Aviation Week & Space Technology* 24 November 2003, p. 36.
60 Pritchard, D.J. (2004), 'Are Federal Tax Laws and State Subsidies for Boeing 7E7 Selling America Short?', *Aviation Week & Space Technology* 12 April 2004, p. 74.

with their development. Millions of people gather in aviation clubs, hundreds thousands visit airshows and aviation museums annually and buy toy models of famous aircraft such as the Boeing 747. Tens of thousands witnessed the final ever landings of the various supersonic Concorde aircraft in 2003 and the first fly-bys of the Airbus A380 in 2005. People admire the elegant flying of aircraft in the sky but they also do become annoyed when they are too noisy. In both cases, emotions are involved.

Figure 1.3   The city of Toulouse thanking Airbus for developing the A380
Source: Ville de Toulouse, with permission.

The development of a new aircraft will therefore surely attract the interest of both people and governments a long time before its appearance in the marketplace. This requires very intense and complex public relations activities by the Communications department of the aircraft manufacturer during the Development phase. This is certainly not required to this extent for most other product developments.

SUMMARY

Commercial aircraft are indeed unique products. Each of their developments represents a complex, multi-cultural and top-of-the-learning curve mega-project exposed to high financial risks and strong public interest.

Other high-technology products come close to this uniqueness but lack some of the criteria described above. Military aircraft, for example, lack the commercial approach by definition. They are also not as international and multi-cultural as commercial aircraft because of technology export sensitivities, which limit the trade of goods and services necessary to develop the aircraft. Commercial space products, such as broadcasting satellites, are certainly complex systems within complex mega-systems too. But the degree of complexity is still far less than for commercial aircraft. They also do not have to comply with the stringent safety requirements requested for commercial aircraft. Safety is an important aspect in the nuclear industry but weight is not a priority there whatsoever. The interface between the structural and systems' functionalities is also less complex compared to aircraft.

Cars provoke a lot of emotions among customers and people, probably even more than aircraft, but they lack the same volumetric complexity. In addition, for a given development project, the financial risk exposure to the car-developing company is relatively small compared to the aircraft company, as there usually are more new car developments in parallel (or almost in parallel). As a result, the development of a new car does not impose a comparable risk to the car company than an aircraft to the aircraft company. In addition, the frequency of new car developments is much higher than for commercial aircraft. Arguably, the same applies for trains and commercial ships. However, unlike for aircraft, other products used for transporting passengers such as cars, trains or ships are not subject to the same intensity of weight reduction measures during their development – justifying only relatively smaller amounts of money spent on them if compared to other important technical principle.

Computers are also generally regarded as high-technology products. But compared to commercial aircraft, they (1) feature significantly less volumetric complexity (a modern aircraft is full of computers but it is still a lot more); (2) the shape of the structural housing of the computer does not really matter except for aspects of aesthetics and ventilation (which means that the interdependence between structural platform and electronic functionalities is not very high); (3) the frequency of new builds is much higher and the financial exposure much smaller. Finally, relatively low-technology products such as washing machines, furniture and so on have no chance to compete with commercial aircraft on the grounds of complexities, financial exposures, weight, safety, and so on.

Commercial aircraft development is unique in the combination of extreme complexity and the need to drive down costs for Management disciplines required to successfully manage it. On the one hand, complexity drives the need for integration of Management disciplines: the more complex a development project is, the higher the need for a common, integrated architecture, on the basis of which Management disciplines can maximise their efficiency and effectiveness. On the other, the strong need to save management costs in commercial projects drives the tendency to concentrate on some essentials of the available Management disciplines. Combining the aspects of integration with the aspect of concentration on Management discipline essentials is therefore the preferred basis for and Management approach to successfully complete a commercial aircraft development project. This is summarised in Figure 1.4.

However, before turning our attention to the more detailed description of this integrative and essential management approach, let us first of all understand how – in brought terms – aircraft development works. This will be explained in two steps: In the subsequent Chapter 2 it will be demonstrated how IT has changed the way aircraft are designed and developed, while in Chapter 3 the different activities, which take place during the various development phases, will be described.

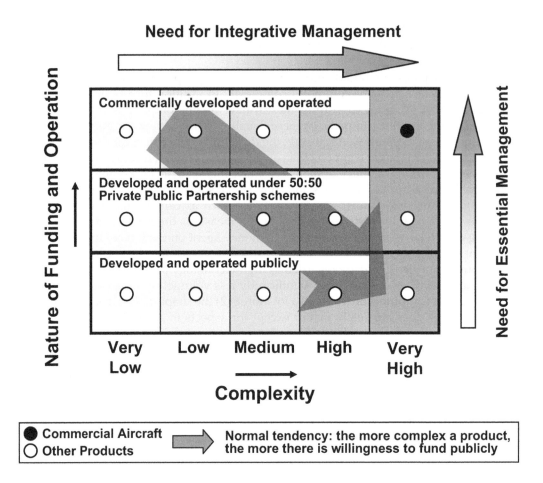

Figure 1.4   Complexity drives the need for integrative management while the commercial nature of a development drives the need for concentration on management essentials (qualitative statement/ figure)

MAIN CONCLUSIONS FROM CHAPTER 1

- Despite some government involvement, modern commercial aircraft development must be regarded as a predominantly commercial venture rather than a governmentally funded one.
- Commercial aircraft developments represent uniquely complex undertakings from a technical, managerial, processes, business and market-situation point of view. They are multi-cultural and top-of-the-learning curve mega-projects exposed to high financial risks and strong public interest.
- There is a fundamental dilemma typical for all commercially funded projects of very high complexity: They require the highest level of Project Management sophistication, yet there is a natural tendency to limit the costs associated with Project Management.
- The necessity to master the extreme complexities associated with commercial aircraft developments drives the need for integration of Project Management disciplines. The strong need to save costs associated with the managerial side of development projects drives the tendency to concentrate on some essentials of such disciplines.
- Therefore, the dilemma can be resolved by concentrating on the *essential* elements of the various Management disciplines necessary to develop complex projects – to keep their principal strengths – and *integrating* them in an intelligent and pragmatic way.

2

How Information Technologies Have Shaped Aircraft Development

INTRODUCTION

Over the last few decades there has been a truly astonishing evolution of Information Technology (IT) systems and tools. Modern tools provide design engineers entirely new ways to approach the design engineering process. They offer techniques which lie far beyond traditional methods and allow engineers much greater freedom to exercise their creativity, offering potentials like:

- high speed in the use of the tool as well as for the communication of data;
- an environment where geographic distances do not matter when exchanging data;
- multi-user functionalities, which allow for an integrated approach;
- solutions for more complex problems, which previously were unavailable; as well as
- three-dimensional geometry visualisation improving problem analyses and communication.

Most industrial sectors have benefited from this, resulting in leaner, more efficient processes. So has the aeronautical sector. Next to improvements in generic business processes like finance, accounting, logistics, supply chain management, and so on, it is primarily in the area of:

- aircraft design;
- factory design;
- parts manufacture;
- marketing and technical publications, as well as
- Project Management;

where improved IT systems have revolutionised the way aircraft are developed.

DIGITAL AIRCRAFT DESIGN

The conceptual and detailed design of aircraft has benefited enormously from what is called Computer Aided Engineering (CAE). It comprises the implementation and application of both, digital *analysis* tools required in the various Engineering disciplines such as aerodynamics (Computational Fluid Dynamics (CFD)) or stress analysis (Finite Element Analysis (FEA))[1] as well as digital *design* tools. For the purpose of this book it is primarily important to look at the evolutions in product *design* tools capabilities rather than in tools for technical *analyses*. This is because there are more intense interfaces between the former on the one hand and Management disciplines on the other. Therefore, while not by any means discrediting the major contributions of digital analyses, the following concentrates on the digital design aspects of aircraft development.

Until the late 1980s aircraft were 'born' on the drawing board, supported by (often full-scale) mock-ups used to correct for the imperfections of managing the enormous engineering complexity of a three-dimensional aircraft with two-dimensional drawings.

> 'Traditionally, new planes had been designed in two dimensions: drawings on paper had been used as a basis for the manufacturing process. But to design a plane entirely this way, with over 100,000 different three-dimensional parts, and then to trust that the two-dimensional drawings had accounted for all the complexities of the three-dimensional airplane would have led to endless unpleasant discoveries at the assembly stage, as a piece designed by one designer arrived at the factory and turned out to be impossible to install because another designer had failed to leave the right amount of space. Furthermore, when it came down to the detail of the plane – the wiring and tubing that ran from one end to the other and required holes to be drilled or cut to allow free passage – the task of accounting for all that in two-dimensional drawings would have been impossible. So the drawings were backed up by what were called mock-ups – successively refined full-scale models of the plane. ... But not good enough. With the inevitable imperfections and the overwhelming complexity of such a hand-crafted object, there were still unpleasant surprises on the shop floor as the first planes were assembled.'[2]

Today, drawings are generated with the help of software tools and computers. Three-dimensional Computer Aided Design (3D CAD) enables the creation of 3D models and derived 2D drawings in a digital format, see Figure 2.1. Drawings in digital format are objects, which offer hugely enlarged potentials for manipulations compared to the old drawings based on blue prints. 3D CAD also provides the basis for photo-realistic visualisation, which is becoming an essential capability for conceptual design (and other applications). It allows engineers to view the products as if they were already produced

---

1   Aircraft require both extremely lightweight constructions and a high safety standard. This in turn requires precise, detailed calculations of structural strengths. The calculations are done using Finite Element Analysis (FEA) defining tensions and deformations in structures under loads. For this purpose, the airframe is divided into a large number of individual elements. The elements are assumed to be connected at node points only. The balance of forces for every element is described as an equation. At node points equal but opposite forces are assumed for the neighbouring elements. The airframe modularised in this way is called the Finite Element Model (FEM), which is the core element of any FEA. The latter results in a set of algebraic equations with a corresponding number of unknowns. The symmetrically linear equation system can be solved using conventional methods. Note, however, that even within the narrow discipline of FEA, there are many specialist disciplines, such as: fatigue analysis, thermal, vibration and magnetic analysis. Plastics, iso-plastics, and composites further complicate the analysis. FEA is performed as part of the standard engineering processes and is one of the more obvious applications to which an existing design is subjected. It reduces significantly the leadtime for the complex iterations needed between stress and design.

2   Sabbagh, K. (1996), *21ˢᵗ Century Jet. The Making of the Boeing 777* (London: MacMillan Publishers), pp. 50–51.

as hardware, and also enables for trying out different variations of the design without the accompanying investment in cost and time that traditional prototyping techniques required. The idea is to completely go away from the generation of 2D drawings, which historically were established as a means of communication between Engineering and Manufacturing. Today, engineers design aircraft parts entirely in 3D resulting in digital data, which can be directly used by Manufacturing.[3]

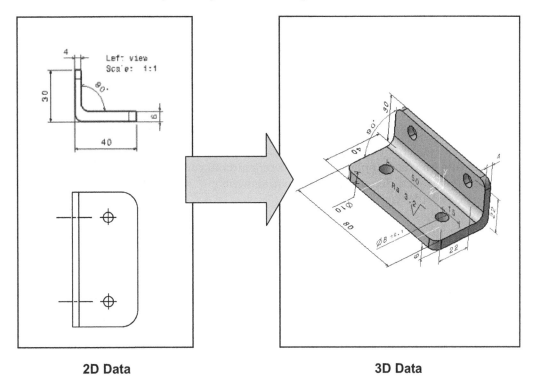

2D Data   **3D Data**

Figure 2.1   **From the drawing board to 3D data**

Source: Airbus, with permission.

Digitally designed workpieces can also be interpreted using what is called Knowledge-based Engineering (KBE). During every new product development, established expertise is combined with new knowledge. KBE supports this process by using established database libraries for norms, tolerances' series and structural elements. In particular, in areas where repetitive structures are to be designed, such as ribs or stringers, KBE is applied with huge success.

'KBE … was used to design and analyze all of the rib feet used in the stretch A340 wing design. The rib feet are flanges used to bolt the wing skin to each rib that runs from the front to the back of the wing and stringer, along its length. Each rib foot is slightly different from its neighbor; using conventional CAD techniques would have taken approximately one full man-year to design and analyze all of the feet. Airbus developed software that itself was able to create the

---

3   There may, however, still be cases where 2D drawings are needed, for example for archiving purposes. In this case, the 2D drawings would come directly off the 3D models.

CAD model needed for the rib feet. The entire rib design for the A340 wing was completed in less than one man-day.'[4]

Modern 3D CAD tools also offer the possibility to automatically adapt parts design to standard parts or user-defined features. For instance, placing a fastener from a catalogue could automatically trigger the creation of the required holes in the joined parts and to position them correctly relative to edges.

In addition to 3D CAD tools, a Product Data Management (PDM) tool is used in aircraft development, which primarily manages:

- the creation and evolution of a product architecture (that is the breakdown, within which the 3D models are organised);
- the digital build-up of an entire aircraft or parts thereof from individual 3D CAD objects;
- the release of 3D objects and associated drawings to Manufacturing;
- the management of change control processes;
- the validation and application of different aircraft configurations (for example, resulting from different customer requirements);
- the access rules for the design teams;
- any confidentiality aspects; as well as
- any design principles.

In most cases, the CAD and PDM tools are, in addition, linked to a vault to manage different models' status such as 'work in progress' or 'released'. However, out of the three of CAD, PDM and vault, the PDM is probably the most important one. It really is the operational and neuralgic engine of aircraft development and definition.

Thanks to the geometric data, which determines the spatial location and orientation of each component in relation to a reference coordinate system, the PDM can compose an entire aircraft from individual 3D CAD component models by positioning them in space. It is therefore possible to view complete sections or zones of the aircraft. This is called the Digital Mock-Up (DMU), see Figure 2.2. It is a very powerful tool used in aircraft development. With the combined strengths of 3D CAD and PDM tools, DMU objects can be aggregated into larger sub-assemblies and assemblies, located on higher and higher levels of the product architecture.

Compared to physical mock-ups, DMUs have the advantage of being remotely consultable. They also can be more easily and quickly updated compared to physical mock-ups. DMUs really are the centre pieces in a concurrent design process to simultaneously study several variants of the same project. Data associated with an object, such as the Bill of Material, masses, weights, moments of inertia and so on can also be managed (aggregated, summed up and so on) in the PDM.

'Boeing started using a computer design system in 1978, for some parts of some aircraft. When they designed a wing strut of the 767 using this system, they found that it halved the number of changes they would have expected to make during manufacture. It was decided that the 777 would be the first plane to be fully built without mock-ups, entirely on the basis of 3D computerized data. Boeing had introduced CATIA, using a Dassault/IBM system,

---

4   Spitz, W., Golaszewski, R., Berardino, F. and Johnson, J. (2001), *Development Cycle Time Simulation for Civil Aircraft*, NASA/CR-2001-210658, p. 3–9.

Figure 2.2   **Using Digital Mock-Ups for presentation and
communication purposes**
Source: Airbus, with permission.

in 1986. Then, when the 777 decision was made, Boeing computing staff devised the add-on program, Electronic Pre-assembly in the CATIA (EPIC), to allow the system to replace mock-ups entirely.'[5]

'In order to be useful for the design of the 777, the CATIA System had to be scaled-up, which was in itself a major engineering effort. Just some numbers help to illustrate the task at hand. Total storage capacity for the overall System reached 3.5 terabytes (the equivalent of 2,500,000 million 3.5-inch high-density disks). … All engineers needed access to all of the computer data. A paperless design meant that instead of waiting for drawings, any engineer working on any part or subassembly could call up all connected parts and subassemblies on any library of the 7,000 workstations that were scattered across 17 time zones. In order to make this possible, Boeing laid dedicated data lines across the Pacific Ocean. About 20 percent of the fuselage structure was being designed and developed by a consortium of Japanese partners including Fuji, Kawasaki and Mitsubishi Heavy Industries. Their engineers had to be logged into the worldwide 777-workstation network.'[6]

There are usually different chapters of the DMU managed by the same PDM, which are used with varying intensity during almost the entire development life-cycle of an aircraft, see Figure 2.3:

- Resulting mainly from the definition of the aerodynamic shape, the Master Geometry represents all the external shapes of the aircraft or sections (wetted surface) and its main geometrical references for frames, rails, ribs, stringers, cut-outs like doors and windows as well as the main interface points.

---

5   Sabbagh, K. (1996), *21ˢᵗ Century Jet. The Making of the Boeing 777* (London: MacMillan Publishers), pp. 55–56.
6   Reprinted by permission of the publisher from: Petroski, H. (1996), *Invention by Design: How Engineers get from Thought to Thing* (Cambridge, MA: Harvard University Press), Copyright ©1996 by Henry Petroski.

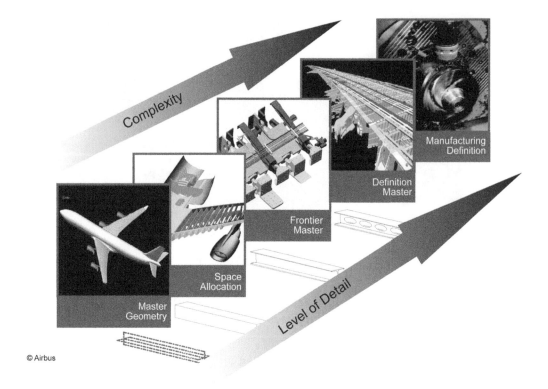

© Airbus

Figure 2.3   Using Digital Mock-Ups for the design and production of commercial aircraft

Source: Airbus, with permission.

The Master Geometry therefore supports the creation, management and interrogation of external surfaces and major structural plane locations. It further includes all coordinate systems necessary to position, for example, a section and it represents a single authoritative source of key data for design, production, and inter-team study work.

For example, the Master Geometry provides the basis for the Stress Design Reference Base (SDRB), which represents the aircraft (and its major assemblies, sub-assemblies and so on) in a schematic way to enable FEA calculations. In a way, the SDRB can be regarded as a tool required for the communication between structure designers and stress engineers.

The Master Geometry is mainly used by Engineering to establish the preliminary definition of the aerostructure and associated design principles. But it is also used by Manufacturing Engineering to have a preliminary definition of the Final Assembly Line (FAL) (operations and building) as well as by Operations engineers to study airport compatibility aspects, that is, the interfaces with airport infrastructures and servicing tasks undertaken at airports.

- The Space Allocation Model (SAM) is composed of 3D objects that initially claim the maximum required space, which physical parts on board the aircraft are expected to eventually consume. They are subsequently refined and detailed to a point where they almost resemble the objects forming the basis for manufacturing (see Definition Master below). The SAM allows to downselect and agree on various configurations as well as to validate the aircraft architecture and industrial processes. Also, maintenance analyses can be performed with the help of the SAM at a very early stage of the development.

Apart from structural design, the SAM is strongly used to lay-out the systems within the aircraft, such as the hydraulics, pneumatics and electrics systems, consisting of equipment items, which are linked by pipes, tubes, harnesses or other. The SAM is therefore of great importance to the system installation engineers performing preliminary installation analyses for their systems and equipment parts. Its purpose is to validate the global aircraft architecture and to serve as a contract between the systems' and structures' assembly specialists. The requirements for the 3D models that make up systems architectures and detailed system/equipment installations are described in documents such as the System Installation Requirement Dossier (SIRD) and the Equipment Installation Requirement Document (EIRD).[7]

- The Frontier Master defines the method of assembling different sections of the aircraft supplied by different manufacturing sites or external companies (such as risk-sharing partners), and to freeze the dimensions and resulting tolerances obtained after joining. It plays an important role in freezing interfaces early in the design process.

- The most detailed DMU is the Definition Master, also called Geometric Reference Model (GRM). It represents all the exact digital 3D repliques (with nominal tolerances) of the parts to be manufactured. Each replique goes through an official 'Release for Production' process and is subject to rigorous configuration control. Once released each replique froms part of the 'Definition Dossier', together with other data such as the Bill of Material, Engineering Change Notes (ECNs), Frontier agreements, stress sheets, standards, specifications and so on. The Definition Dossier represents a basis for aircraft certification, that is, it is used to demonstrate to airworthiness authorities that the design is based on sound assumptions having taken all requirements, formal airworthiness regulations and safety aspects into account.

- Elementary part drawings as well as assembly drawings released from the Definition Master for the purpose of manufacturing are used for machine programming, assembly processes and inspections, thus creating the Manufacturing Definition. The latter contains all data required by Manufacturing to build the aircraft according to the design. This also includes routings, which describe in detail how to prepare, perform and finish the actual assembly or installation on the shop floor.

In addition to the aforementioned major advantages offered by the use of IT systems, there are numerous other advantages in using DMUs during aircraft design and development. For example, where there are supposed to be parts' interfaces (structure/

---

7   The System Installation Requirement Dossier (SIRD) represents the contract between systems architects and system installation designers. It contains segregation rules, installation constraints and maintenance aspects, among others.

    The Equipment Installation Requirement Document (EIRD) describes the maximum space in which the final equipment geometry will have to be inserted. It also generates and provides information on any installation constraints, maintenance aspects and interfaces.

structure, systems/structure or system/system) engineers can check whether interfaces are as they should be. As a Boeing source stated: 'Studies at Boeing show that part interference … and difficulty in properly fitting parts together in aircraft final assembly are the most pervasive problems in manufacturing airplanes'.[8]

> 'We've had a situation in the wing trailing edge where all the simple interferences were worked. We had a torque tube that fitted within the wing trailing edge and everything was OK. And then we found out through further analysis that we couldn't get the torque tube in, and we couldn't get it out. So we've had to go back and look at the swept volume – the space that this torque tube needs around it so that it can be removed. It's really saved the company much money. Usually you don't find these things out until the first airplanes are built or till somebody tries to service them.'[9]

Interference checking using the DMU during the space analysis process can today be semi-automated using established knowledge rules. By taking into account specific clearance requirements, tolerated contacts or even limited overlaps, software tools can reduce significantly the time required to review interference and other packaging issues.

Any component can be viewed on the computer screen without having to climb around a large wooden mock-up. Pictures of the digital aircraft can be extracted and can, for example, be sent to distant teams not directly linked to the DMU, thus greatly easing communication. The fact that interfaces can be validated with the help of the DMU prior to release of drawings considerably reduces the amount of rework when the actual assembly and systems installation, respectively, take place.[10]

However, what remains challenging is the management of tolerances of large assemblies in the DMU. A digital component model can, of course, be attributed with data describing acceptable tolerances to be respected by Manufacturing. But in the case of larger assemblies components may clash because of (cumulated) tolerance overlaps. On the shop floor, clashes could occur despite individual components being manufactured within their respective tolerance bands. In these cases DMUs still are not able to provide a warning signal to the design engineers. As aircraft are composed of large assemblies, this is a particular problem for the development of aircraft compared to smaller products.

Similarly, man/machine interfaces can be analysed during dedicated DMU-based ergonomics studies, investigating any area in the aircraft where people will be working, such as cockpit or crew rest compartments.

Another key application of the DMU is to confirm that unintended interferences do indeed not occur. This is called clash detection. There are clash detection analyses for static cases as well as kinematic cases. The latter include, for example, interferences between the fuselage skin and the inner flap during extraction as well as the landing gear deployment and retraction kinematics. As part of what is called Particular Risk Analysis (PRA), it can also be checked which parts of the aircraft would be affected by potentially hazardous bursts of engine rotors or tyres. This gives the designers a chance to re-design or better protect individual zones of the aircraft.

---

8   http://www.boeing.com/commercial/777family/pf/pf_computing.html, accessed 28 April 2004.
9   Steve Johnson, one of the designers responsible for parts of the 777 wing, mentioned in: Sabbagh, K. (1996), *21st Century Jet. The Making of the Boeing 777* (London: MacMillan Publishers), pp. 57–58.
10 There is a lot more to releasing drawings than just the checking of interfaces. Interference analysis, structure design, mass properties, adherence to safety and/or corporate standards and imposition of local codes and regulations are additional requirements for a design to be accepted. Design generally must pass these analyses before it can be considered for manufacturing.

In summary, designing in 3D is not only about managing pure geometries. It is mainly the way to ensure that parts are designed in a holistic approach, which integrates and manages functional requirements and constraints (such as interfaces, holes, clearances, parametric definitions). Clearly, this can be envisaged for all product or part families of an aircraft, such as sheet metal, machined parts, composites, piping, tubing and electricity.

DIGITAL MANUFACTURING

Manufacturing Simulation: The Digital Factory

When Manufacturing Simulation and Manufacturing Preparation – two of the three Manufacturing Engineering disciplines[11] – started introducing modern IT tools to allow for virtual Manufacturing, they boosted their capabilities in as much the same way as Design. For Manufacturing Simulation purposes, the main role of a modern product DMU is to support communication and to aid visualisation of the aircraft components in order to:

- have an awareness of spatial constraints of aircraft components and assemblies;
- view dynamically a given configuration of the aircraft components and assemblies;
- view clashes between the aircraft and industrial facilities;
- evaluate technical solutions for manufacturing operations;
- check the feasibility of assembly and disassembly (repair);
- check accessibilities for operators and tools; as well as to
- check zone inspection feasibilities.

However, where aircraft DMUs reach their limits, CAD-based 3D manufacturing simulation tools can extend their capabilities to analyse in more detail the interactions between the product (the entire aircraft or smaller sections of it) on the one hand and industrial facilities on the other. With the combined support of the aircraft DMU and any additional 3D models of a planned assembly site (such as the FAL), a Digital Factory can be created. Figure 2.4 provides an example. A Digital Factory allows analysis of the different assembly scenarios, logistics dynamics and complex kinematics required, for example, for the transport of aircraft sections within the factory. This helps significantly to optimise factory layout as well as assembly processes, thus reducing leadtimes and cost.

In particular, jigs & tools concepts can be studied in 3D. This includes validating compatibility of the jigs & tools' interfaces with the corresponding aircraft components. It also includes the associated jigs & tools kinematics and operations, leading to an optimised usage of resources and capacity. When considering space and accessibility aspects, aircraft DMUs can also be used to analyse the aspect of producibility of sections using large and complex Numerically Controlled (NC) machines. Of particular interest are questions of process improvements in an attempt to reduce production costs. Finally, IT also offers the possibility to design Manufacturing facilities and jigs & tools concurrently with the product design, thus, significantly reducing development leadtimes.

---

11 The third Manufacturing Engineering (ME) discipline, ME Operations, will be discussed later. Its main tasks are to prepare Manufacturing work by issuing work orders and to provide production support to Engineering during Series Production.

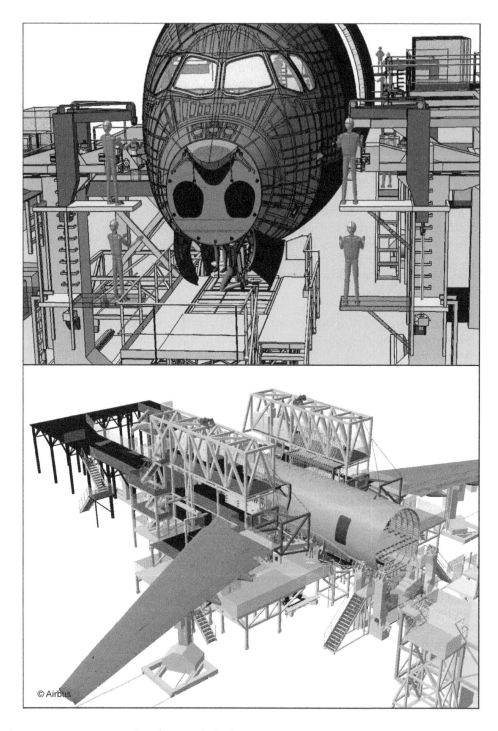

Figure 2.4 Examples for a Digital Factory
Top: Jigs & Tools for Section Build-up
Bottom: Final Assembly

Source: Airbus, with permission.

As all these simulations help define the complete manufacturing and assembly process of the aircraft. It is therefore possible to establish the classical Manufacturing Plan much earlier than in the past. This in turn allows engineers to shape the product architecture in the PDM in a way, which takes the manufacturing and assembly processes into account. In Chapter 11 we will come back to this important point.

Manufacturing Preparation: Computer Aided Manufacturing

Manufacturing Preparation, the second discipline of Manufacturing Engineering, uses Computer Aided Manufacturing (CAM) tools and techniques to simulate and implement aspects such as pattern nesting,[12] tool design, fixture design, sheet metal development, manufacturing quality control analyses, training as well as the actual NC programming itself. With regards to the latter, the DMU can be used once again: the basic process of programming NC machines and robotics software starts with the reading of the workpiece geometry data directly off the DMU.

> 'With the EPIC system [used by Boeing for the 777 programme], there could now be a direct link between the computer description of the design of a component and the instructions that a machine tool would need in order to make it. ... For the first time in a Boeing plane, there was to be a direct – electronic – connection between the decisions made by the designers about the dimensions of a component and the data that the tool would need in order to put the pieces together.'[13]

The complete product process from first pre-design concepts in the design office to detailed programming of the numerical controlled machine for the manufacture of basic parts can nowadays be supported digitally. An example of how this can be done is provided in Figure 2.5.

However, for CAM to really be supporting Manufacturing Preparation, manufacturing aspects must be allowed to influence the design of components and assemblies. Examples for this include:

- the sequencing of machining operations, ensuring standardisation of methods and techniques as defined by the company know-how;
- location of clamps to hold parts while they are machined;
- machine-to-fit tolerances given the practical availability of real machine tools;
- the validated toolpaths; as well as
- recognition of user-defined feature types and attributes and creation of rules based on them – to automate operations on these or associated parts.

Without taking manufacturing aspects into account during design, released 3D models and drawings are often impractical to manufacture. Setup costs, non-compliance with existing manufacturing methods or excessively complicated operations may preclude the consideration of an otherwise good design, causing that design to be modified.

---

12 Nesting uses algorithms to optimise the use of raw material. For example, in the case of parts to be cut from sheet metal, nesting uses algorithms to position part geometries on the metal sheet in such a way, that material waste at the end of the production process is minimal. See also Figure 2.5.
13 Sabbagh, K. (1996), *21st Century Jet. The Making of the Boeing 777* (London: MacMillan Publishers), p. 59.

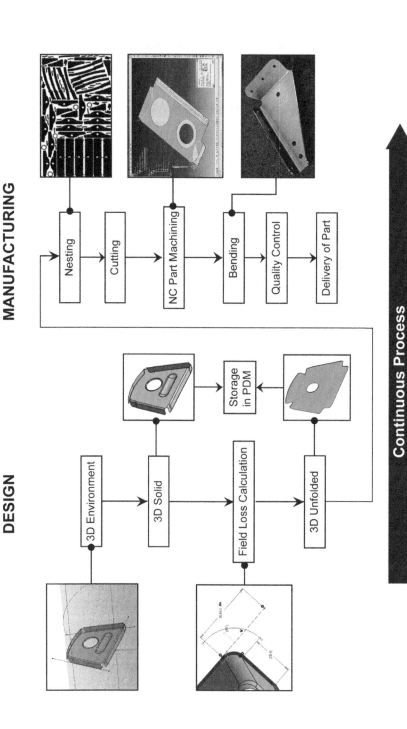

Figure 2.5   Using digital data for producing sheet metal parts.
Note that both, the 3D model for the '3D Solid' as well as for the '3D unfolded' are stored in the
Product Data Management (PDM)

Source: Airbus, with permission.

However, before machines can actually be programmed, they still require some degree of user interaction to enter technological data such as feed, cutting speed, way of clamping and so on. In particular where tolerances must be considered, the interaction of the human expert is necessary. Determination of these process parameters is largely based on empirical knowledge. Attaining specific shape and position tolerances is in most cases not predictable and the selection of cutting parameters will often be based on recommendations of the cutting tool manufacturers. Thus, initial NC programming is still a semiautomatic production process and the use of geometry data is often only the simplest aspect of the NC process after all.

After the clamping situations have been examined and applicable technologies determined, the final step in the NC programming process consists of generating the NC code required for manufacturing of the part. It defines the toolpath whereby the translation of the geometric toolpath descriptions into a language, which each machine tool understands, is an art in itself. With the simulation of the toolpath one can see the form of the finished part that will come out of the machining process, thus allowing to correct early enough any problems which may arise.

Manufacturing Operations Benefiting from the Digital Mock-Up

Manufacturing Operations can now compare the geometries of a part as determined by the DMU with the actually measured dimensions of the machined part. This can be done online so that concessions[14] can be recorded while the quality check is going on. Similarly, actual versus DMU comparisons can be done for quality checks on assembled aircraft sections.

The DMU can also be used for training shop floor staff prior to start of assembly in an attempt to gain an in-depth understanding of assembly sequences. This speeds up the staff's learning curve, which in turn allows for a steeper industrial ramp-up. If technical solutions vary from one aircraft to another (for example as a result of customisation), the DMU enables the shop floor to view the applicable configuration well in advance.

However, where the DMU is intended to be used in Manufacturing Operations to provide to the shop floor virtual images illustrating the production sequence, there may still be some challenges preventing full implementation on the shop floor. Examples include the following:

- For aircraft development, the DMU usually represents the aircraft in its final condition and in flight. Individual assembly steps to come to that final condition are not represented. The DMU may, for example, only show images of fully installed harnesses. If so, then in cases where fully equipped fuselage sections need to be transported to the FAL with the end of the harnesses temporarily hanging loose (as they are not yet connected to the corresponding harnesses of the adjacent section), a DMU image of the fuselage section would not show the correct condition of supply to the FAL.
- If a delayed component needs to be installed in the aircraft at a different Assembly Stage (AS) during the production sequence than originally foreseen, the DMU

---

14 A concession is a documented deviation from the intended aircraft configuration, for example, as a result of errors in Engineering or Manufacturing, with a description of the implemented repair solution, if any.

image illustrating the production status at that stage would usually not show the missing part. It would assume it to have been installed at the foreseen stage, unless the IT system, which is managing missing parts, would be linked to the DMU. However, this is beyond the capability of most companies today.

There is still a long way to go before shop floor staff will be able to load data on their monitors in order to get a virtual image of the *actual* 'as-is' production status.

OTHER DISCIPLINES BENEFITING FROM THE DIGITAL MOCK-UP

Other company disciplines also take advantage of the capabilities offered by the DMU. Marketing, for example, gets a huge push as DMUs can be used to support cabin layout studies. Early visibility inspires customers to work 'through their cabin'. This motivates them, stimulates their creativity and binds them closer to the aircraft under development. In addition, the leadtimes required to respond to customers requesting additional systems or equipment to be installed in the aircraft can be reduced. This is because interfaces of the new elements with the existing aircraft structure or systems can be checked well in advance, thus avoiding downstream problems.

Maintenance and repair aspects of the aircraft, such as parts exchange, can also be simulated using the DMU. This helps to verify inspection requirements as well as accessibility options for maintenance staff, see Figure 2.6. In addition, maintenance process sequences can be determined. Finally, the customers' future maintenance costs can be optimised during the early design phases, and any weaknesses with regard to maintainability and repairability of the aircraft can be identified and removed.

Interfaces of the aircraft with on-ground infrastructure (such as airport terminal gates as well as ground handling and servicing vehicles for cargo loading, catering, cleaning and de-icing) can also be analysed. This leads to an optimisation of the aircraft turn-around time, which allows the aircraft operator to reduce cost. At the same time, it also increases operator revenues as the aircraft can be better utilised.

Illustrations needed for documents and reports to customers, airworthiness authorities and others can be directly extracted from the DMU. Typically, this is done in support of the Illustrated Parts Catalogue (IPC), the Aircraft Maintenance Manual (AMM), the Structural Repair Manual (SRM) and many other technical publications, leading to an improved and earlier understanding of potential issues. Among other, this also contributes to a reduction of Aircraft on Ground (AOG) times, where an aircraft is not authorised to take-off due to technical problems.

Finally, the customer can be provided with a CD-ROM containing the complete DMU of the aircraft, showing its exact configuration – even if this deviates from the target configuration. In case of problems, this allows the customer to 'dive' directly to the problem area of the aircraft and to have a better understanding of what might have gone wrong.

Figure 2.6   **Using digital data to check maintenance processes
Here: operator in clash with ducts and pipes**
Source: Airbus, with permission.

PROJECT MANAGEMENT INFORMATION TECHNOLOGY TOOLS

Modern project management tools represent significant advances to project control. Solutions bring together in a single environment all relevant project and resource data for display, modification, processing and reporting. This applies equally to structured data (dates, resources, costs and so on) as well as unstructured data (documents, e-mails, reminders and so on).

Today it is possible to combine multi-user project planning, resource allocation and scheduling, cost control and graphical reporting capabilities, thereby enabling more effective planning and scheduling of core resources. It is also possible to maintain in parallel different versions of the same project plan, such as the baseline plan for a reference purpose, the current status version, versions used for 'what-if' analyses and so on. For large projects like a commercial aircraft development it is also possible to establish and maintain inter-project relationships to control the interdependencies between, say, fuselage and wing.

Tools are centred around a Work Breakdown Structure (WBS), Cost Breakdown Structure (CBS) and an Organisation Breakdown Structure (OBS). These are different architectures than the one used by the PDM, but are also structures traditionally used during aircraft development. In Chapter 11 a suggestion will be made about how to integrate these architectures into a single one.

A major element of any project control tool is the time recording system. Individuals working on projects book the times spent on the project against project codes. Today's tools are effective web-based or client server timesheet systems designed to provide a comprehensive understanding of project and resource activities across all collaborating units. Office-based users (for example via a Local Area Network) and remote users (via Internet or intranet) can access, edit, approve and update their timesheets from across the hall or around the world. This helps to collect timely and accurate activity information, thereby offering increased visibility of project, resource and schedule performance. Information can be consolidated and reported in a meaningful and organised manner. This offers the means needed by Project Managers to understand where resources are being assigned, determines the level of project and resource efficiency, and ensures that expended effort is aligned with overall business objectives. Managers can set realistic project goals by identifying key resources and activities, updating project performance and assessing an accurate view of staff progress and performance.

In order to analyse project status, modern tools allow for easy and fully graphics-supported navigating, analysing and reporting on project cost, resource and schedule information. Individuals throughout the organisation can consolidate multiple project data and can easily drill-down through relevant data dimensions to make informed business decisions. Also, database querying became significantly faster, simpler and more effective over the past years.

FINAL REMARKS

IT tools have changed dramatically the way aircraft are designed, developed, built and project managed. With the help of web-browsers it is today possible to link collaborating units or individuals from all over the world together in real time. Data from the DMU, Finite Element Model (FEM), integrated schedule plans, resource information and so on can be exchanged online, and frozen snapshots of this data can be exchanged quickly and without geographic limits using the Internet, intranets and e-mails.

As a result, IT also offers the chance to involve customers as well as suppliers very early in the design phase and in much more depth than traditionally possible. Interfaces with supplier parts can be managed and checked, supplier know-how can be integrated into the contractor's design where deemed advantageous, supplied equipment can be easily incorporated digitally to evaluate fit, form and function, and supplier schedule plans can be integrated into the aircraft developer's plan. Customers can 'see' the product before they buy it, and all this significantly before release of drawings to Manufacturing. Thanks to IT, individual aircraft development processes have become faster and the first aircraft built during a development project is generally more mature compared to the past.

But above all, development processes became considerably more integrated and concurrent. Functions like Engineering, Manufacturing, Procurement, Marketing, Quality, Product Support and so on can all use the same information. The existence of a DMU has, in particular, vastly improved communication between Design and Manufacturing Engineering (ME). Any design changes can be rapidly communicated to the ME department. Any issues with the production process or tooling can be highlighted and resolved earlier, thus mitigating risks, which otherwise could materialise at later stages of the aircraft programme. Functions can feed their requirements into the design – thus influencing an area, which traditionally was an Engineering domain only – while having constant visibility on the design evolutions. Functions can concurrently develop the product definition, particularly structures, on-board systems, overall aircraft analyses as well as manufacturing plans and tool designs.

> 'Today's full capabilities of design and development IT tools, packaged as what is called a Product Lifecycle Management (PLM) software, was introduced by Boeing for the development of the 787. It included modules dedicated to life-cycle application, collaborative data management as well as digital manufacturing. [Boeing] wants not only a virtual prototype, but a virtual rollout of the aircraft – to have it digitally designed, digitally simulated, digitally produced – before any physical parts are made. … PLM will provide a fully distributed, worldwide collaborative workspace integrating … [787] partners into a single, seamless community – a requirement dictated by globalisation that increasingly drives the aerospace industry. …
>   One aspect of the PLM approach is that a part can be designed in the context of how it relates to other parts, subsystems and systems around it, so that if a change is made, the surrounding parts and subsystems can automatically adjust for it, avoiding the possibility of a collision during production and virtually eliminating rework costs. … Another key to the concept is that it ensures not only spatial but temporal control of the product cycle. It covers not only the part, but the materials, tooling and processes to make it, the systems necessary to test it, the space necessary to install it and the manuals to operate and maintain it. …
>   It gives … the entire movie of everything [in the manufacturing process] even before the physical job has started. Instead of getting a picture of a particular airplane, with all its parts, PLM generates a series of pictures of all stages of that airplane throughout its life – from conception, detail design and parts production through final assembly, testing and operation. … This allows Engineering, Manufacturing and Maintenance bills of material to be done in parallel rather than in sequence, speeding up the process and giving engineers full control of change and configuration management. … Later, it enables mechanics on the airline shop floor to access the precise tasks they need online and execute them accurately, minimizing computer capacity requirement and reducing errors. … For the first time, all design, production, maintenance and other relevant information are accommodated from day one in a single database … and the up-to-date configuration known and controlled at all times, in real time.'[15]

However, digitally supported development projects require drastically improved levels of direct communication between people working on the project in order to ensure *faster* communication, yet one which concentrates on the exchange of *relevant information*. This may look like a paradox but is due to the fact that, for example, digitally managed design changes can be implemented so much quicker compared to some 20 years ago. Significantly improved ways of communication are therefore a mandatory prerequisite for complex projects seeking rapid product development. The problem here is not so much the speed of sending information, but the rate at which people can absorb it.

---

15 Hughes, D. and Taverna, M.A. (2004), 'Expanding the Digital Envelope', *Aviation Week & Space Technology* 10 May 2004, pp. 50–52.

Unfortunately, different IT tools for different applications, which have been developed over the years, are often not compatible with each other. Their capabilities to interoperate are sometimes poor. This only too often generates significant additional manual work to transfer data from one tool to another compared to the ideal situation of full interoperability between tools. As this is usually not a one-off exercise but often needs to be done on a regular basis, the huge advantages IT potentially offers are somewhat offset by increased manual efforts. This is in particular a problem for the relationship between the aircraft-developing company and its suppliers, which very often use different tools.

Also, the demand for IT hardware capacity and speed on computers and data highways seems always to be bigger than what is available. It still takes too long to work concurrently with a full DMU in all its details. While hardware continues to improve, advances have also been made to pin-down the real needs of DMU users. The latter usually only need a detailed DMU in some aircraft zones at a given point in time but not in others. Where details are not needed, algorithms have been developed to approximate the DMU zones by thousands of simple shapes such as triangles, which can be calculated much faster. But the overall situation is still far from being satisfying.

Finally, over-reliance on computers and digital design processes may increase the risk of error as classical cross-checks are no longer undertaken at the same level of thoroughness if compared to pre-digital times. Some caution needs to be applied here.

> 'May be computer usage has passed the point of enhancing productivity. How else can we explain why the fully computerised 777 ... took longer to develop than the 747.'[16]

However, for commercial aircraft development, full exploitation of today's capabilities of IT is about the only way to deal satisfactorily with the complexities embedded in such a project. Given this, the fundamental question is then whether a company's organisation, processes and management cultures are adequately set up to provide the best possible support to exploit all the capabilities of modern IT environment. After all, the real challenge of applying computing tools is in the cultural and organisational changes that must come along in order to take advantage of IT supported processes.

How should modern commercial aircraft development be managed and organised to fully exploit the advantages offered by modern IT? What would the required architectures be looking like to integrate DMUs, teams, schedule plans and so on? Is there an architecture, which could serve as a 'communication bus' to which individual IT tools can dock on so that interoperability can be ensured? However, before providing some answers to these questions, it is helpful to first understand how a generic aircraft development process looks like. Experienced readers may wish to proceed directly to Chapter 4.

---

16 Runkel, M.A. (2009), '747 was the Mold-Breaker', *Aviation Week & Space Technology* 10 August 2009, p. 8.

MAIN CONCLUSIONS FROM CHAPTER 2

- During the last two decades or so, information technology tools have dramatically changed the way aircraft are designed, developed, built and project managed.
- Thanks to information technology, aircraft development processes became considerably more integrated and concurrent, which in particular has helped to bridge the classical gap between Engineering and Manufacturing.
- Information technology also offers the chance to involve customers and suppliers much earlier in the design process and involve them much more deeply than traditionally possible.
- For a commercial aircraft development, full exploitation of today's capabilities of information technology is about the only way to satisfactorily manage the complexities embedded in such a project.
- However, to fully exploit the advantages of information technology, there are cultural and organisational changes, which must come along. After all, this is the real challenge of applying computing tools.

3

Developing a Commercial Airplane

THE MAJOR MILESTONES

The typical life-cycle of a commercial aircraft programme consists of individual phases, from Research, Development, Production and Operation until Retirement and Disposal, see Figure 3.1. Each of these phases features its individual activities, challenges and objectives. For example, during the Production phase the major challenges centre on ensuring the guaranteed high quality of the product, on-time delivery as well as continuous cost reduction. For the Operations phase, during which the aircraft will be operated by the customer, the major challenges lie in ensuring lowest possible operations costs and highest possible dispatch reliability as well as maintaining a high asset value.

The Development phase, however, is somewhat special as it consists of distinctively different activities compared to other phases. During this phase:

- Engineering plays a major role;
- compared to earlier phases, many more people are involved;
- activities are highly interlinked;
- the complexity of interdependencies between different skillsets is extremely high;
- various iterative steps and design loops have to be performed; and
- the final outcome of the phase is not exactly known at its beginning.

The time for the Development phase until delivery of the first customer aircraft typically spans from six to eight years for entirely new designs, whereas for derivative products it spans from 28 to 40 months.

Because of its complexity, the Development phase – starting from the product idea and ending with the termination of any kind of design modifications long after the aircraft's Entry into Service (EIS) – is broken down into sequential sub-phases, following a process called Phased Project Planning (PPP). In this process a project sequentially passes through checkpoint milestones to ensure that all the deliverables required at that milestone are available and at the right level of maturity. However, in reality sub-phases do overlap as a result of Concurrent Engineering (to be introduced below) and other Design methodologies.

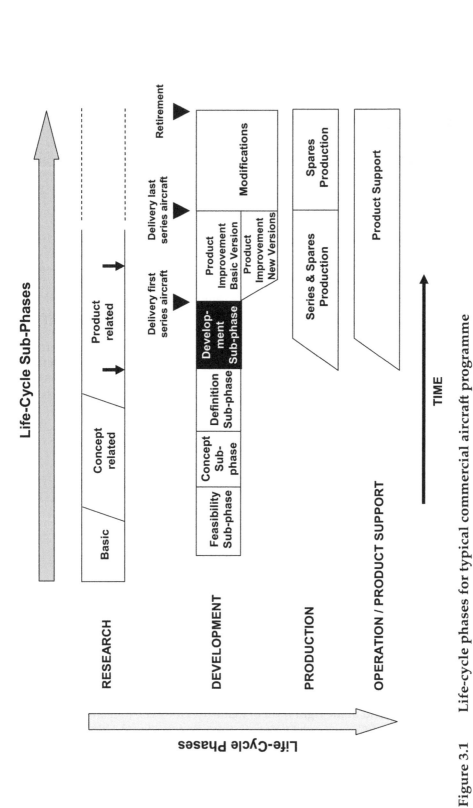

Figure 3.1   Life-cycle phases for typical commercial aircraft programme
This book mainly deals with the management challenges related to the Development sub-phase.

Based on: Schmitt, D. (1999), 'A New Perspective on Aeronautics R&T for Europe'; in: Lawrence, P.K. and Braddon, D.L. (eds) (1999), *Strategic Issues in European Aerospace* (Aldershot: Ashgate Publishing), p. 214, with permission.

Different breakdowns for the Development phase have been proposed and introduced. Typical breakdowns for Airbus and Boeing can be seen in Figure 3.2.[1] Airbus, for example, called for the five sub-phases Feasibility, Concept, Definition, Development and Post Entry-into-Service for the development of the A380. For its 777 development, Boeing called its phases Concept Definition, Configuration Definition, Product Definition, Freeze Period and Production. Note that the terminology 'Development' can encompass an entire total life-cycle phase as well as a sub-phase (see Figure 3.1 and Figure 3.2). Sub-phases might be further broken down into sub-sub-phases. However, hereafter a phase or sub-phase or sub-sub-phase shall always be called 'phase' for reasons of simplicity, except where a differentiation is regarded necessary for the understanding of the context.

However, whatever the terminologies involved or the level of phases under consideration, if applying PPP a phase will only be regarded as completed if the associated termination milestone is achieved. In fact, the key principle of phased developments is that no new Development phase should be initiated until the previous one is complete. In theory, thus, the aircraft development will not proceed at all beyond the end of any such phase if the conditions to complete the corresponding milestone are not met, that is, if the milestone is missed.

In reality, though, it still will proceed as a result of the pressures emanating from the necessity to protect the programme.[2] Thus, in case deliverables are not available or are not at the expected level of maturity at that milestone, the checkpoint cannot be declared as passed. But the project proceeds and continues towards the next checkpoint, while declaration of former checkpoint milestone achievement has to wait until all issues are tackled and solved. While already commencing with a new phase, incomplete activities will therefore have to be completed in parallel and as soon as possible. From a Project Management perspective, this represents a significant additional challenge.

It is important to note, that with the advent of Concurrent Engineering technologies to be described in Chapter 4 and 14, respectively, the application of PPP, in particular the stringent breakdown of the Development phase into smaller periods, is no longer adequate. If therefore the PPP philosophy is still applied, it must be regarded as a way to focus the minds of the involved people towards the achievement of checkpoint milestones. It no longer does describe development reality as a strict series of sequential process' steps separated by Go/No-Go decision points. Keeping this in mind, in this book we will apply a simplified PPP breakdown, which is also represented in Figure 3.2.

There are, however, three milestones which are fundamental and which could indeed bring the development to a halt if milestone objectives or deliverables are missed. The first one is the Authorisation to Offer (ATO) milestone. It is at this point where the company and its shareholders allow the team in charge of the aircraft development to offer the aircraft to potential customers. In order to be able to do so, the aircraft design must have reached a level of maturity which allows performance values to be guaranteed to the potential customer, such as range, weights, payload capacity, noise levels, turn-around times and operating costs.

---

1   For Airbus data see again Schmitt, D. (1999), 'A New Perspective on Aeronautics R&T for Europe'; in: Lawrence, P.K. and Braddon, D.L. (eds) (1999), *Strategic Issues in European Aerospace* (Aldershot: Ashgate Publishing), p. 214. For a detailed description of the Boeing 777 development cycle, see Spitz, W., Golaszewski, R., Berardino, F. and Johnson, J. (2001), *Development Cycle Time Simulation for Civil Aircraft*, NASA/CR-2001-210658, pp. 3–13.

2   More about these pressures will be explained in subsequent chapters.

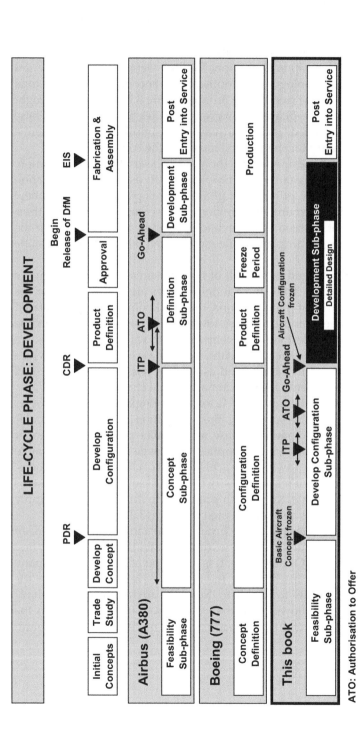

LIFE-CYCLE PHASE: DEVELOPMENT

	PDR		CDR		Begin Release of DfM	EIS
Initial Concepts	Trade Study	Develop Concept	Develop Configuration	Product Definition	Approval	Fabrication & Assembly

Airbus (A380)

| | | | ITP | ATO | Go-Ahead | |
| Feasibility Sub-phase | Concept Sub-phase | | | Definition Sub-phase | Development Sub-phase | Post Entry into Service |

Boeing (777)

| Concept Definition | Configuration Definition | ITP ATO Go-Ahead Develop Configuration Sub-phase | Product Definition | Freeze Period | Production |

This book

Basic Aircraft Concept frozen

Aircraft Configuration frozen

| Feasibility Sub-phase | ITP ATO Go-Ahead Develop Configuration Sub-phase | Development Sub-phase — Detailed Design | Post Entry into Service |

TIME →

Note: length of bars not representative for periods of time

ATO: Authorisation to Offer
CDR: Critical Design Review
DfM: Data for Manufacture
EIS: Entry into Service
ITP: Instruction to Proceed
PDR: Preliminary Design Review

Figure 3.2   Phased Project Planning used for Development phase for commercial aircraft

Note that ITP and ATP are commercial decision milestones, which are often driven by market opportunities and therefore not necessarily linked to the status of technical milestones. This is why ITP and ATP milestones can be regarded as moveable milestones. This book mainly deals with the management challenges related to the Development sub-phase.

Based on following sources: for Airbus: Schmitt, D. (1999), 'A New Perspective on Aeronautics R&T for Europe'; in: Lawrence, P.K. and Braddon, D.L. (eds) (1999), Strategic Issues in European Aerospace (Aldershot: Ashgate Publishing), p. 214, with permission; for Boeing: Breuhaus, R.S., Fowler, K.R. and Zanatta, J.J. (1996),

Very important too, at this point in time, is the company's capability to explain in detail the cabin layout, as well as the potential it offers to customers to differentiate themselves against competitors. The team should also be able to inform potential customers about the aircraft's systems and their performances. In addition, the price and cost of the aircraft must be known, that is, the Business Case for the aircraft programme must be very mature. All potential showstoppers, which could bring about a high risk to the aircraft ever coming into existence, must have been removed. This includes risks associated with the application of new technologies as well as any incompatibilities with existing aircraft, airports or airworthiness requirements.

Between ATO and the next fundamental milestone called 'Go-Ahead', Marketing is playing a very active roll in trying to offer the aircraft to customers. But one needs to take care that this phase does not already implement some first problematic seeds into the programme: Marketing may well be tempted to sell a better aircraft performance to the customers compared to what can in the end be achieved. A closed feedback loop between Marketing and Engineering must be secured during this period.

If then sufficient customers order the aircraft on the basis of what is known about its performances and features, that is, if the aircraft programme's Business Case had a good start, then the official 'Go-Ahead' could be given. However, usually the 'Go-Ahead' will not be given on the basis of a number of aircraft ordered alone, but the ordering airlines need to have the right reputation, financial strength and geographic distribution too. The rationale for this is: if 'high calibre' airlines in North-America, Europe and Asia order the aircraft than others are more likely to follow.

In a way, ATO and 'Go-Ahead' are very much the same kind of milestone from an Engineering perspective (although, of course, the time lag between the two milestones is used to further define the aircraft, for example by incorporating more customer inputs). But at 'Go-Ahead', the company has a lot more confidence in the success of the aircraft programme as the market has responded positively. It can be certain now that the product meets market demand and that the expected EIS complies well with the customers' own business planning to acquire new aircraft.

With the 'Go-Ahead' the company and its shareholders express their confidence in the Business Case and are prepared to accept the (financial) risks associated with it. The team in charge of the aircraft development has successfully convinced potential customers, shareholders and other providers of financial means as well as its own top Management that the new product will be an interesting case for business. It can be proud about this achievement. 'Go-Ahead', which comes about six to 15 months after ATO,[3] kicks off the Development sub-phase.

A third critical milestone is the granting of the Type Certification by the airworthiness authorities just prior to EIS. No customer would accept an uncertified aircraft. Passing this milestone smoothly requires a tremendous amount of preparatory work throughout the development. To ensure this work is always on track, it is good practice to involve authorities in the design and during the development from very early stages onwards.

---

3   Schmitt, D. (1999), 'A New Perspective on Aeronautics R&T for Europe'; in: Lawrence, P.K. and Braddon, D.L. (eds) (1999), *Strategic Issues in European Aerospace* (Aldershot: Ashgate Publishing), p. 223.

THE PRE-AUTHORISATION TO OFFER ACTIVITIES

Achieving the Basic Aircraft Concept

At start of the aircraft development, Marketing intensifies its market research in order to analyse the needs of the market in the targeted aircraft payload/range segment. Based on these activities, it derives some first top-level requirements for the future aircraft, such as:

- range;
- speed;
- number of passengers;
- direct operating costs;
- turn-around times;
- legal requirements (such as allowable noise levels);
- cargo capacity; as well as
- reliability.

Engineers (for example from the Future Projects department) will then conceive a first set of design concepts, which look promising to fulfil the market needs. Should the aircraft have two, three or four engines? Should it have one deck or two? What type of cargo containers should it be capable of carrying and how many? Should it be an overwing aircraft where the fuselage is above the wing or an underwing? What would be the shape of the empennage? It is this kind of questions an aircraft design concept should be capable of answering.

In order to be able to eventually narrow down the list of alternative concepts, a profound understanding of the relationships between the following design areas must be developed:

- basic geometry;
- aerodynamic and other loads data;
- airframe structure;
- systems;
- landing gear; as well as the
- flight control units interfering with fuselage and wing.

Also some understanding of the design principles[4] for joints as well as manufacturing and repair principles need to be established.

---

4   Design principles are established in order to:

- standardise design solutions throughout the aircraft;
- harmonise interfaces;
- formalise, present and demonstrate technical solutions;
- share the design rationale;
- provide a basis in order to exchange and capitalise knowledge among all stakeholders;
- provide directives or constraints needed to model the elementary parts in solid 3D.

Design principles for in-house use often include the rationale for the principles, while design principles provided to supplies only contain rules and recommendations which must be respected by the suppliers, but no rationale.

This necessitates the generation of 2D and 3D schemes[5] to develop and understand the various constraints imposed upon the design by these options. As the down-selection process will not be completed at this stage, several concepts will be required to ascertain their viability. Consequently, drawings are initially produced for all potential concepts. These drawings, called General Arrangements, indicate the size, the outline and the position of all major external components including fuselage, wing, tail, powerplants and landing gear, see Figure 3.3.

Tools to facilitate the concept down-selection process include the Master Geometry (MG) and Space Allocation Model (SAM) introduced in Chapter 2.

As far as aircraft systems are concerned at this stage, drawings and models are generated to show the functions of, for example, hydraulics, pneumatics, fuel and other systems in a diagrammatic form. They are drawn primarily for the use by designers and operators, which deal with aircraft operations.

Eventually, an aircraft concept is chosen which meets the overall aircraft requirements.[6] At this stage none of the detailed aircraft geometries are frozen but the overall layout of the aircraft – that is, its structural and systems architectures (such as under-fuselage wing, T-tail and so on) – as well as its principal mold line would no longer be expected to change considerably. Thus, there is now a basic aircraft concept available. The number of passengers for the aircraft, its Maximum Weight Empty (MWE), Maximum Take-Off Weight (MTOW), cruise altitude and Mach-number as well as the principal cabin layout, among others, are all defined now. This concept is described and documented in the Data Basis for Design (DBD) document. It provides evidence that the major market requirements identified earlier by Marketing can be met.

Selecting the basic aircraft concept[7] is a very important milestone in the development process indeed. It concludes upon some strategic decisions such as:

- key technologies to be implemented;
- major Life-Cycle Costs (LCC);
- the transport principles for major components; as well as
- manufacturing and tooling concepts, which may necessitate larger investments in new buildings and facilities.

These decisions should preferably later not be altered any more as this would cause very significant repercussions to the programme. The generation of the DBD marks the milestone which completes the Feasibility phase[8] as a feasible basic aircraft concept can be presented (see again Figure 3.2).

---

5 A scheme is a method of presenting design principles or methods in either a 2D and/or 3D form depending on the project requirements. These schemes may not be geometrically to scale but will include the general layout of the aircraft or section showing basic aircraft configuration, structure and systems to meet the requirements.
6 However, the General Arrangement drawings subsequently still get further refined as the selected concept evolves.
7 If not a single aircraft is to be developed, but an entire aircraft family, the selected concept would have to cover all members of that family.
8 Despite the fact that 'Feasibility' is shown as a sub-phase in Figure 3.1 and 3.2, it will hereafter be called a 'phase' for reasons of simplicity.

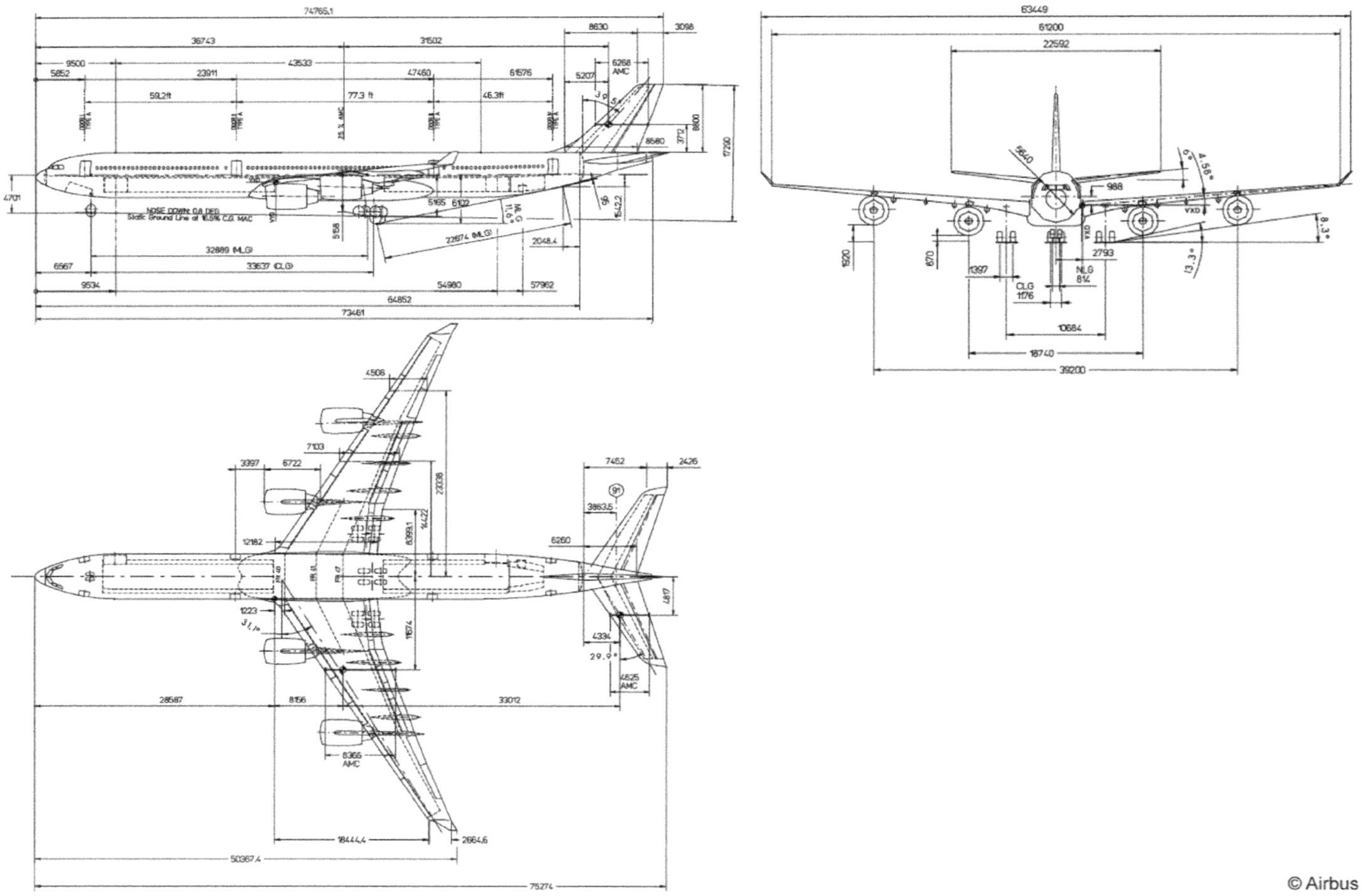

© Airbus

Figure 3.3   Example for a General Arrangement drawing

Source: Airbus, with permission.

Achieving an Aircraft Configuration Which Meets Customer Expectations

During the subsequent Develop Configuration phase more detailed analysis and optimisation is undertaken to confirm the viability of the chosen concept. The objective of the Develop Configuration phase is to identify the optimum solution from a large number of potential configurations, which are all possible within the chosen concept. There is, however, a fundamental difference to the previous phase with regards to the down-selection process: while during the Feasibility phase different basic concepts were analysed *in parallel*, during the Develop Configuration phase an initial configuration evolves through a series of iterative steps to a final configuration. The iterative configuration evolution process is supported by, among others:

* windtunnel tests to confirm the high-speed aerodynamics of the aircraft during cruise as well as its low-speed aerodynamics during take-off and landing;
* continuous feedback from potentially interested airlines;
* the involvement of potential suppliers and risk-sharing partners;
* the Cost Estimation department to refine the development and production costs of the aircraft and to demonstrate that the aircraft project represents a viable Business Case;
* engine suppliers providing data on thrust and noise levels;
* airport authorities contributing with their know-how on airport compatibility;
* and so on.

The schemes established during the Feasibility phase are now progressed to a standard, which is to scale, incorporating some initial sizing with improved loads data. Each iteration implies that the size and position of major components are amended and refined to achieve a new acceptable configuration that meets both performance and cost targets. The existing General Arrangement drawings are updated with each new status of the aircraft and act as a fundamental reference source for configuration studies. A number of new supportive 3D models and drawings are also produced including wing, fin and tailplane General Arrangements. These contain extra geometric information indicating, for example, the size and position of leading and trailing edge devices for the wing. These 3D models and drawings are maintained and updated with each new status in the same manner as the aircraft level drawings. In addition to the referential General Arrangements, several intermediate drawings will be produced in support of the many trade-off studies undertaken during the Develop Configuration phase.

Other activities during the Develop Configuration phase include the following:

* the SAM created during the Feasibility phase is refined to a stage where it becomes the master Digital Mock-Up (DMU);
* all the moveable component motions (for example for flaps and slats) are calculated and systems are routed;
* more detailed manufacturing principles such as data on the required Conditions of Supply (CoS)[9] are incorporated into the major components' and assemblies' schemes;

---

9 CoS determine the form, fit and function aspects of deliverables to be supplied in order to ease transport, handling and assembly tasks.

- structural configurations as well as provisions for the systems are progressing to a stage where design of major jigs, tools, facilities and buildings can commence;
- the top-level requirements are further refined.

Eventually, one configuration will emerge which satisfies the customer needs, ensures the aircraft will be in the market on time, and which represents a viable Business Case for the aircraft company. The strategic, economical and technical aspects of the programme have all been assessed. The manufacturing processes have been put on a technically and economically sound basis. The detailed geometries of, for example, the landing gear position or the engine or wing planform are essentially fixed but will still require final sizing. In addition, the fuselage cross-section, which is a key factor for commercial success of the aircraft programme,[10] is frozen. Also, the structure architecture is frozen, including frame orientations and location of floors and doors, as well as the architecture for the systems. The definition of the critical interfaces is very mature and a concept for the final assembly is established. The basis, on which certification will eventually be granted, is agreed with the authorities.

Finally, the worksharing among parties interested in participating in the development programme has been agreed. Finding the right worksharing is a tremendous task on its own. Many conflicting interests – as described in Chapter 1 – have to be balanced out. In addition, all potential implications of managing a complex supply chain have been carefully analysed.

> 'Our design philosophy called for all ... [fuselage] doors to operate alike so that cabin crews would not be confronted with different ways of opening and closing them or arming and disabling their automatically deploying emergency slides. Standardization of this sort is important for intuitive use. Unfortunately, things went wrong when many suppliers with different detail design responsibilities took different design routes. Worse still, they farmed out some of their Boeing work to second-tier suppliers who likewise employed a wide mix of design methodologies.
>
> The upshot was that not one of those five similarly sized 747 doors had its detail design performed by the same people. My people had provided overall design guidance for such aspects as the type of linkage, locks, slides, and so on; we thus expected similar doors. But when they began arriving, we were dismayed to find that these cabin doors were all different in just about every regard. Just as troubling, many of their features did not work well. ...'[11]

For example, the workshare for the Boeing 787 was well defined before ATO. It granted the right to Japanese companies to take 35 per cent of the design-build share, mainly for the wing. Other companies were granted 30 per cent, with only some 35 per cent remaining for Boeing itself.[12] Never before did Boeing allow such a large volume to be outsourced.[13]

---

10 For example, the success of the first Airbus, the A300, was largely due to the choice of the fuselage cross-section, which allowed an eight-abreast accommodation and the loading of industry-standard LD3 containers in the cargo hold. The market success of the Boeing 777 also illustrates the importance of a successful harmony of fuselage cross-section, capacity and customer comfort. It is interesting to note that Airbus has only three basic body cross-section designs, while for Boeing each aircraft is different.

11 Sutter, J. (2006), *747: Creating the World's first Jumbo Jet and other Adventures from a Life in Aviation* (New York: Smithsonian Books, HarperCollins Publishers), pp. 132–3.

12 Boeing's own workshare further reduced significantly by the outcarving of its Wichita, KS. and Tulsa, OA., plants in 2005, thus forming the independent aerostructures company Spirit Aerosystems in charge to build the 787 fuselage's front sections and cockpit as well as the leading edge.

13 — (2003), *Boeing Announces Work Share for 7E7 Structures Team*, (Everett: Boeing News Release), 20 November 2003, http://www.boeing.com/news/releases/2003/q4/nr_031120g.html, accessed 10 October 2009.

With the optimised configuration at hand, Management has at this point in time sufficient information available to identify whether it wants to proceed or not. However, certain circumstances may have changed in the meantime: as some years have gone by since the beginning of the aircraft programme, the prospects of the future market should again be looked at. But also the *current* market situation may have changed. May be there was a sudden industry downturn and the aircraft company is suddenly struggling to lever the necessary funds for the Development phase. Or some analysed technologies turn out to be too risky after all. There are certainly many reasons why an otherwise promising development project could come to an end at this point. Management will therefore carefully analyse the situation. If it concludes that the proposed aircraft configuration meets the needs of the market and stands on viable business grounds, it will grant the Instruction to Proceed (ITP).

What remains to be done, apart from further technical refinements[14] before the aircraft can be officially offered to the airlines, is the proper preparation of the Development sub-phase. This is not only necessary to have a sound basis for a successful management of the development but also to raise the level of confidence with potential customers.

So far, all work was done on aircraft level or on the level of the larger assembly components (such as fuselage, wing, and so on). It was done by a relatively small number of people. More detailed analyses, to be launched subsequently, will require more and more individuals to work concurrently. Therefore, the question of organisation becomes more and more important. Shortly after the ITP has been granted, organisational and way of working principles have to be established, roles and responsibilities need to be sorted out and resource ramp-ups have to be planned.

As a result, a first high-level Organisation Breakdown Structure (OBS) will be established and individual deliverables are identified for each unit of this OBS. This is captured in a document called the Project Plan. It describes how the aircraft development should be run, who will be in charge of what, what processes are to be used, and so on. For the first time, a rather detailed schedule plan for all phases until EIS is established. It shows all major milestones and serves as a master schedule reference for all subsequent planning activities. Other plans, like the Manufacturing Plan, the Quality Plan, and so on are also going to be established.

At the point where the technical refinement has culminated in the generation of the product specification (to be described in more detail in Chapter 8), and with the Project Plans at hand, Management can now grant the ATO the aircraft to potential customers. However, the precise timing of the ATO will, of course, depend on the current market situation.

During the sales campaigns, the configuration will be further refined, in particular as a result of more intense contacts with the airline customers. Once Management has confirmed that its launch conditions have been met, the industrial development of the aircraft programme can go ahead. This start means that the company's Management, the shareholders, the engine manufacturers, other important suppliers and risk-sharing partners as well as the company's own Functions like Engineering, Manufacturing

---

14 For example, the Master Geometry needs to be fully frozen as well as the detailed geometries for the high lift system (important for the climb and descent phases in flight) and the wingbox (important for calculation of available fuel and therefore range). Also, the finalisation of the interfaces of the systems with the structure, defined in System Installation Requirements Documents (SIRD), needs to take place as well as the writing of specifications for the systems' equipment. Work on the Equipment Installation Requirements Documents (EIRD) can subsequently commence too.

Procurement, and Finance all declare their full support for the new programme. The precise timing of 'Go-Ahead' therefore not only depends on the market situation but also on a relatively small window of opportunity where approval of all stakeholders can be gained. The programme is now becoming a top priority for the aircraft company. From now on, the Management in charge of the aircraft development project is authorised to spend a significant amount of money, resulting in resources' ramp-ups and rapidly increasing cash consumptions.

POST GO-AHEAD: THE DEVELOPMENT SUB-PHASE

Industrial launch usually takes place a few years after Marketing has communicated the initial market opportunities. A relatively small team has worked very hard to get to that stage. Work was mainly focused on trade-offs with high impact on configuration and architectures. But all this represents a relatively small effort compared to what still lies ahead. The period from 'Go-Ahead' until EIS is what is called in this book the Development sub-phase of the total aircraft programme,[15] see Figure 3.2. It is during this Development sub-phase that the majority of the total Development phase' cost will be spent and the aspect of volume (in terms of number of people involved, 3D models and drawings to be released and so on) becomes the predominant feature. As an example, Figure 3.4 sketches the required ramp-up of engineers for the development of the A380.

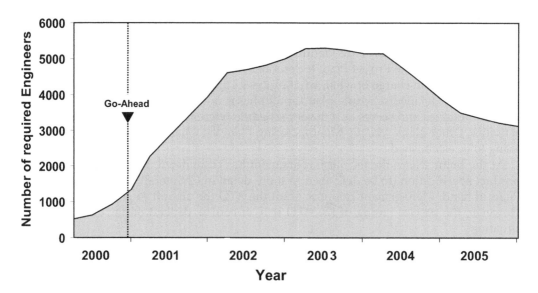

Figure 3.4   **Required ramp-up of Engineering resources at Airbus for the development of the A380**

Data taken from: Thomas, J. (2001), *Die A380 – Eine große technologische und wirtschaftliche Herausforderung für Europa* (Berlin: presentation to German MEPs), 29 May 2001, p. 19, with permission.

---

15 Neglecting for a moment all the relatively minor development activities following the Entry into Service.

The 'Go-Ahead' kicks-off a tremendous series of activities. Although the aircraft concept and configuration is frozen on the aircraft level, on lower level, for example on the level of major components, many concepts and configurations still need to be analysed. In principle, the same basic process with parallel analyses of concepts, down-selection of one concept and iterative configuration evolutions, which we saw on aircraft level, is now repeated on lower levels. However, the number of concepts and configurations per design area can be much higher.

Concepts are the basic solutions generated by the Engineering department to meet top-level requirements. Thus, whenever solutions are to be frozen by a concept design review, it is checked whether the requirements can actually be verified by the design. Longlead items, such as forgings, and detailed interface configurations are analysed first. Eventually, they are designed to a level, which allows to:

- kick-off their production; as well as to
- freeze interfaces in detail, ensuring largely independent design flows for structure and systems, respectively.

Then, design concepts for other detailed areas are established, each one confirmed in design reviews. Once applicable requirements have been verified during concept design reviews, the design can move to the next step, that is, into detailed design. Any remaining degrees of design freedom are now limited to local zones of the aircraft. The link between Engineering and Manufacturing intensifies and the need for sophisticated Configuration Management increases dramatically.

Detailed design, which kicks-off now, encompasses all of the design work needed to go from a mold line to cutting metal. The primary driver in this phase is the creation of manufacturing drawings and 3D data sets, which entail detailed structural analysis. Thousands of detailed drawings, which are to scale and where the sizing of structural components is carried out to a standard commensurate with the available loads information, need to be established.[16] Part drawings, assembly drawings, interface drawings, installation drawings, Interchangeability Drawings (ICY) and frontier drawings: all need to be generated and released. Assembly and installation drawings physically bring together various assemblies under a specific function or location (join-ups). ICY drawings describe parts (and their attachment tolerances), which can be interchanged by replacing them by new ones, such as a damaged flap. Frontier drawings are used to define the method of assembling different sections of the aircraft supplied by different companies (for example risk-sharing partners), and to freeze the dimensions and resulting tolerances obtained after joining. It also serves as the contract of responsibilities between the partners involved in a frontier.

Compared with the Feasibility and Develop Configuration phase, the period of Detailed Design is not a very creative one. Instead, a significant volume of 'execution' work at highest possible quality level has to be processed. The 747, for example, required about 75,000 individual drawings to specify[17] and it took Airbus about 79,000 drawings to design the A380.[18] Each drawing needs to be carefully checked and approved by the company's

---

16 Loads include aerodynamic loads and ground loads.
17 Petrosky, H. (1996), *Invention by Design: How Engineers get from Thought to Thing* (Cambridge: Harvard University Press).
18 Leichter, S., Airbus, private communication.

appropriate signatory authorities prior to release. Despite the advancements of computer systems this still takes a considerable amount of time.

Once part drawings are released, the information contained therein needs to be converted into formats, which can be used by the machines in Manufacturing. The latter is called Data for Manufacture (DfM). Note that in some companies DfM is regarded as an inherent part of the detailed design activities while in other companies it is a part of the Manufacturing steps following the release of detailed drawings. With the technological progress of Information Technology (IT) tools this distinction gradually becomes an historic one. As was said in Chapter 2, 3D models can nowadays be directly converted into codes for Numerically Controlled (NC) programming. In this book, therefore, DfM is regarded as a part of the Detailed Design phase, see Figure 3.2.

With the release of the first detailed design data, manufacture of parts can start. The associated milestone is called First Metal Cut (FMC). This used to be a very important milestone at times where Computer Aided Design (CAD) and NC machines were not abundant. Today, it is more a symbolic public relations event. While in former times FMC essentially marked the end of the Engineering activities, that is, after the engineers had 'thrown the drawings over the wall to Manufacturing', it is today a milestone, which is celebrated to provide confidence to the customers that the company is serious about the programme.

As far as assembly or installation drawings are concerned, they carry the information of the Bill of Material comprising all those parts, for which the drawing describes the join-up. This includes the basic components (for example, machined pieces, fasteners and other standard parts) as well as already assembled parts. Manufacturing converts this information into generic routings. The latter describe in detail the specific activities, which need to be done (and in which sequence, using which jigs & tools, and so on) in order to prepare, perform and finish the actual assembly or installation on the shop floor.

Routings typically do not carry dates on them, but durations. However, once the overall schedule planning for the aircraft deliveries is mature, start and finish dates of assembly processes for individual aircraft can be calculated. It is then possible to allocate the generic routings to specific aircraft and to convert the durations carried by generic routings into individual work orders, which carry dates.

It is also possible now to identify the precise dates when components need to be available at each assembly station. However, at the beginning of the industrial ramp-up, parts should always be planned to arrive on site with sufficient margins as there will always be some of them arriving late. Thus, for a development project – quite different to Series Production – a controlled Just in Time approach may not be a wise approach. The planning for all the parts belonging to all the work orders of an individual aircraft build will generate what is called the 'demand' for all items to be supplied. Demand information contains all relevant data related to delivery dates, part numbers, quantities and so on, to be forwarded to the suppliers. Once the parts arrive on site, they are 'goods-received' and a First Article Inspection (FAI) is performed.

To ensure, during this intense phase, that all structural assemblies, the system installation parts like pipes, brackets and so on, as well as all equipment are supplied on time represents a major management challenge. While adequate control of thousands of individual parts to be delivered by internal and external suppliers can only be executed in reasonable time thanks to the support of modern IT systems, it is absolutely essential that project leaders and team members also keep very close personal ties with their supplier counterparts to ensure delivery of the items at their scheduled dates. Upfront personal presence of members of the project team on the suppliers' premises can also reduce the number of failed FAIs. It is very frustrating for a team in charge of installing parts or assemblies to send them back after they have already been goods-received.

The effort to ensure that all drawings for the thousands of parts are released, that the routings are created, the assembly process schedules are established, the parts arrive on time, are inspected, goods-received and fit for purpose, is tremendous. This period of the Development sub-phase is most difficult to manage, even if there were no design errors or changes. But, unfortunately, changes repeatedly occur during this charged period, too, because:

- Engineering may be obliged to change the design for well justified reasons;[19]
- Manufacturing does not produce or assemble as designed, resulting in concessions to be approved by Engineering;
- parts arrive late and therefore the assembly processes must be constantly revised; and because
- parts, although arriving on time, are not in line with the specification and need to be replaced. (This can often only be done at a later stage as soon as parts with the right quality become available.)

Each time a design change is requested, a new 3D model and part drawing has to be created, resulting in a new assembly drawing, a new routing, new demand for a part and so on. As there are many changes during an aircraft development, it is absolutely essential that the company manages the process from 3D model and drawing release to arrival of the changed components at the assembly site as slickly and smoothly as possible. The better this process is managed the less time will be needed for the development cycle overall. Managing this process well – apart from minimising the amount of changes in the first place – is therefore a core competency of an aircraft-developing company.

During the Development sub-phase, not only will the first few aircraft be built, but also some important test specimen:

- Static test – Static tests verify the data calculated by the Design and Stress departments and contribute towards the further development of mathematical models. Loads are simulated as generated by forces, mass forces or transverse forces, taking into account environmental influences such as temperature, air pressure and humidity. Computers regulate, monitor, measure, record, evaluate and document all systems and test parameters.
- Fatigue test – Since the mysterious accidents with the De Havilland 'Comet' type aircraft in the 1950s, which proved to have been caused by fatigue, the airworthiness authorities have required fatigue tests to be performed on the airframe. Certification evidence is required that the airframe can withstand not only the largest load that can occur, but also the quantity of loads and stresses which can occur over its service life. The fatigue tests allow authoritative statements to be made about the degree of material fatigue.
- Iron Bird – The iron bird is a testbench of actual flight worthy electrics and hydraulics systems as well as flight control actuators operating under controlled

---

19 Typical reasons for change include safety aspects, loads changes, customer requests, and requirements changes.

conditions in response to inputs from flight control computers. It is used for initial component development, system development and test. It is also used to solve problems, aiming at developing mature systems.

- Cabin test facility – With the cabin test facility the effects of aircraft interior conditions, including noise, vibration, temperature, and humidity on cabin and cockpit crew and passengers are examined. It considers the health and comfort effects of the aircraft environment with an aim to improving environmental comfort while working within parameters that will not adversely affect factors such as fuel consumption, fire safety and weight. Human factors such as personal perceptions and individual comfort levels will also come into play.
- Landing Gear test facility – In the landing gear test facility, forces similar to a fully loaded aircraft can be applied in predetermined sequences to simulate extended periods of operational use. With the addition of rigs to perform landing gear door movements and other sub-systems to landing gear deployment and retraction, entire landing gear systems can be tested. In addition, critical strength and fatigue tests are performed. The landing gear is then subjected to loads well in excess of those likely to be experienced in real life.
- Engine testbeds – Whenever a new engine is developed it has to be extensively tested for various parameters like power, fuel consumption, vibration and leaks prior to fitting it on the aircraft. It is necessary to run an engine 'live' on the testbed facility. Usually, there are engine testbeds on ground as well as flying testbeds attached to the wing of an aircraft. The creation of an engine testbed facility also allows to locally change specific modules or systems/sub-systems in the engine during development.
- and many others, for example for materials and individual systems testing.[20]

Test facilities are needed to validate the data calculated by the various departments (such as Stress, Design and so on). All of these test specimen need aircraft and non-aircraft components to be designed, produced and delivered more or less in parallel with the components for the first aircraft intended for flight.

When a major component assembly is finished, for example a wing, it is transported to the Final Assembly Line (FAL). Prior to shipment, however, FAL people will inspect the component. They will identify the quality of the component and the level of outstanding work, that is, work which could not get completed at the major component's assembly site. The mainstream final assembly process, which usually is a complex set of activities, should at best not be disturbed by outstanding work. However, if outstanding work needs to be performed at the FAL, planners will identify slots in the FAL activities' schedule plan where outstanding work can be performed, either by FAL staff or by staff from the major component assembly site. As this staff cannot work continuously but rather during an irregular and interrupted sequence of working time slots only, outstanding work in the FAL takes much longer and is more expensive than an equal amount of work at the home site. A factor between 3 and 6 is common. All management effort must therefore be directed towards minimising the amount of outstanding work.

However, eventually the aircraft is fully assembled and systems can be switched on. This milestone is called 'Power On'. Soon afterwards it is ready for 'Roll-out' of the hangar,

---

20 For the development of its 777, Boeing tested 57 major systems supplied by 241 companies. The associated test facilities consisted of 46 stand-alone test facilities as well as three major super laboratories. From: Norris, G. and Wagner, M. (2001), *Boeing 777 – The Technology Marvel* (Osceola, WI: MBI Publishing).

which is another major public relations event. 'Roll-out' is followed by a large number of ground tests culminating in the declaration of flight worthiness. Then the day of the first flight approaches. For all the people involved in the aircraft development project this is the milestone which generates the most emotions. Bringing such a complex product successfully into the air still represents – after all – an old dream of mankind coming true.

Obviously, at this point the flight is not for pleasure but already represents the first in a series of many flights, during which additional tests are performed, such as the confirmation of:

- the flight characteristics and flight control laws;
- the climbing and descent characteristics;
- the landing behaviours at touch down and during braking; as well as
- the confirmation of the calculated loads.

Flight testing is very time consuming and can use up as much as two years of the entire Development sub-phase of the first aircraft. In fact, the actual number of required flight test hours has grown over time. This is a result of the continued growth in aircraft complexity. The latter in turn is partially due to increasing customer specifications and partially due to the increased stringency of certification requirements issued by the Federal Aviation Administration (FAA) and the European Aviation Safety Agency (EASA) – to name but the two most important and influential ones.

> 'Flight testing depends importantly on whether the design is new or derivative. For example, the Boeing 777 program included an objective of achieving ETOPS certification concurrently with FAA certification. This required dramatically more test flight activity than would otherwise have been the case. Sources indicate that ETOPS added approximately 2,000 flight hours to the test flight program.'[21]

Finally, the certification process, which over the entire period of the Development phase focused on the safety aspects of the aircraft as determined by FAA and EASA, and as spelled out in documents such as Federal Aviation Regulations (FAR) and Certification Specifications (CS), respectively, will result in the certification being granted by these authorities. The certification includes the certification for the type of aircraft ('Type Certification' as opposed to the certification of an individual aircraft), assurance that the applied manufacturing processes produce aircraft that conform to technical drawings, as well as the statement of airworthiness.

From a safety point of view the aircraft is now authorised to fly and to transport people. But before entering into service it still has to demonstrate that it can deliver the guaranteed performances in quasi-operational circumstances. The aircraft is therefore subject to a series of route proving tests. In particular the turn-around times, dispatch reliability, maintenance aspects and way of working in the cabin are tested. Eventually, the aircraft can be delivered and handed-over to the customer.

In reality, however, the first aircraft is usually not delivered to a customer. It will be used by the aircraft company for continued testing of, for example, modifications prior to

---

21 Spitz, W., Golaszewski, R., Berardino, F. and Johnson, J. (2001), *Development Cycle Time Simulation for Civil Aircraft*, NASA/CR-2001-210658, p. 3–6.

implementing them to Series Production aircraft. Furthermore, it can be used for future research activities (for example for new high-lift devices) or as a flying testbed for new engines.

The next few aircraft may not be delivered immediately to customers either until a much later stage. This is because they are needed for tests which cannot be performed in a timely manner using the first aircraft only. For example, the aircraft's newly developed cabin may need to be tested for a considerable amount of time. In order to save overall development cycle time, cabin tests should therefore preferably be performed in parallel to the tests performed on the first aircraft. A certain number of additional dedicated test aircraft are therefore required to avoid disruptions to the rest of the programme schedule. So the first aircraft to be delivered directly to an airline customer may only be the seventh or eighth aircraft or thereabout.

'A typical build of test airplanes may look like the following:
- Airframe No. 1 – flight testing
- Airframe No. 2 – flight testing
- Airframe No. 3 – static testing
- Airframe No. 4 – flight testing [first cabin]
- Airframe No. 5 – fatigue testing
- Airframe No. 6 – flight testing
- Airframe No. 7 – flight testing.'[22]

EIS finishes the Development sub-phase. But before closing this chapter it is worth having a closer look at two technical design processes which an aircraft manufacturer must be capable of mastering in any case, but which have a high potential for adversely affecting development cycle time:[23]

- the Loads-Design-Loop (LDL), fundamental for aircraft structural design; as well as
- the design of systems and of their installation on the aircraft.

The LDL is an iterative design process comprising a certain number of loops. The problem with regards to adherence to planned cycle times is that the number of required loops cannot be exactly forecasted. The design and development of systems and of their installations on the aircraft is not only very complex on its own, it also drives the design of brackets, of which there are thousands on an aircraft and which are needed to hold the systems' equipment units and associated interconnecting elements. As simple as they might be, brackets are the last structural components to be designed, following the design of the systems and their parts. However, on the shop floor they must be installed relatively early during the aircraft's assembly process. Thus, schedule problems associated with the design and delivery of brackets can severely disrupt the assembly process, adversely affecting cost and leadtime objectives. Project Managers must regard them as symptoms for big risks embedded in the design of systems and their installations, requesting their highest management attention.

---

22 Spitz, W., Golaszewski, R., Berardino, F. and Johnson, J. (2001), *Development Cycle Time Simulation for Civil Aircraft*, NASA/CR-2001-210658, pp. 3–11.

23 A third major process, the engineering of systems, will be described in the following chapter.

THE LOADS-DESIGN-LOOP

In a way, aircraft structural development can be regarded as managing a design through a number of loads loops. The ability to manage LDL is therefore a primary core competency of any aircraft company. An overview is shown in Figure 3.5 for the example of an aircraft's wing design iteration. External forces and moments – generally understood as loads – act on the aircraft, such as aerodynamic loads or ground loads introduced during the taxiing. The Stress department looks into how these external loads propagate through the aircraft as stress to the structure. However, in order to do so, it needs to assume a certain aircraft design. The Design department delivers this input as an initial step to the calculation loop. The structural design of the aircraft is then modelled by the Stress Department using the Finite Element Model (FEM) introduced in Chapter 2.

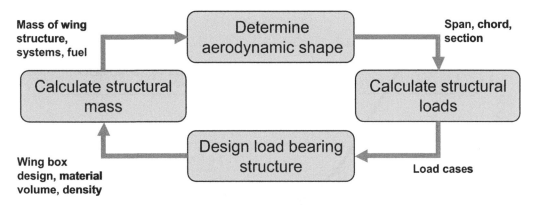

Figure 3.5   The Loads-Design-Loop

Idea adopted from: Lawrence, P. and Scanlan, J. (2007), 'Planning in the Dark: Why Major Engineering Projects Fail to Achieve Key Goals', *Journal of Technology Assessment and Strategic Management* Vol.19 No.4, pp. 509–25, with permission

The FEM not only takes the properties of the materials selected for the components into account but also the actual structural design: a stiffly designed component will propagate the loads more than a component with more inherent flexibility. The stress analysis will then reveal if and where the design has to be adapted to match stress levels, for example by changing the materials used or the design itself. Eventually, a new FEM based on new materials – if any – and changed design will be established. However, in the meantime, the loads usually have changed too. This comes as a result of better knowledge of the aircraft configuration, its mission and the application of more detailed calculations. The loop starts again.

What makes this process even more complex is the fact, that at interfaces between major component assemblies – often involving different design teams with different design philosophies and tools – interface loads have to be calculated. For example, the team designing the flaps not only has to convert the (external) aerodynamic loads on the flaps into stress levels *within* the flaps but also into loads acting on the interface points (at the trailing edge of the wing), to which the flaps are attached.

Each time a loads loop is completed, changes in the stress levels will result in changes in design of components. Design changes are time consuming and generate cost. Thus, the loads loop process has to be carefully project managed in much the same way as the design of the components itself. For example, it should be communicated to all teams when new loads and stress data can be expected and schedule plans must include the milestones where the new loads will be delivered.

Note that the calculation of a complete loads loop could take months. The whole process is usually repeated a few times between design freeze and Type Certification. If managed well, the loads changes will rapidly become smaller with each loop, and design converges into the final configuration, which is compatible with the loads to be expected, see Figure 3.6. After the first aircraft has been built, ground and flight tests will confirm – or not – the previously calculated load and stress levels. They generate a final load loop, perhaps resulting in final design changes to ensure that the aircraft is fit for certification by the authorities.

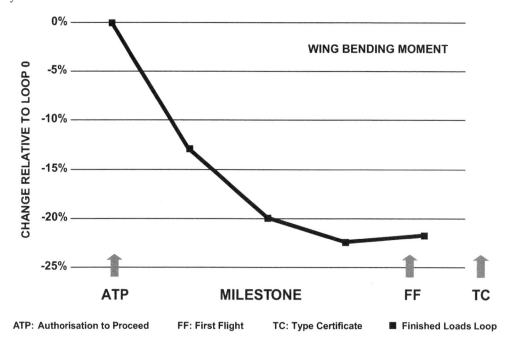

Figure 3.6   Typical evolution of load analysis with time
             Here: Airbus A380 wing root bending moment

Source: Anon. (2006), *Flight Physics World* (Toulouse: Airbus), Vol. 1 July 2006, p. 15, with permission.

A big trade-off to be made is the amount of margin in the design beyond what is required to ensure safety. If design margins are reduced early in the design process it will be easier to achieve the weight targets of the aircraft. However, components will in this case be more often susceptible to design changes as changes in the stress levels cannot be absorbed by the margins.[24]

---

24 In this context it is worthwhile to look at some important definitions:
     Limit Load (LL):   Maximum load that an aircraft could ever experience during its operational life
     Safety Factor: Factor of 1.5 on top of LL, imposed by airworthiness authorities

If, on the other hand, margins remain too generous, it may not be possible to take out sufficient weight in order to achieve a viable aircraft weight early in the design process. Everyone would get very nervous in this situation and pressure from Management to do something about the overweight would increase. Unfortunately, overreactions leading to premature decisions can at this point not be excluded. Because of this, and because the design margin has direct consequences for the management of the development project, the margin policy must be discussed within the entire team leading the project. This fundamental trade-off should be well discussed, should involve experienced senior experts and should be thoroughly documented. In any case the decisions related to the margin policy should not be left to Engineering alone.

THE DESIGN OF SYSTEMS AND OF THEIR INSTALLATION ON THE AIRCRAFT

Of particular complexity is the development of the various systems as classified by the Air Transport Association (ATA) and their installations on board the aircraft. For each system, a Systems Requirements Document (SRD) is generated, outlining the functional requirements to the system as well as providing any constraints to its design. On the basis of this document, system engineers nowadays develop a computer model of the system, which simulates:

- a systems architecture consisting of modules like equipment units and their interconnecting elements;
- the intended functionalities for each module; as well as
- the required properties of the modules necessary to fulfill the functionalities.

The modularity of the system and the properties of the modules are modified as many times as necessary until the simulation provides evidence that the system as a whole will achieve the desired functionalities. With this approach, 'to-be' properties of both the equipment units as well as their interconnecting elements are determined at this point. They are subsequently translated into individual requirements for:

- equipment units;
- mechanical interconnecting elements (pipes, tubes and so on);
- harnesses;
- interface elements such as brackets; as well as
- other elements like protections, labels, decorations and so on.

On the equipment unit side, system engineers lay down these requirements in Equipment Requirements Documents (ERD) before the design and development of any equipment unit can start. ERDs are of significant commercial importance as aircraft manufacturers

---

Ultimate Load (UL):   Limit Load multiplied by Safety Factor, i.e. UL= 1.5 × LL
Allowable Load (AL):   Load which the aircraft is required to sustain.
    As a rule, the Reserve Factor (RF) – calculated as AL/UL – has to be equal or greater to 1. In other words, the aircraft structure is to be designed so strong, that it can sustain at least the Ultimate Load. If it can sustain more, then there is design margin.

tend to outsource most of the equipment units' design and development to external suppliers. In this case, the ERDs serve as a basis for the establishment of specifications to suppliers. What is important to note is that for the development of equipment there is very often a concurrent development of hardware and software, with the latter having in many cases longer development cycle times than the former.

When the equipment engineers (for example of a supplier company) do the detailed design for the equipment, they also establish their requirements for the later installation of their deliverable on the aircraft. This is documented in the Equipment Installation Requirements Document (EIRD). It not only describes the maximum space in which the final equipment geometry will be inserted, but also generates and provides information on any installation constraints, maintenance aspects and interfaces. Among others, it also specifies the interfaces to the interconnecting elements.

As far as interconnecting pipes and tubes are concerned, system engineers formulate their properties – such as the diameter, weight, type of material, pressure resistance, thermal and other environmental constraints – in a document called the System Installation Requirements Document (SIRD). It represents the contract between systems architects on the one hand and system installation designers on the other. It also contains segregation rules, installation constraints and maintenance aspects, among other.

For the electrical interconnecting elements such as wires, cables and harnesses,[25] another approach needs to be applied, see Figure 3.7. It is based on what is called Principle Diagrams (PDs), which are designed for *each* system (or sub-system). PDs are drawings established on the basis of both, the SIRDs and EIRDs, which:

- represent the overall electrical circuit needed for a system;
- specify the grouping of wires into cables;
- determine the cables' gauges; and
- describe the interfaces between cables and equipment units (for example by specifying electrical connectors).

An example for a PD is depicted in Figure 3.8.

But cables (or wires) can and should be grouped together to bundles wherever possible, ideally even if the cables belong to different systems. This is to save attachment points and brackets, thus reducing weight and complexity. In order to do so, engineers firstly define potential pathways[26] (that is, their volume, geometry and attachment points) within the aircraft's DMU. This is a complex task but less complex than it might seem as there usually is a lot of experience from previous aircraft programmes about the typical pathway routes.

Each pathway is identifiable by an individual denomination. Engineers then allocate the individual wires and cables from all the different PDs to any of the potential pathways. Thus, for each system it is specified, which cable will belong to which pathway. Precisely which cables of a system-specific PD are supposed to be fastened together to form bundles or harnesses, and how this grouping is achieved (for example through the use of tie-raps, common insulating surrounds, etc.), is then specified in Wiring Diagrams (WD).

---

25 While simple wires are single conductors surrounded by an insulating material, cables consist of two or more wires contained in a common covering or twisted or moulded together without common covering. A harness is an assembly of any number of wires or cables with termination that is designed and fabricated so as to allow for installation and removal on the aircraft as a unit.

26 An electrical pathway is a physical route between electrical components.

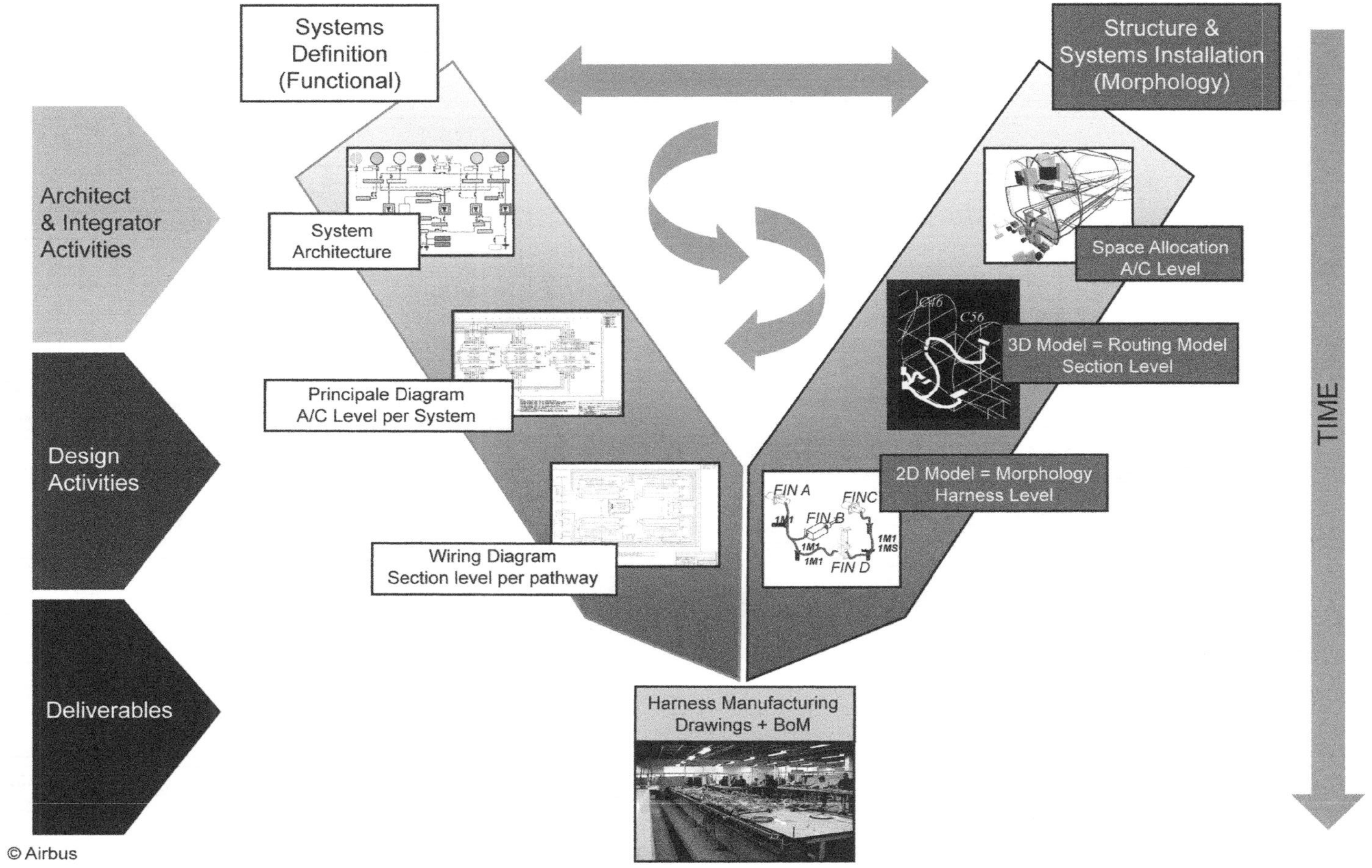

Figure 3.7   Electrics design approach

Source: Airbus, with permission.

© Airbus

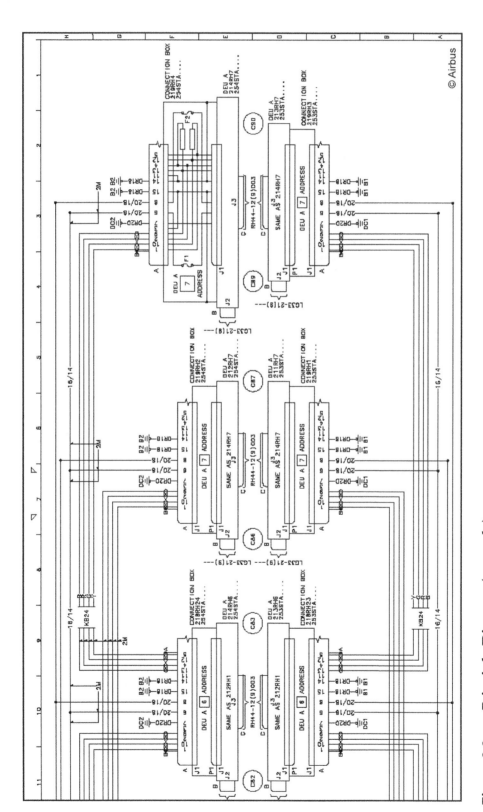

Figure 3.8　Principle Diagram (example)

Source: Airbus, with permission.

In a way, a WD is a more detailed PD but it is specific to a given pathway. For a PD of a given system, a WD only captures this system's cables foreseen for a specific pathway.

It is worth noting that the necessity to reduce aircraft weight wherever possible is the main reason and driver for the complexities associated with the design of a harness. It is therefore an excellent example of how aviation's imperative to save weight can increase the complexity of a product.

Basically, each ATA-system on board the aircraft is initially designed more or less independently from the other systems. However, there clearly are many interfaces between the systems, which need to be looked at and defined. For example, system engineers need to know the available electrical power as an input for dimensioning individual equipment units. At the same time this information is an output of yet another system, i.e. the power supply system. Every attempt must be made to freeze such interfaces as soon as possible, definitely before start of the Development sub-phase. The necessity of an early interface freeze can in fact not be over-emphasized: many aircraft development projects have suffered from significant delays and cost overruns because of too late a freeze of:

- definition of the systems;
- the interfaces between the systems;
- the interfaces of the equipment units with their interconnecting elements; as well as
- the interfaces between the equipment units and their interconnecting elements on the one hand and the airframe platform on the other.

Only if these aspects can be frozen early enough, system design teams can indeed work independently from each other for some time. This must be regarded as a major enabler for achieving the challenging cycle times imposed on commercial aircraft development projects.

However, all equipment units and interconnecting elements have eventually to be accommodated in the various aircraft zones. Thus, there is a moment in time where the system-by-system design approach is no longer adequate and must be replaced by a zonal approach. The zonal approach is necessary to find out whether the elements of *all* systems can be accommodated in any given zone.

This is the point where the use of a DMU adds again significant value to the design process as the accommodation arrangement for all types of interconnecting elements passing through individual zones can be designed in 3D. Usually, this is done for the pipes and tubes first as they are made of less flexible material than electrical cables where it is easier to twist them around obstacles. It is, in fact, the PD (and WDs), which provide the required information to enable a 3D zonal view for the electrical harnesses. As an example, Figure 3.9 provides a 3D image for the electrics installed in the front part of an aircraft's fuselage.

Using the DMU, some fundamental questions can be analysed: Are interconnecting elements supposed to be accommodated in a particular zone? Are there any other interconnecting elements planned to be penetrating through the same zone, possibly of other systems? Are there any clashes between the elements of the various systems? How can all of them be accommodated using as few attachment points as possible while respecting all installation requirements? What do the attachment points look like in detail? Can these attachments be standardised? Can available standard brackets be used? Any problems identified might force a redesign of a system's installation!

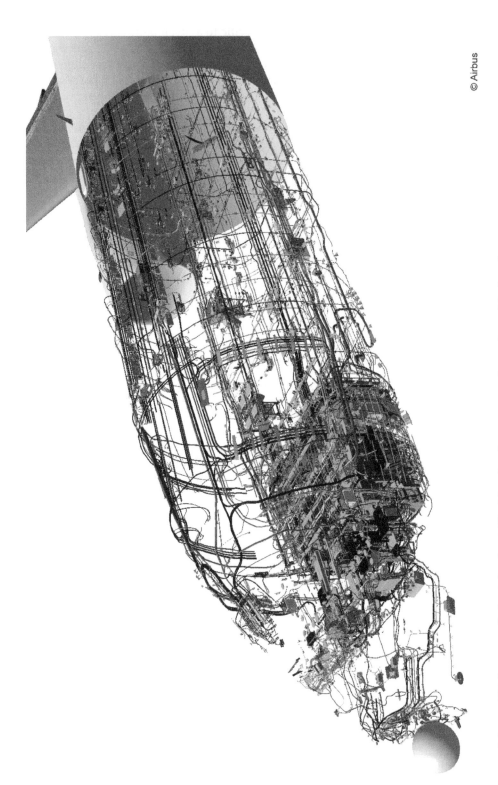

Figure 3.9   **3D image of electrics installed in the front part of an aircraft's fuselage**

Source: Airbus, with permission.

Using the DMU, one can also determine the approximate lengths[27] of individual bundles, which is the last missing information before detailed harness drawings and harness installation drawings can be established. They contain all necessary information to allow production, installation and removal of a bundle as a unit. This includes the design of the conduit fittings for bundle protection (if any), sleeves as well as terminations such as connectors or lugs, to name but two.

Finally, with the bundles and their pathways fully determined, appropriate brackets can be designed. While it is possible to design the brackets for installation of equipment units as well as pipes and tubes earlier, the detailed design and production of the brackets for electrical harness' attachments can only start now. This is quite late in the development process, considering that brackets for harnesses need to be installed before bundles can be attached to them.

Clearly, the zonal approach requires people who pull the relevant information from the various Systems Engineers in order to integrate all systems in that particular zone. In this context it is important from a process and leadtime point of view that the PDs are all ready and frozen at their planned milestones. If they are not, zone integrators will not be able to deliver their data on time either. Imagine, for example, that individual cables need to be changed as a result of design changes to equipment units resulting from changed customer requirements. This could modify the grouping of cables into bundles. As a result, the overall length of the bundle could change as well, and the original brackets might have to be redesigned, reproduced and replaced in the aircraft. All this would consume a significant amount of time and cost, not to mention the frustration of the people involved. It is therefore of utmost importance to design the PDs as robust as possible. To this end:

- the aircraft's development schedule plan should entail some margins around the PDs' delivery milestones to compensate for possible delays;
- it should be investigated whether the bundles for the test aircraft – where the weight targets applicable to series aircraft may not have to be achieved as these aircraft are not or not yet delivered to customers – could carry extra wires so that design changes do not automatically lead to changed harnesses and brackets;
- Contractual Definition Freeze (CDF) milestones, after which airline customers are not supposed to change their specifications any more, should not be allowed to shift to the right. This would cause disturbances in the design and production process resulting in higher overall cost.

In conclusion, it is worth noting that not only the complex design and development of airframes and on-board systems, about which more will be said in subsequent chapters, represent a challenge for the project's management team. It is also the design processes for the installation of equipment units and their interconnecting elements, in particular for the electrical harnesses, which need high management attention. Objectives will be successfully achieved only if proven design processes are followed and deadlines are kept.

As stated above, the aircraft's EIS finishes the Development sub-phase. However, further development work for design improvements or refurbishments may still need to be done. But the overall Project Management challenges for the subsequent life-cyle

---

27 The length of electrical harnesses cannot be determined very precisely. This is because different variables influence the equation, such as: wire properties, number of bendings along a pathway, sagging, degree of wire twist, wire and cable diameter and so on. The interdependencies between these variables is still poorly understood.

phases are by far not as big as the ones during the Development sub-phase.[28] This book therefore concentrates on the Development sub-phase only, as it is during this period where the managerial challenges are biggest and where the risks of not meeting the project objectives are highest. It is therefore now time to have a look at some of the managerial disciplines, which contribute to integrative management.

MAIN CONCLUSIONS FROM CHAPTER 3

- The life-cycle of a commercial aircraft is broken down into sequential phases.
- One of the top life-cycle phases of any aircraft programme is usually called 'Development'. It can be further sub-divided into a variety of sub-phases, one of which is denoted here as 'Development sub-phase'. From a Project Management point of view it is this sub-phase which is the most demanding period of all life-cycle phases and sub-phases.
- Individual sub-phases are separated by checkpoint milestones where it is checked whether all the deliverables required at that milestone are available and at the right level of maturity. If a checkpoint milestone is not achieved as planned, this does not mean that the entire development project comes to a standstill. Instead, the checkpoint milestone formally remains at 'not achieved' until all non-conformances are removed, while overall development activities continue. Removal of non-conformances must happen as soon as possible after the checkpoint milestone has taken place. This requires additional Project Management attention.
- There are, however, three milestones, which are so fundamental that they could indeed bring the development to a halt if milestone objectives or deliverables are missed: Authorisation to Offer, Go-Ahead and Type Certification.
- The development of a commercial airplane encompasses not only a variety of test aircraft to be build, but also a considerable number of larger and smaller test specimen and test installations – all of which to be project managed during the Development sub-phase.
- There are two technical design processes with a high potential for adversely affecting development cycle time: the Loads-Design-Loop as well as the design of systems and of their installation on the aircraft. From an aircraft development project management perspective, it is in particular these two processes which require significant management attention.

---

28 Note that this is except for major non-achievements during the Development sub-phase – possibly requesting a significant continued Project Management activity beyond Entry into Service.

4

Overview on Essentials for Integrative Project Management

As was claimed earlier, the development of a commercial aircraft project demands a Management approach (as opposed to a technical approach, which is not the subject of this book), which is different to the one applied to most other projects. The proposed approach is integrative and essential at the same time. It is integrative because it integrates state-of-the-art Management disciplines on the basis of an integrated project architecture. These disciplines comprise:

- Systems Engineering;
- Life-Cycle Costing;
- Rapid Development;
- Project Management;
- Total Quality Management; as well as
- Multi-Cultural Management.

However, at the same time, it limits the application of these Management disciplines to some fundamental essential elements. This is because a commercially-driven development cannot afford the application of the above disciplines with the same level of intensity and sophistication which are commonly applied by government-funded projects (for example military aircraft, large space projects). Cost reductions can be achieved through concentration on the essential strengths of such disciplines while combining them in an intelligent and pragmatic way. Let us explore the essential elements of some of the state-of-the-art Management disciplines for use in commercial aircraft development.

SYSTEMS ENGINEERING

Introduction

Systems Engineering is the branch of the Engineering discipline concerned with the development of systems. A system is generally described as an assembly or combination of interrelated elements such as people, complete products and/or product parts, services and processes that provide a capability to satisfy a stated need or objective. The system's elements are understood to be working together perfectly to achieve a common objective. According to this definition, modern aircraft can be described as systems, even as highly complex systems. More precisely now, Systems Engineering can be described as a structured process of bringing together a variety of (possibly disparate) functional and/or hardware elements into a larger system that satisfies the customer's needs and requirements.

It is the increasingly complex and interdependent design evolution which has required a more explicit application of Systems Engineering on modern aircraft compared to the past. Gone are the days where a few very talented individuals were able to keep track in their minds of the complete set of functionalities and its implementation in the airplane's systems. It is Systems Engineering techniques which today establish the confidence that the development of a system will be accomplished in a sufficiently disciplined manner to limit the likelihood of errors that could impact customer requirements and aircraft safety. As T.J. Petersen and P.L. Sutcliffe of Boeing's Commercial Airplane Group point out – based on their experience with the application of Systems Engineering techniques to the Boeing 777 – 'there are no major drawbacks with this approach, only advantages – provided the systematic approach is started early enough and is followed by all concerned.'[1] According to S. Jackson, 'The basic point of Systems Engineering...[is] that the functions, performance requirements, and all the constraints will have been so well defined that the design will ... be much easier.'[2] As can be seen in Figure 4.1, Systems Engineering extends well beyond Product Engineering.

Market and Mission Analyses

In order to satisfy customer needs, Systems Engineering requires first of all the careful analysis of to-level goals or objectives to be fulfilled by the system (here: aircraft) under development. The key element of this analysis is to identify customer and product needs and it must include direct communication with potential customers for proper understanding of their demands.

> 'It may seem unsurprising that a manufacturer would seek input from its customers before deciding on a new product, but for decades Boeing policy had been to dream up a new plane, design it, make it, and then sit around and hope that enough people would buy it. But the costs and uncertainties, helped by intense competition in the market, had necessitated a new approach – Boeing had to find out what the customers really wanted.'[3]

> 'There is an important interaction between customer airline needs and the manufacturer's product lineup. Both Boeing and Airbus have a lineup of products that reflect the large up-front capital costs involved in designing, testing, and manufacturing a jet aircraft. Because of these large expenses, a small number of designs form the basis for the derivative aircraft types offered by both manufacturers. Both companies offer a large number of variants from these base designs with differentiated product characteristics (range, speed, fuel economy, seat and cabin space, etc). The primary purpose of the market analysis phase is to identify ... whether [the needs of potential customer airlines] ... can be met with current products, variants of these products,

---

1   Petersen, T.J. and Sutcliffe, P.L. (Boeing Commercial Airplane Group, Seattle, Washington) (1992), 'Systems Engineering as applied to the Boeing 777', *Aerospace Design Conference* (Irvine, California: AIAA 92-1010), 3–6 February 1992, p. 2 (reprinted by permission of the American Institute of Aeronautics and Astronautics, Inc.).
2   Jackson, S. (1997), *Systems Engineering for Commercial Aircraft* (Aldershot: Ashgate Publishing), p. 79.
3   Sabbagh, K. (1996), *21st Century Jet. The Making of the Boeing 777* (London: MacMillan Publishers), p. 13.

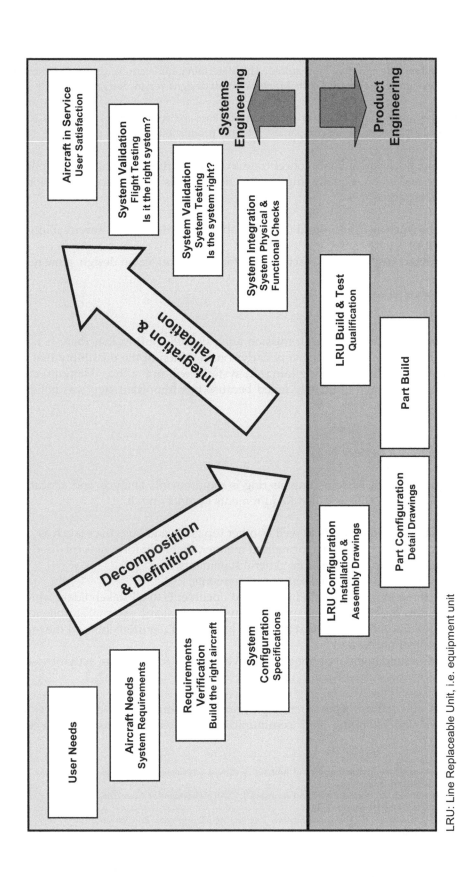

LRU: Line Replaceable Unit, i.e. equipment unit

Figure 4.1   Systems Engineering versus Product Engineering

Based on: Jackson, S. (1997), Systems Engineering for Commercial Aircraft (Aldershot: Ashgate Publishing), with permission.

or if a new aircraft design may be required. This determination is crucial due to the extreme variation in design cycle time and costs for new aircraft designs versus derivative ones.'[4]

'An increasingly important aspect of this analysis phase for both Boeing and Airbus is the competitive analysis of the other companies' potential product offers and/or plans.'[5]

Analyses should, however, not only concentrate on the aircraft as such with its capabilities of speed, range and so on. Equally important is the analysis of the aircraft's mission(s), which includes aspects of:

- operations (such as turn-around times, reliability, maintainability, serviceability, availability);
- human factors (such as man/machine interfaces, comfort, cabin design, crew rest areas);
- maintenance; as well as
- disposal;

to name but a few. Only thorough mission analysis will confirm that there is a real need for a new aircraft. 'It forces the very early consideration of the possibility that, for example, minor upgrades of existing [aircraft] systems will not suffice. Many [aircraft] system developments have ultimately failed because this important step was not fully explored.'[6]

Requirements Engineering

The next major aspect of Systems Engineering is requirements analysis and allocation. The corresponding discipline is called Requirements Engineering. It:

- captures the customer needs as well as other top goals and objectives (such as Business Case objectives) in a formalised manner, that is, in form of formalised requirements as opposed to more general statements;
- converts them into individual sets of requirements;
- cascades those in structured and formalised documents to all stakeholders which need to know them (such as as specifications to internal and external suppliers);
- ensures full traceability between all issued lower-level requirements and the top-level customer needs; and
- represents a fundamental prerequisite for Verification & Validation activities (see below).

The success and failure of an aircraft programme depends to a very large extent on the proper understanding and communication of requirements. This is why

4   Spitz, W., Golaszewski, R., Berardino, F. and Johnson, J. (2001), *Development Cycle Time Simulation for Civil Aircraft*, NASA/CR-2001-210658, pp. 3–2.
5   Spitz, W., Golaszewski, R., Berardino, F. and Johnson, J. (2001), *Development Cycle Time Simulation for Civil Aircraft*, NASA/CR-2001-210658, pp. 3–3.
6   Eisner, H. (2002), *Essentials of Project and Systems Engineering Management*, 2nd Edition, (New York: John Wiley & Sons), p. 195.

professional Requirements Engineering is so important during the early stages of aircraft development.

Systems Architecture

For requirements flow-down to work, it is a key Systems Engineering principle to view a commercial aircraft as a hierarchy, in which lower level elements, such as sub-systems like flight control, water/waste and so on, are subordinate to the aircraft, see again Figure 1.1. Remember that the actual aircraft itself is a subordinate to a higher-level aircraft system.

The decomposition of a system into modules (or units) with defined interfaces to other modules is what is called the system's architecture. It is a very important aspect of Systems Engineering.

'The main reason for showing an aircraft architecture is to illustrate its hierarchical nature ... However, the hierarchy is not a hardware description. It is merely a set of 'buckets' into which requirements [and functionalities] can be placed. When the aircraft hierarchy is defined, one of the first steps ... towards aircraft system synthesis will have begun. The hardware selection is [but] the final step in system synthesis ... Studying the aircraft architecture helps us understand what is really included in the aircraft 'system'... .'[7]

The necessity of decomposing complex systems into smaller units stems directly from the fact that they are just too complex to be managed by the human brain. The complexity needs to be 'cracked' into levels of complexity, which can be understood and managed. Unfortunately modularity introduces interfaces and interfaces generate additional complexity and cost. Adding only one more module roughly doubles the total number of potential interfaces, a fact not always appreciated by engineers and managers.[8]

In doing so, Systems Engineers start with an analysis of functionalities (called 'functional analysis'), which respond to requirements derived from customer needs. In aircraft design, functionalities are assured via dedicated systems or systems' equipment like avionics, landing gear, hydraulics system and so on. As a functionality can contribute to the fulfilment of more than one requirement and as it may take more than one functionality to respond to a requirement, analysis of functionalities inevitably leads to a decomposition of functionalities into higher and lower-level functionalities. This is not an easy task. For example, the requirement of retractable landing gears leads to static, dynamic and kinematics functionalities with lower-level functionalities such as provision of forces and moments, control and sensor electronics, mechanisms to hold the gear in certain positions, and so on. These functionalities need to be allocated to hardware units (structures and systems with their equipment), such as the hydraulic system, which makes

---

7   Jackson, S. (1997), *Systems Engineering for Commercial Aircraft* (Aldershot: Ashgate Publishing), p. 9.
8   Breaking a complex, physical system down into many virtual modules to deal with them as representations of the physical world (that is, in computer systems, mock-ups or simply on paper), can generate new problems as all tools to manage representations have their limits too. For example, the potential number of interactions N among modules n approximately doubles with every new module added. The equation is given by

$$N = \sum_{i=2}^{n} \frac{n!}{[(n-i)!i!]}$$

This is why digital models become slower when they increase in size, given the same hardware capacities.

use of hydraulic pressure to retract the gear. However, the same hydraulic system is used to extract the flaps and other flight control units. Remember that the aircraft's structure has to assure important functionalities too (see Chapter 1).

One can learn from this example, that the decomposition of an aircraft into lower-level hardware modules such as structural elements or equipment units – called the product breakdown – is heavily linked to the decomposition of the functionalities. Fortunately, the link is not completely rigid: the analysis of the functional decomposition is more important for the understanding of the aircraft's systems, while the analysis of the aircraft's product breakdown is more important on the structural side.

Interface Management

Within the chosen architecture there are many different functional and physical interfaces between modules. Efficient management of interfaces is therefore another major element of Systems Engineering. Life will be much easier for everyone during the Development phase if interfaces remain as stable as possible. Stability of interfaces is so important as they represent the key external constraint of a module. Changing the interface could mean changing the entire module and the repercussions of this could be very expensive and time consuming. Freezing interfaces early and for a defined and communicated period of time is therefore good Systems Engineering practice.[9]

Aircraft design is typically suffering from the problem of interface instability. Final design of systems and equipment in particular tends to be late because this is the area where engineers will try hardest to introduce the very best, latest technologies until as late as they possibly can. For example, as a result of changes to the pneumatics system, the interface points to the supporting aerostructure often change too. If these changes occur at a point in time of the development cycle where the structure has already been built, the only way to be able to accommodate the system changes would be through retrofitting and this could turn out to be very expensive indeed. However, if things need to get changed, the change could perhaps be limited to certain modules only – instead of changing the entire architecture. Therefore, proper design of the system's architecture is a key task, which deserves a significant amount of management attention already early in the design process.

Physical Integration

The definition of interfaces does, however, not only depend on the functional and structural architectures. To a large extent it also depends on the aspect of assembly and physical integration which defines *how* the individual modules are brought together. Interfaces generated between modules must reflect the assembly processes needed, that is, the build sequence of architectural modules, in connecting pieces of hardware and software to make larger builds. Analyses of assembly and integration processes may well change the original layout of the product architecture. In other words, the intended assembly

---

9   Note that interfaces can be changed at a later stage when the rest of the design is more mature. This may for example be necessary to save aircraft weight.

processes provide a feedback to the design of the product architecture via the selected interfaces.

To be able to interoperate in the intended manner, successful physical integration requires that *all* the interfaces are correct. Architectural design is therefore an iterative process trying to find the optimum solution for functional *and* structural breakdown *as well as* assembly and integration. Experience shows that many projects have failed because of lack of proper integration.

Verification & Validation

Once the design is more mature, it needs to be checked whether it fulfils the requirements. In Systems Engineering this is done by a process called Verification & Validation (V&V). Verification attempts to confirm that the design meets the requirements, while Validation checks – for example through simulation, tests and evaluation – whether the final product meets the customer needs. In combination with Requirements Engineering, V&V contributes significantly to achieving a safe and mature product of high quality. It also contributes to cycle time and cost reduction as non-compliances are detected early and can still be corrected at relatively low cost.

Configuration Management

Commercial aircraft usually are products with a predefined number of configurations from which to choose or are built to order.[10] Aircraft 'options', 'variants' and 'derivatives' are all examples of different product configurations based on a mostly similar underlying product design.

However, aircraft manufacturers face a huge challenge to keep track of the complex network of component design change information (including versions and revisions), or other engineering information (such as design verification at each change level) that goes into building the aircraft.

> 'This may seem very basic, yet in 1997 ... Boeing completely lost control of the aircraft configuration process. In a huge and costly production meltdown the 747 production line had to be shut down because of out of sequence assembly.'[11]

With so many different variations, control can only be exerted by applying structured processes in a disciplined manner. The technique of Configuration Management offers just that. As a dedicated discipline within Systems Engineering, it is defined by the renounced Institute of Configuration Management (ICM) as 'the process of managing products, facilities and processes by managing the information about them, including changes, and ensuring they are what they are supposed to be in every case'.[12]

---

10 Building to order in theory involves a very high number of variations. However, in reality it is more often based on a combination of predefined underlying configurations with some added customisation.
11 Lawrence, P. and Scanlan, J. (2007), 'Planning in the Dark: Why Major Engineering Projects Fail to Achieve Key Goals', *Journal of Technology Assessment and Strategic Management* Vol.19 N°4, pp. 509–25.
12 http://www.icmhq.com, accessed 10 October 2009.

Very often Top Management, and indeed also many engineers, underestimate the fundamental contribution of Configuration Management in achieving high maturity. If, for example, the company cannot document the precise aircraft configuration for the customer airline, it will be difficult for the airline to do repairs efficiently. Also, safety might be at stake if not every single item assembled and used on the aircraft is known. Airworthiness authorities therefore rightly insist on high standards of Configuration Management within aircraft-producing companies.

Development-related requirements, documents, methods, processes, to name a few, are all subject to change too. Most of these changes will eventually lead to changes in the aircraft configuration. In other words, Configuration Management represents the culmination of the wider and more upfront aspects of change control. But even the more upfront changes need to happen in a controlled manner. To ensure this, changes need to be communicated to everyone concerned and potential repercussions of the change need to be investigated. Controlling upfront change is a prerequisite for good Configuration Management.

LIFE-CYCLE COSTING

Introduction

Life-Cycle Costing (LCC) encompasses all those costing techniques that take into account both the initial costs as well as the future costs and benefits of a programme over its entire life-cycle. In fact, NASA describes life-cycle costs as 'the total direct, indirect, recurring, non-recurring, and other related costs incurred, or estimated to be incurred, in the design, development, production, operation, maintenance, support, and retirement over the planned life span of a project'.[13]

LCC was originally introduced by the US Government to manage the life-cycle costs of military programmes. It distinguishes four main life-cycle cost categories, which correspond to four main life-cycle phases:

1. Research, Development, and Test and Evaluation (RDT&E);
2. Acquisition (or Procurement);
3. Operations and Maintenance (O&M); as well as
4. Retirement and Disposal.

Whereby all costs of these phases are to be funded by the US Government as the customer of military programmes, it is nevertheless important to analyse the interdependencies between the phases: How can, for example, the costs of one phase be reduced by investing more during another? What does this mean for the total life-cycle costs taking financing costs into account? For any given programme, is the distribution of life-cycle costs over the years compatible with the annual budgetary process? In the case that a government has reached the spending limits of its fiscal year(s) budget, can costs be shifted at relatively short notice from one life-cycle phase to another? It is to answer these types of questions where the management of life-cycle costs has become very important indeed.

---

13 Shishko, R. et al. (1995), *NASA Systems Engineering Handbook* (Pasadena: Jet Propulsion Laboratory), NASA SP-6105.

Applying Life-Cycle Costing to Commercial Aircraft Programmes

Nowadays, the principles of LCC are also applied to commercial aircraft programmes. Here (1) is called Non-Recurring Costs (NRC),[14] (2) unit price, (3) operating costs and (4) retirement costs. But besides different terminologies there are fundamental differences to the application of LCC when comparing commercial with military aircraft programmes: the NRC of the former are initially to be borne by the aircraft company and can only be later amortised if a given number of aircraft has successfully been sold to airline customers. While therefore in military programmes the unit price essentially consists of Recurring Costs (RC) – representing the unit production costs plus the profit for the aircraft company – in commercial programmes the unit price is further increased by the cost of NRC amortisation.[15] In other words, life-cycle cost categories and phases do no longer correspond to each other.

Whilst military projects normally have only one customer (that is, the government), the commercial aircraft business is confronted with some competition between aircraft manufacturers on the one side and with customers (airlines) requesting different types of aircraft on the other. The aircraft company can therefore realistically not expect to receive any additional funding from customer airlines to finance its development cost[16] even if this would reduce purchase price or airline operating costs. Instead, it is confronted with a strong aircraft sales price battle to win market shares and orders. Thus, in reality no obvious link exists between the costs for development and production of the aircraft on the one hand and its sales price (market price) on the other.

Finally, the aircraft company is also expected to guarantee direct operating costs to the customer, among others, while the same customer could not care less about the risks associated with the development of the aircraft. The risk to launch a new aircraft lies entirely with the aircraft manufacturer in competition with other aircraft manufacturers, while the risk to operate the aircraft lies with the airline in competition to other airlines.

Thus, life-cycle cost management in the military domain primarily serves the customer needs for fiscal year budget planning, while in the commercial aircraft business it is a necessity to analyse the commercial viability of a new aircraft.

Despite these differences, some of the experiences gained during military programmes are also applicable to commercial aircraft. For example, it turned out that the total life-cycle unit cost of an aircraft is to the largest extent defined already during the Development phase.[17] This can be seen in Figure 4.2. Note in particular that already at 'Go-Ahead' around 80 per cent of the total life-cycle costs are defined whereas only around 20 per cent of the NRC are spent.

After investigating 2,000 components, Rolls-Royce found that '80 percent of their production cost was attributable to design decisions.'[18]

---

14 NRC depend very much on whether an entirely new aircraft is to be developed or a derivate of an existing model. '[Non-recurring] costs of a major derivative can incur NRC of 75 percent of those of a new design while for a relatively minor derivative, the costs can be from ten to 25 percent of a new design.' Source: Spitz, W., Golaszewski, R., Berardino, F. and Johnson, J. (2001), *Development Cycle Time Simulation for Civil Aircraft*, NASA/CR-2001-210658, pp. 3–6.
15 The cost of NRC amortisation usually accounts for 5–10 per cent of the list price of a commercial aircraft.
16 This is except any down payments made by airline customers when purchasing a new aircraft.
17 The total life-cycle unit cost of an aircraft can be approximated as the cost of (1) divided by the number of aircraft produced, plus the sum of cost of (2), (3) and (4) per aircraft ((1) to (4): life-cycle phases as introduced in this chapter).
18 Smith, P.G. and Reinertsen, D.G. (1998), *Developing Products in Half the Time*, 2nd Edition, (New York: John Wiley & Sons), p. 243.

Slightly different, 'a British Aerospace study has reported that 85 percent of a product's manufacturing cost will depend on choices made in the early stages of its design.'[19]

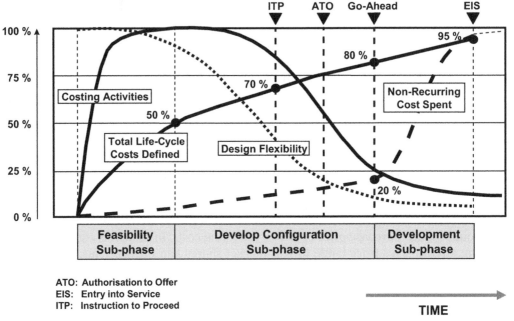

Figure 4.2   Typical levels of defined and spent costs for commercial aircraft development

Whether in the end one believes in a 20/80 or rather a 15/85 law does not matter so much here. What does matter though is the conclusion that any attempt made during the early stages of design to reduce costs for later life-cycle stages could provide a definite competitive advantage.

What can also be seen in Figure 4.2 is how the design flexibility, which could principally be used to reduce life-cycle costs, quickly loses impact over time. At Go-Ahead, not much design flexibility is left.[20] Finally, Figure 4.2 shows the relative amount of activities by the Costing or Cost Estimation department for costing all life-cycle phases. Not surprisingly, costing activities are mainly required before Go-Ahead.

The Business Case

In order to minimise the financial risks associated with commercial aircraft programmes, the launch of an aircraft development should always be based upon an extremely well-prepared Business Case. This Business Case defines the expected cost and revenue streams

---

19 Smith, P.G. and Reinertsen, D.G. (1998), *Developing Products in Half the Time,* 2nd Edition, (New York: John Wiley & Sons), p. 243.
20 This represents a strong message for Chapter 7 where the setting of priorities for the Development sub-phase will be discussed.

as well as the selected funding strategy. Every aspect affecting LCC – such as application of new technologies, material maturity and prices, sub-contracting opportunities, design capabilities, skills, capacities, certification requirements – to name but a few, should be analysed and considered in the Business Case.

The completed and fully documented Business Case serves as a proof for the commercial viability of the programme. Shareholders as well as banks base their financial decisions on it. It is therefore a very important piece of work indeed. When at last the aircraft development is launched and the Business Case is then sealed, key objectives, which stem from the Business Case, are well known on aircraft level:

- the performance characteristics of the aircraft to be guaranteed to the customers such as weight, speed, range, fuel consumption, operating costs, turn-around times, maturity levels and so on;
- the well selected Entry into Service (EIS) date to gain as much a competitive advantage as possible;
- the Return on Investment (RoI) promised to all those who provide the funding, such as shareholders, banks and so on.

After launch, constant monitoring of these objectives is necessary to take rational decisions.

The establishment of the Business Case leads to an in-depth understanding of the interdependencies between development costs, unit production cost and operating costs. The knowledge of these interdependencies will later help to establish a pragmatic cost control. For example, an unfavourable variance to the development cost may be justified if it can be demonstrated that unit production cost can be reduced, for example, through some kind of expensive Design to Cost (DtC) activity. This, however, requires the constant monitoring of all life-cycle costs. In any case does the Business Case represent the reference, against which cost control must be performed. It also defines the overall level of contingency, which the company foresees for the aircraft programme. For more details about Business Case set-up, please refer to Appendix A1.

For the purpose of this book it is also important to note that because the Business Case sets stringent limits to the (non-recurring) development costs it also prescribes an upper limit for what the aircraft development company can spend on management and programme control: it has to be just about enough to ensure overall programme success. This limits Management to implementing some pragmatic essentials rather than the full-blown management methods and tools required – and paid for – by governmental customers.

RAPID DEVELOPMENT

Introduction

The time it takes to develop a product until it is mature enough to be introduced to the market is called the development time, cycle time or development leadtime. The longer the development time the higher the risk of failure in the market place. This is because a product idea usually is borne out of analyses of future market needs. Essentially, the product ideas are therefore based on forecasts. They inherently have the tendency to become more unreliable the longer the forecast period lasts. It is a lot easier to judge the market behaviour and needs for, say, the next year, than to forecast what the market will need in five or seven years time. Aircraft developments take several years and the risks of not meeting customer demand are high. Shortening the development leadtime is therefore an essential element of risk reduction for the aircraft manufacturer.

The classic example for this was the Anglo-French Concorde, a supersonic aircraft with a very high fuel consumption. Initially, customers did not worry about high fuel consumptions as fuel prices were low and speed was the perceived top priority. However, EIS was already much delayed when shortly afterwards fuel prices rose dramatically during the early 1970s as a result of what became known as the 'oil shock'. Operating the Concorde became so expensive that all airlines except two cancelled their orders. All of a sudden airlines preferred fuel-efficient aircraft. Had the Concorde been available on the market earlier, a lot more aircraft could have been in operation.

Apart from achieving short leadtimes in the first place, keeping leadtimes as projected is also important to keep costs under control. For example, aircraft performances contractually guaranteed to potential customers nowadays often include guaranteed EIS dates. If the latter cannot be met, the corresponding contractual clause could turn out to be very expensive for the aircraft company. Even worse, this would come on top of the already generated unplanned costs associated with keeping the 'marching army' of engineers running for longer than originally scheduled.

But despite shorter development cycles, things have to be done right. This is because of the 20/80 rule described earlier: late changes are particularly expensive and must be avoided as much as possible. Too ambitious a development cycle reduction would increase the risk of not having mature design solutions, resulting in late changes, and consequently delays. Unfortunately, recovery of a delayed project becomes increasingly more expensive for each succeeding project phase. The cost increase in fact follows an exponential law. Thus, development cycles need to be challenging but must remain realistic.

It is important to note that cycle time reduction is not only vital for the Development sub-phase of a product but also for the costs and responsiveness to customer needs during Series Production.

> 'In his "1999 Vision", Ron Woodard, [Boeing's] Renton Division vice-president and general manager, has declared that cycle time for the [Boeing] 737 and 757 programs, from definitive agreement to delivery, will be reduced to six months [the 757, for example, was by then integrated over 315 manufacturing days]. Boeing Commercial Airplane Group Executive Vice-President Bob Dryden has stated that "by the year 2000, we must reduce our cycle time well over fifty percent, and possibly seventy-five percent, in order to retain our market share and competitive edge."'[21]

Leadtime reduction opportunities during the Production phase always have the potential to lower the cost of borrowing, simultaneously increase cash flow and therefore lower the overall capital risk to the producing company. However, for the purpose of this book it is sufficient to consider Rapid Development techniques for the Development sub-phase only. Some of the most important Rapid Development techniques for this sub-phase include:

---

21 Blair, S.A. and Premselaar, T (1993), 'Cycle-Time Reduction – Second Stage of the Quality Revolution', *Manager* January-February 1993, p. 11.

- Concurrent Engineering based on commonly shared data as well as overlapping;
- stable interfaces (again!) between aircraft components, defined and frozen as early as possible;
- improved communication through co-location of multi-functional teams;
- efficient Risk Management in order to compensate for some of the deficiencies of the aforementioned; as well as
- full exploitation of the capabilities of modern Information Technologies as a prerequisite of the aforementioned.

Concurrent Engineering

Concurrent Engineering or Concurrent Product Definition (CPD) as used in aeronautics is a systematic approach to the integrated and simultaneous design, analysis and planning of a series of interrelated airplane parts, as well as the creation of the supporting tool designs and Numerically Controlled (NC) programmes for those parts. Its one major cornerstone is a common database, which is fed, continuously updated and shared by all stakeholders involved in the aircraft development. Figure 4.3 demonstrates this principle of Concurrent Engineering. Note that a common database should not only be used for the development of aircraft structures and systems but should also include the concurrent development of manufacturing concepts, facilities and jigs & tools.

Individual design and production teams often feel that *their* share of the overall development effort is of greatest importance and that the headaches in design are only due to the requirements of the other 'less important' teams. It is the integrated working using the same common database which forces people to overcome this type of silo mentality with all its undesired effects on integrated working.

The use of a common database also allows for a reduction of what otherwise still is a major reason for late engineering change: errors and reworks due to bad part interfaces and fit up. Significantly reduced numbers of changes after initial drawing releases to Manufacturing result in considerable cost reductions, further reduced development leadtimes and an earlier production ramp-up. Applying a commonly shared database therefore contributes enormously in gaining confidence that eventually the completed airplane will represent a valid compromise of the knowledge, experience and desires of *all* stakeholders involved.

While using the common base of design and development data, Concurrent Engineering also means that the different Functions like Engineering, Manufacturing, Procurement, Quality and so on are working *in parallel* and not in a timely sequence. To make this possible every project activity should start as early as possible and should not wait to start until another is completely finished.

Parallel, concurrent working is represented in a schedule plan by what is called 'overlapping'. It is the other major cornerstone of Concurrent Engineering. Overlapping is possible wherever an overall interdependency between a (internal or external) customer and (internal or external) supplier can be broken down into lower-level interdependency steps distributed over time. For example, instead of waiting with the manufacture of a part until the final drawing is released, the rough shape of the part could already much earlier be milled from a billet, provided a preliminary drawing could be released, which would describe the required rough dimensions. Issuing a series of drawings with increasing levels of dimensional details can kick-off a stepped manufacturing process, which can start *and* finish earlier than compared to the classical case where Manufacturing waits for the final drawing. The total duration of

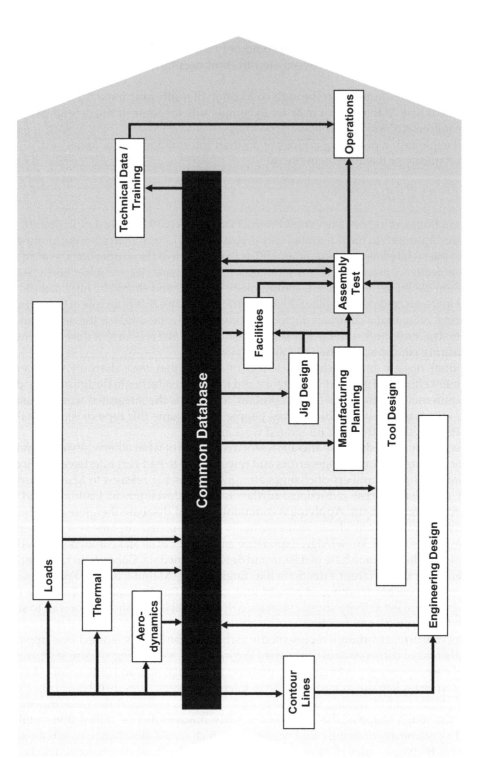

Figure 4.3   The principle of Concurrent Engineering

the stepped manufacturing process may be longer, but manufacture can be finished earlier. In view of Rapid Development, this is what counts in the end.

Another example: tooling must be designed and manufactured prior to the start of manufacture of aircraft components. Traditionally, tool design and manufacture started after release of final drawings. In order to save cycle time, however, the design of the tooling is preferably performed during largely the same period where the Engineering design of the parts is taking place and not afterwards. Using Concurrent Engineering techniques this would be possible.

Frozen Interfaces

Even more cycle time reduction can be achieved if the modularity introduced by Systems Engineering is combined with stable interfaces. This would allow for Concurrent Engineering not to be applied to the aircraft as a whole, but individually to aircraft sub-assemblies (at least the larger ones such as the fuselage or the wing). For this to work, there must be stable interfaces between the components, so that design changes affecting one component do not propagate across the interface to another. Thus, the stability of the interfaces must be such that the design of interfacing components can be performed more or less independently from each other. The full perimeter of what is depicted in Figure 4.3 would then be applied to each component individually, but in parallel for all components.

Early seeking for and defining and freezing of stable interface opportunities is therefore not only very beneficial for the application of the Systems Engineering methodology, but also is another major contributor to achieve Rapid Developments. In fact, stable interfaces depend very much on the clever architectural breakdown of an aircraft system into smaller modules. One therefore has to confess that Systems Engineering in general and architectural design in particular are of pivotal importance to development leadtime reduction too. This fact is very often underestimated, if not completely ignored.

As the number of interdependencies significantly increases as a result of the application of Concurrent Engineering, professional Interdependency Management is required in addition to interface management in order to avoid a drastic increase in project risks. Note that interdependency and interface management as prerequisites to effective Rapid Development techniques in turn require excellent standards of communication as well as stakeholders, which are committed to stick to agreements.

Co-located, Multi-functional, Integrated and Integrating Teams

As Concurrent Engineering represents a simultaneous, multi-functional approach requiring improved levels of communication, why not ensure co-locating the various Functions into Integrated Product Teams (IPTs)? Established early in the Development phase, that is, already during the Feasibility phase, IPTs bring together all the people who have an influence on how the aircraft is to be developed and placed in the market, such as Engineering, Manufacturing, Marketing, Sales, Contracts, Procurement and so on.[22]

'IPTs are one area that has significantly altered the way large systems [such as aircraft] are developed and built. What IPT means is that parts of the aircraft are developed as 'products',

---

22 At times, IPTs are also referred to as cross-functional teams or multi-disciplinary teams.

such as the wing, nose, empennage, etc., rather than as functional areas, such as avionics, mechanical, etc. For example, the nose ... [IPT] would all work together, including the electrical, mechanical, avionics, etc. personnel. In addition, manufacturing and other personnel would be members of the nose IPT. This arrangement helps communications enormously.

A key aspect of the IPT is the principle of requirements ownership. That is, the IPT owns all aspects of requirements. For example, the IPT is responsible for implementing requirements allocated from higher level IPTs. ... Secondly, the IPT is responsible for developing and allocating requirements within its own domain. Finally, the IPT is responsible for every aspect of implementing its requirements. This includes design, procurement, fabrication, installation, and verification.'[23]

Once the Development sub-phase has started, IPTs evolve into multi-functional Design-Build Teams (DBTs) – concentrating on *design* and *build* of individual components rather than on conceptual studies on aircraft or major assembly level.[24] DBTs are the primary approach to implement the design-build process. They operate on a principle of partnership between Engineering, Manufacturing, Procurement, Materials and Processes, Customer Services, and other Functions as specific cases demand.

The design-build process is a commitment to work together to develop a more producible, more error-free design and is implemented by secondment of functional staff into DBTs. Engineering is responsible for the design and will incorporate Manufacturing's producibility knowledge to the maximum degree possible. It will define the systems and structure concurrently, and will make sure all parts are defined and fit together. Manufacturing will verify the producibility by accomplishing sufficient planning and tool design to provide valid knowledge to the Engineering design. Multi-functional DBTs are therefore working on the basis of a continuous flow of information; they are information-based teams. As put by R.Holman et al.: 'Instead of using a linear approach to collect information, make a decision, and then base other decisions on the first one, information-based teams solve problems continually and combine their findings frequently.'[25]

In modern aircraft development projects, co-location is a fundamental principle to make the integrated, multi-functional and concurrent approach a reality.

'I would love to have a building in which the entire organization was within fifty feet of each other. With ten thousand people that turns out to be really hard. So you start devising other tools to allow you to achieve that – the design-build team. You break the airplane down and bring Manufacturing, Tooling, Planning, Engineering, Finance and Materials all together in that little [co-located] group. And they are effectively doing what those old design organizations did on their bit of the airplane.'[26]

But co-location comes at a price: co-locating individuals from different Functions, which most likely have never worked together before, requires a good deal of team building effort. Without a good team spirit and clear understanding of each other's roles and responsibilities, a co-located multi-functional team will not operate properly. The team building effort must be even higher if multi-cultural aspects are involved. As no

23 Jackson, S. (1997), *Systems Engineering for Commercial Aircraft* (Aldershot: Ashgate Publishing), p. 142.
24 Some companies just keep the name IPT which would then include the design and build aspect, others change the name, for example to DBT, to mark the change in focus and responsibility.
25 Holman, R., Kaas, H.-W. and Keeling, D. (2003), 'The Future of Product Development', *The McKinsey Quarterly* 3, p. 34.
26 Boeing's Phil Condit quoted in: Sabbagh, K. (1996), *21st Century Jet. The Making of the Boeing 777* (London: MacMillan Publishers), p. 64.

manufacturer develops commercial aircraft any more within its national borders alone, this is most likely the case.

Leadership is another important aspect of managing co-located teams. One criteria for good leadership, which we shall look at in more detail in Chapter 7, is the setting of the right priorities at the right time. Typically, priorities include schedule, cost and quality, but in commercial aeronautics priorities like safety and weight are even more important. Setting the right priorities at the right time is a challenging task, which requires excellency in leadership.

Risk Management

Shortening development leadtimes can be regarded as a huge mitigation plan to reduce the risk of missing market needs. However, essential methods of leadtime reduction, such as early interface freeze and overlapping, generate new risks. For example, early interface freeze based on generous space margins around physical interfaces consumes weight, that is, it puts the overall weight target at risk. Another risk is represented by those components which have been machined very early during the Development phase based on rough dimensions and which could end up as scrapped parts because of design changes. In a sequential world, where Manufacturing only starts machining the parts once it has received the final drawings, this risk is much reduced.

These risks – and certainly many others – need to be addressed. Risk Management is the discipline which deals with them, for example by braking down the top critical and difficult to manage risks into a larger number of smaller risks. The latter can be more easily addressed by implementation of risk mitigation plans. The management of risks is a necessary ingredient of Rapid Development.

PROJECT MANAGEMENT

Introduction

Project Management is a structured and disciplined management approach to ensure that project objectives are achieved within not-to-exceed time and budget limits. With the words of R.D. Archibald, the objectives of Project Management can be described as follows:

> 'to assure that the programs and projects when initially conceived and approved contain acceptable risks regarding their target objectives: technical, cost, and schedule;to effectively plan, control, and lead each project simultaneously with all other programs and projects so that each will achieve its approved objectives: producing the specified results on schedule and within budget.
>
> ... Too frequently, project failures can be traced directly to unrealistic technical, cost, or schedule targets, and inadequate risk analysis and management.'[27]

---

27 Archibald, R.D. (1992), *Managing High-Technology Programs and Projects*, 2nd Edition, (New York: John Wiley & Sons), pp. 4–5.

The evolution of Project Management as a Management discipline received a tremendous push in the 1960s, mainly driven by larger military aerospace projects as well as the Apollo space programme. In all cases, project complexities had significantly increased compared to projects of the 1950s and management tools available then were no longer adequate.

'The forces pushing us toward project management emanate from the inexorable evolution of technology. These fundamental forces are not likely to diminish within the foreseeable future.'[28]

In an attempt to reduce project complexities, the fundamental approach of Project Management is to provide a structure to both the people working on a project as well as the tasks which need to be performed. In addition, tasks are individually schedule and resource planned and monitored. In this way, Project Management primarily helps to achieve cost and schedule targets. Only indirectly it supports the achievement of quality, maturity and performance objectives.

However, effective Project Management not only requires having good control methods and procedures in place in areas like planning, scheduling, estimating, budgeting, work authorisation, monitoring and reporting, but also requires that Project Managers understand and actually use these methods and procedures. The existing company culture therefore plays an important role in defining how Project Management is introduced.

'Implementing or improving Project Management itself requires the use of effective Project Management practices, and must be viewed from a long-term perspective. There is no one best answer that fits all situations. The concepts of Project Management must be tailored to the situation and culture, including the cultural mix of the project team.'[29]

It is to be remembered that commercial aircraft developments represent dynamic projects with many unknowns and 'unknown unknowns' (see Chapter 1). They require adaptation of the classical Project Management methodologies as defined by the various existing Project Management bodies of knowledge.[30] To this end, it is in fact important to note that adequate leadership styles are required to assure that on the one hand the Project Management planning and control functions are in place and are properly used and that on the other there is enough flexibility to deal with the challenges of the dynamic nature of commercial aircraft development projects.

Work Breakdown

The starting point for the Project Management process usually is the decomposition of the total work to be done into manageable work packages. In aircraft design, work packages include, for example, loads calculation, stress analyses, design activities, manufacture and assembly activities, flight testing and many more. Each work package is described

---

28 Archibald, R.D. (1992), *Managing High-Technology Programs and Projects*, 2nd Edition, (New York: John Wiley & Sons), p. 15.
29 Archibald, R.D. (1992), *Managing High-Technology Programs and Projects*, 2nd Edition, (New York: John Wiley & Sons), p. 135.
30 A good overview for project management approaches for dynamic environments is provided by: Collyer, S. and Warren, C.M.J. (2009), 'Project Management Approaches for Dynamic Environments', *International Journal of Project Management* Vol. 27, pp. 355–64.

in a Work Package Description (WPD) covering the content of the task, inputs required, outputs delivered, resources required, as well as the leadtimes needed.[31]

> 'If I could wish but one thing for every project, it would be a comprehensive and detailed WBS. The lack of a good WBS probably results in more inefficiency, schedule slippage, and cost overruns on projects than any other single cause. When a consultant is brought in to perform the role of "project doctor", invariably there has been no WBS developed. No one knows what work has been done, nor what work remains to be done. The first thing to do is assemble the planning team and teach them how to create a WBS.'[32]

Organisation Breakdown

The next questions to be addressed is then: Who is going to be in charge of individual work packages? To answer this question, it is necessary to have a clear idea of the kind of organisation which is selected to support the project. Should it, for example, be a matrix organisation or a classical line organisation or yet another type of organisation? Once that is sorted out, the organisation needs to be broken down into organisational units – resulting in an Organisation Breakdown Structure (OBS) – to which the work packages can be allocated. For each organisational unit one has to define the roles and responsibilities of key individuals belonging to the unit, the way of working between all stakeholders as well as the applicable lines of reporting. Any leadership principles to be applied need to be sorted out too.

Task Responsibility Matrix

If the aircraft is to be developed using co-located DBTs, they are most likely the organisational units, to which work packages need to be allocated. The resulting allocation is documented in what is called the Task Responsibility Matrix. This is the single reference where it is described who does what. It is an important document and may even have contractual and/or legal significance if teams of other companies, sometimes even foreign companies, are involved. The more complex the organisation, the more international and multi-cultural the teams, the more care has to be taken to generate a high-quality Task Responsibility Matrix.

Planning

Each work package needs to be planned. This comprises on the one hand the schedule planning with start and finish dates of individual activities necessary to complete the work package as well as its associated milestones. The logical interlinkages of different

---

31 A Work Package Description (WPD) is also sometimes called a Statement of Work (SoW). The difficulty with the latter terminology is that in the commercial aircraft industry it often represents a mixed bag of requirements, solutions, and tasks. It is good Systems Engineering and Project Management practice to ensure that a SoWs only function is to describe tasks. If requirements are included and not captured in dedicated requirements documents, the risk is that they are ignored and not respected for the design. To avoid confusion, the terminology 'Work Package Description' is used in this book.

32 Devaux, S.A. (1999), Total Project Control: A Manager's Guide to Integrated Planning, Measuring and Tracking (New York: John Wiley & Sons).

activities representing inputs/outputs to each other define potential routes for the plan's Critical Path as well as any margins embedded in the plan. On the other hand, activities are resourced in terms of estimated workload (manhours or number of people) to do the job. This estimate is converted into budgets for the work packages.

In dynamic projects such as commercial aircraft development, the quantity of change resulting from 'unknown unknowns' makes too detailed plans very difficult to maintain. Plans with excessive details are therefore often found to be misleading and are abandoned in favour of a higher-level rolling wave approach. Thus, the planning detail is progressively developed as the project progresses and as more is learned.

Cost Control

Costs relevant to the achievement of the project are usually monitored on the level of individual activities and/or work packages. They are compared with the released work package budgets in order to allow for analyses of variances. The challenging aspect of project cost control is that it not only needs to satisfy the project, which is structured according to the various architectures mentioned before, but also the wider needs of the aircraft programme's Business Case. In addition, cost control needs to comply with the needs of the company's financial accounting structures.

Earned Value

If progress of the work is also measured, it can be compared with the Cost Variances (CVs). One measure of doing this is called Earned Value: it is the value of the work, which has been earned until a given date, expressed in units of cost (such as $ or €). Earned Value represents a better performance indicator than the analysis of pure CVs. If, for example, cost is below budget, it is not clear with CV analyses alone whether this is due to the performed work being cheaper than expected or just less work performed. If actual Earned Value is below budget there can be no doubt that the project is in trouble.

TOTAL QUALITY MANAGEMENT

In the commercial aircraft business, the number of entirely new products (that is, aircraft) generated over time is very small compared to other industries. Thus, failing to meet the customer's quality expectations on one aircraft programme will not only lead to severe loss of reputation but it can also not be quickly corrected by doing things better on other programmes. In fact, it may by then be too late altogether and the company may no longer exist. Thus, for every new aircraft programme, quality must be right with the first aircraft delivered.

However, there are very different definitions for what quality is all about. For example, some regard quality as the degree of conformity with requirements. Others describe quality only as 'fitness for use'. And again others disregard the existence of an absolute standard for the definition of quality all together. They rather believe quality to be the *perception* of it by customers relative to the perceived quality standard of the competitor. In international projects, quality is often judged subjectively leading to different opinions about the level of achieved quality. This is because different cultural values define different standards of quality. This could become an issue if manufacturer and customer belong to different cultures, a situation quite common in commercial aviation. In any case is quality a question of mindset for both the one who manufactures a product as well as for the customer.

However, the company or project organisation which believes that the traditional quality control techniques will suffice to achieve high-quality standards, will fail. Employing more inspectors, tightening up standards, developing corrections, repair and rework teams is not enough to promote quality. In order to meet all the features and characteristics of the aircraft – and its related services – which make it fit for purpose and satisfy the customer's stated or implied needs, quality should in fact apply to every aspect of the business.

Traditionally, quality has been regarded as the sole responsibility of the Quality department. Total Quality Management (TQM) requests instead that the mindset of competitive quality penetrates the entire organisation while keeping the cost down at the same time. TQM can be regarded as a continuous effort to meet the agreed requirements of the customer at lowest possible cost through the full involvement of all employees. It focuses on mindset change and process improvement – in addition to the traditional inspections – and requires that:

- the cost of non-quality due to, say, scrap and rework, is fully considered;
- everybody is involved and contributes; and that
- a climate of mindset change, as well as knowledge of processes capable of producing competitive quality outcomes, is empowered and enabled through adequate leadership behaviours.

Thus, in TQM, employees are the one key element, with those processes being the other which begin with the external customers' statement of requirements and return back a product or service to this source. In a way, there are many things in common with Systems Engineering. In fact, the NASA Systems Engineering Handbook describes TQM as 'the application of Systems Engineering to the work environment'.[33]

For the Development sub-phase, multi-functional teams are the right forum where cross-functional processes can be applied by employees. If they work in structured coordination, with agreed objectives, common methods, sharing complementary roles, they become the teams upon which the Total Quality process depends to achieve its results. Total Quality will therefore be most effective in organisations where teamwork supplants individual work and where leadership is confidently empowering, that is, 'freeing' people to perform to the best of their abilities through fast communication and good training.

MULTI-CULTURAL MANAGEMENT

As was explained earlier, commercial aircraft development projects are very international undertakings. Internationality in projects stems from the appearance of different cultures and languages among the people working for them.

Individuals, who have gained project experience, have learned strategies of how to deal with the typical project issues such as objectives, tasks, information, responsibilities and

33 Shishko, R. et al. (1995), *NASA Systems Engineering Handbook* (Pasadena: Jet Propulsion Laboratory), NASA SP-6105.

so on. These strategies are influenced by different values, norms and history. If individuals have only learned about these strategies in uni-culturally-led projects, then multi-cultural projects will pose for them significantly increased challenges, because the diversity of values, norms and histories is much higher in international projects. Without additional training, there is a much higher risk for conflicts arising from cultural misunderstandings and misinterpretations.

Even worse, due to lack of capability to interpret the signals and behaviours of members of other cultures, one's own proven behaviours may not be constructive any more for the working together within an international project. In addition, the others may not see their project and/or team role expectations fulfilled when looking at them from their cultural background. All of this generates frustrations among the people involved.

Problems occurring in international projects are often of higher magnitude than in national projects. In international projects it is therefore wise to assume more coordination effort compared to national projects before reaching a decision. On the other hand, the success potential of international projects can be much higher than for national projects too.

Cultural differences can influence international projects enormously and the organisation and steering of projects in a trans-national context becomes very complex. However, there are a multitude of measures, means and paths available, depending on subject, task and function, and project staff will need additional guidance to minimise the risk of cultural conflicts.

As more and more international projects have emerged over the recent decades, the demand for intercultural awareness and management has also increased. This has stimulated significant research in this area and different models to explain the relations between rules and norms of a culture, and to compare cultures with each other, have been developed. As with every model they do not represent reality in all its facets but concentrate on certain cultural aspects only, thereby representing a simplification and/or approximation of reality.[34] However, knowing these models nevertheless helps to understand complex cultural relations and to manage international projects.

As an example, the model of an onion with three shells is often used to explain more specifically what is actually meant by the term 'culture' in the context of international projects, see Figure 4.4. The inner shell represents the social culture, which is determined by:

- the social background;
- the ethnic group;
- the religious beliefs; as well as
- all the other cultural determinants of a country or region;

to which someone is exposed during childhood. In other words: the social culture depends on the social context during one's early years.

---

34 When generating models to understand cultures it is important to differentiate between stereotypes and prejudices. If a model states that, for example, reliability is important in a specific culture, it does not say that every member of that culture is reliable. Even in that culture there will always be people who are not reliable at all. Thus, there is an accepted deviation from the generalised statement. The model only describes a standard mean, also called a stereotype. A prejudice emerges if the stereotype of a value is assumed to be applicable to all members of a culture. Thus, there is a clear difference between the statement 'All members of culture A are unreliable' (prejudice) and the statement 'Reliability is important to culture A' (stereotype).

Figure 4.4   The onion-shells of culture

The second shell represents the cultural properties of a profession. It originates from the period of professional qualification (for example during the course of studies) and it therefore happens that individuals with different social cultures still enjoy cultural similarities if they have enjoyed the same professional qualifications. It is, for example, quite common that engineers from different countries have less of a problem of working together than, for example, engineers and lawyers belonging to the same social culture.

The third layer represents company-specific norms and habits. It is the shell of the corporate culture, which one adopts during one's career at a company.

Any of the three shells features rules and norms to decide about the acceptability, politeness or morality of behaviours. The term 'culture', in fact, encompasses any of the three shells or a combination thereof. However, to succeed in international projects one should know well about the social culture shell. But it is also the most difficult to deal with. The rules and values which are internalised during the youth of a person – and thereby becoming a set of personal and individual convictions – lies deepest within the onion. They are the most protected from outside influences. Accessing them in order to understand and deal with them requires peeling away the overlying shells of professional and corporate rules, norms and convictions with great sensitivity and empathy.

The problem with understanding a culture – and the social culture in particular – is the fact that the culture-specific learning process is rarely appreciated *consciously* by members of a different culture. When individuals with different cultural backgrounds encounter each other, it is a lot more common that each of them *unconsciously* assumes that the others orientate themselves along the same set of rules and values. Clearly, misunderstandings and conflicts can easily result from this fundamentally wrong assumption. It is therefore important to understand the rules and values of the other culture.

Many other cultural models have been developed (such as by E.T. Hall, E. Schein, F. Trompenaars & C. Hampton-Turner, see Multi-Cultural Management: Further Reading section at the end of this chapter) but a particularly interesting one is the one developed by Geert Hofstede. What makes his model so useable for the analysis of multi-cultural problems in businesses and, thus, projects, is the fact that it is based on employee attitude surveys covering different subsidiaries in 66 countries of a single company (that is, IBM) comprising 88,000 employees.[35] Hofstede and Hofstede distinguish different cultural dimensions, which they define as follows:

- **Power Distance:** 'The extent to which the less powerful members of institutions and organizations within a country expect and accept that power is distributed unequally.'[36] [37]
- **Uncertainty Avoidance:** 'The extent to which the members of a culture feel threatened by ambiguous or unknown situations.'[38] [39]
- **Individualism versus Collectivism:** 'Individualism pertains to societies in which the ties between individuals are loose: everyone is expected to look after himself or herself and his or her immediate family. Collectivism as its opposite pertains to societies in which people from birth onward are integrated into strong, cohesive in-groups, which throughout people's lifetimes continue to protect them in exchange for unquestioning loyalty.'[40] [41]
- **Masculinity versus Femininity:** 'A society is called masculine when emotional gender roles are clearly distinct: men are supposed to be assertive, tough, and focused on material success, whereas women are supposed to be more modest, tender, and concerned with the quality of life. A society is called feminine when

---

35 The data is about 30 years old, but cultures change slowly.
36 Hofstede, G. and Hofstede, G.J. (2005), *Cultures and Organizations: Software of the Mind*, 2nd Edition, (New York: McGraw-Hill), p. 46.
37 In a project, Power Distance describes the emotional distance between boss and subordinate. Western cultures like in the US, UK, North and Central Europe regard the boss as being equal with equal rights. Thus, the power distance is low. The opposite is true for Latin-American, Asian and African cultures, where there is huge respect on the sides of subordinates for their chiefs. From: Hoffmann, H.-E., Schoper, Y.-G. and Fitzsimons, C.J. (2004), *Internationales Projektmanagement* (München: Beck-Wirtschaftsberater im Deutschen Taschenbuch Verlag), pp. 26–7, free translation by the author.
38 Hofstede, G. and Hofstede, G.J. (2005), *Cultures and Organizations: Software of the Mind*, 2nd Edition, (New York: McGraw-Hill), p. 167.
39 Uncertainty Avoidance describes the degree to which the members of a culture feel themselves threatened by uncertain or unknown situations. Members of cultures with a high degree of uncertainty avoidance (such as Japan, South Europe, German-speaking countries) try to avoid unclear situations, they are looking for rules or structures, with which results can be interpreted and predicted. Members of cultures with a low degree of uncertainty avoidance (Scandinavia, Anglo-Saxon countries, Asia) do not like formal rules. Insecurity is accepted as a fact of life, strange things generate curiosity more than fear. Spontaneous, creative and innovative solutions to problems are preferred. From: Hoffmann, H.-E., Schoper, Y.-G. and Fitzsimons, C.J. (2004), *Internationales Projektmanagement* (München: Beck-Wirtschaftsberater im Deutschen Taschenbuch Verlag), pp. 27–8, free translation by the author.
40 Hofstede, G. and Hofstede, G.J. (2005), *Cultures and Organizations: Software of the Mind*, 2nd Edition, (New York: McGraw-Hill), p. 76.
41 Members of individualistic countries (Anglo-Saxon countries, Scandinavia, Germany, etc.) regard individuality as the core of their identity. Members of collectivistic cultures (Latin-America, Asia, Arabic countries, Africa) draw their identity from the membership to a group, family, clan, company, or ethnic group. Which of those groups is key differs from one collectivistic culture to another. In Japan, the company as a group is more important than the family. In Africa, the contrary is often true. From: Hoffmann, H.-E., Schoper, Y.-G. and Fitzsimons, C.J. (2004), *Internationales Projektmanagement* (München: Beck-Wirtschaftsberater im Deutschen Taschenbuch Verlag), p. 27, free translation by the author.

emotional gender roles overlap: both men and women are supposed to be modest,
tender, and concerned with the quality of life.'[42, 43]

- **Long-Term Orientation:** 'The fostering of virtues oriented towards future rewards –
in particular, perseverance and thrift.'[44]

Hofstede and Hofstede use their model to characterise cultures and to gauge cultural
dimensions in relative terms. An example can be seen in Figure 4.5. Here, a typical US
American's cultural dimensions are compared to eight other cultures.

Notice that, for example, the British and Canadian culture are very close to the US
American culture, while the French culture has a higher Uncertainty Avoidance and the
German culture a weaker Individualism than the US American one. The Chinese culture
exhibits a significantly larger Long Term Orientation compared to the US American one.
As Hofstede and Hofstede point out, 'People from cultures very dissimilar on the national
culture dimensions of power distance, individualism, masculinity, uncertainty avoidance,
and long-term orientation … can cooperate fruitfully. Yet people from some cultures will
cooperate more easily than others with foreigners. The most problematic are nations
and groups within nations that score very high on uncertainty avoidance and thus feel
that what is different is dangerous. Also problematic is the cooperation with nations and
groups scoring very high on power distance, because such cooperation depends on the
whims of powerful individuals. In a world kept together by intercultural cooperation,
such cultural groups will certainly not be forerunners.'[45] The model by Hofstede and
Hofstede will be used later in this book as an exemplary model to generate awareness for
multi-cultural, project-related problems and to understand how to deal with them.

DETERMING ESSENTIAL AND INTEGRATIVE
MANAGEMENT ELEMENTS

The foregoing explanations lead to some conclusions about which essential and integrative
elements of the various Management disciplines should be used for commercial aircraft
development with its specific characteristics. Besides the full exploitation of modern IT
capabilities as well as robust Business Cases, we should have a closer look to the following
for the Development sub-phase:

- As good communication is always a fundamental success criteria we will look into
the details of co-locating multi-functional teams. As was said earlier, this supports
Rapid Development thanks to Concurrent Engineering and leads to better quality
and maturity of the product.

---

42 Hofstede, G. and Hofstede, G.J. (2005), *Cultures and Organizations: Software of the Mind*, 2nd Edition, (New
York: McGraw-Hill), p. 120.

43 Masculine cultures distinguish strictly between the roles of man and woman. Men are regarded as assertive, tough,
and concerned with material success, whereas women are more modest, tender, and interested in the quality of
life. In feminine cultures the roles are not so strictly separated. Conflicts are solved through negotiation and
compromise. From: Hoffmann, H.-E., Schoper, Y.-G. and Fitzsimons, C.J. (2004), *Internationales Projektmanagement*
(München: Beck-Wirtschaftsberater im Deutschen Taschenbuch Verlag), pp. 28–9, free translation by the author.

44 Hofstede, G. and Hofstede, G.J. (2005), *Cultures and Organizations: Software of the Mind*, 2nd Edition, (New
York: McGraw-Hill), p. 210.

45 Hofstede, G. and Hofstede, G.J. (2005), *Cultures and Organizations: Software of the Mind*, 2nd Edition, (New
York: McGraw-Hill), p. 366.

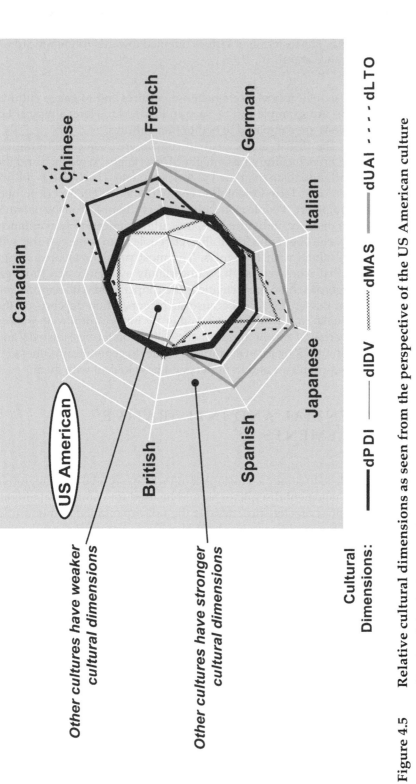

Other cultures have weaker cultural dimensions

Other cultures have stronger cultural dimensions

Cultural
Dimensions: ——dPDI ——dIDV ·······dMAS — – –dUAI - - - -dLTO

Figure 4.5   Relative cultural dimensions as seen from the perspective of the US American culture

dPDI: delta in Power Distance
dIDV: delta in Individualism
dMAS: delta in Masculinity
dUAI: delta in Uncertainty Avoidance
dLTO: delta in Long Term Orientation

- As very often team members in a co-located environment have never worked together before, there is the important aspect of team building and leadership, without which even the most advanced Concurrent Engineering will not work.
- Team building also includes the aspect of early involvement of both customers and suppliers in the development process, for example, by allowing them to participate directly in the co-located teams.
- We will have to look at the aspects of Multi-Cultural Management as not only sub-contractors and suppliers will usually come from all over the world but co-located teams may well be of a multi-cultural nature too.
- There are many aspects of leadership, but an important one is about setting the right priorities to teams at the right time. Typical for aircraft development, weight and safety play a fundamental role in setting priorities.
- Creating DBTs focuses the minds of team members towards a deliverable, such as a fuselage section, with the advantages of ensuring leadtimes and producibility, among others. We will look into the interrelations between multi-functionality and a deliverable-oriented design-build philosophy. We will also have to explain how the downside of this philosophy, that is, the more complex management of 'distributed' sub-systems (such as electrics), can be overcome.
- Having a good understanding of the market and customer needs on all levels of the programme organisation is cardinal for success. We therefore will have to look into the methods and processes of Requirements Engineering. We will extend this to the aspects of verification and validation to ensure that requirements have been met.
- Both Systems Engineering and Project Management are looking for some kind of architecture. While Systems Engineering is seeking an architecture for functionalities and product breakdown, Project Management requests a breakdown of organisation and work.[46] Obviously, there are strong links between these architectures. In fact, making use of these links will result in synergies. As architectural aspects in an environment dominated by Information Technologies are very essential indeed, a good part of this book will be dedicated to them.
- While architectures describe the breakdowns of product, functionalities, work, organisation and others, changes to individual modules or their content or to the architecture itself need to be carefully monitored in order not to lose control. This is why change control in general and Configuration Management in particular is so important. We therefore will look at them as another set of our essential elements.
- It goes almost without saying that schedule and resource planning are essential features of every project. However, we need to explore the links between planning and architecture and we will discover some major synergies in this area. Also, the subject of using schedule (and other) plans for communication purposes will need to be addressed. Among others, this requires to aggregate planning information into something, which can be easily digested by Management. Communication based on schedules' aggregation is important in every development environments with thousands of collaborating individuals on different hierarchical levels.

---

46 The analysis of functionalities responding to requirements cascaded to the development teams is an important prerequisite for a well-structured product architecture. Functional analysis has to be initiated very early in the development process in order to better integrate transverse functions.

- We will have to look into ways of how to reduce development cycle time as much as possible since only faster Time to Market offers the chance to increase market share and reduce production costs of the aircraft. These factors all lead to higher corporate profits.[47]
- We will in addition have to look to the major enablers of Concurrent Engineering, such as interface control enabling simultaneous engineering of structures and systems as well as professional Interdependency Management. Both have a strong link to schedule planning and Risk Management.
- But Risk Management does not only cover the risks associated with Concurrent Engineering and Overlapping. All kinds of other risks can be covered. The repercussions which would materialise if a risk were to become a reality (that is, an issue) can be expressed in terms of variances of status indicators such as leadtime, cost, quality and performance. We therefore need to look into the possibility of using these repercussions for establishing the outlook of the key indicators, for example, for the financial outlook. Next to schedule planning, Risk Management is another powerful tool forcing teams to look ahead and to do something about identified risks. We will dedicate an individual chapter to the management of risks.
- We have not yet mentioned cost control, which in Project Management is heavily linked to the control of work progress via the method of Earned Value. But in addition we need to control the costs for production, operations and refurbishment to be able to maintain the Business Case. We need to have the entire life-cycle costs in mind. The challenging aspect of cost control in our context is that it not only needs to satisfy the needs of the project – for example because we would like to monitor costs for each work package – but also the wider needs of the aircraft programme's Business Case as well as of the company's financial accounting procedures. We will have to look into this aspect to derive a suitable Cost Breakdown Structure (CBS), among others.
- Having looked at all these aspects, in particular the aspect of architectures, we are now able to identify the needs and possibilities for monitoring, control and reporting in a wider sense. Communication of status information and variances is obviously Project Management basics. We therefore should not neglect this aspect. However, as mega-projects such as aircraft development tend to lead to micro-management by company executives, we will also have to find easy ways of aggregating data while allowing for rapid responses to specific queries.

This is the list of (project) management essentials suggested for commercial aircraft development. When implementing all the items on this list one is entering the world of integrative management essentials. One can achieve best assurance for successful project integration by systematically examining all interfaces between the items on the list. In the next chapters each of the management essentials will be described in more detail and some pragmatic applications will be suggested. Let us start with the co-location of multi-functional teams.

---

47 Note, that this represents a fundamental strategic shift compared to the past: today it is economies of time instead of economies of scale, which in the past represented the classical barrier to market entry and the protector of profits in the aircraft industry.

MAIN CONCLUSIONS FROM CHAPTER 4

- Essential and integrative elements of classical Management disciplines
  to be applied for the Development sub-phase of a commercial aircraft
  programme with its specific characteristics need to be selected from the
  following:
    - multi-functional Design-Build Teams, which are co-located;
    - leadership and team building;
    - Multi-Cultural Management;
    - priorities sequencing;
    - Requirements Engineering;
    - integrated project architecture, satisfying the needs of Systems
      Engineering, change control, Project Management and communication
      management;
    - Rapid Development techniques such as Concurrent Engineering,
      overlapping, as well as interface and Interdependency Management and
      control;
    - risk and opportunity management;
    - Earned Value Management; as well as
    - lean monitoring, control and reporting.

PART II

Integrative People Management

5
Multi-Functional Design-Build Teams

THE RATIONALE FOR MULTI-FUNCTIONAL DESIGN-BUILD TEAMS

Thousands of decisions need to be made during the development of a commercial aircraft. But poor communication within such a complex project can lead to delays in the decision process or in poor decisions in the first place. Both can result later – and very often do – in unnecessary rework. If this happens to one decision it will usually not have serious consequences. But the aggregate effect of myriads of delayed or poor decisions is staggering. It results in unwanted consequences with regard to aircraft quality, maturity, development cycle time and cost. It must therefore be ensured that:

- the *right decisions* can be taken;
- in a *timely manner*;
- at the *right level*.

This is an organisational challenge.

The *'right decision'* requirement is fulfilled by the multi-functional approach. It ensures that all stakeholders, that is, Functions like Engineering, Manufacturing and so on as well as customers, suppliers and authorities are not only jointly represented during the design and development process but are encouraged to actively influence the latter. It must be heard what Engineering *and* Manufacturing *and* Procurement *and* Product Support *and* others *together* have to say before taking a decision. Only improved intra- and cross-functional communication will ensure highest standards of quality and maturity of the aircraft when entering service, as well as adherence to leadtimes and budgets.

'We can view our design choices as having to meet the needs of both Engineering and Manufacturing. When a design fails to meet the needs of either of these groups, rework is required. The key difference in rework that comes from manufacturing issues is that it is expensive and late in comparison to rework originating from engineering issues. … The problem with most design processes is that they first design to meet the needs of Engineering and then assess manufacturability. This means that the needs of Manufacturing are poorly served until late in the design process. Instead, we do time-consuming rework of the design, frequently on the critical path where it can do the most damage to our cycle time. A better approach is to address the needs of Manufacturing and Engineering simultaneously. This reduces the number of design decisions that fail to meet the needs of Manufacturing.'[1]

As Henry Shomber puts it, who was brought in very early on the [Boeing] 777 programme to help develop new ways of designing planes to avoid some of the problems of the past:

'Prior to the 777, the release of an engineering drawing was entirely up to Engineering to decide. There was a signature from the designer who prepared it, the one who checked it, the stress man who certified that it was strong enough, and the group's supervisor that was responsible for the part. There may also have been a Boeing materials-technology signature on it. But in addition we have now added a signature from the manufacturing engineer who planned how the part will be made. Now his signature means something different from the Engineering signature. The Engineering signature is a statement about engineering: that the engineering is correct. The Manufacturing Engineering signature is a statement that it is *producible* – i.e. that we have incorporated to the degree we could the producibility suggestions that they've brought forward that we hope will make the airplane easier to build. This has been unheard of within Boeing, and in many ways was one of the most uncomfortable aspects when first proposed. Engineering's view was "I'm responsible", and yet what we're trying to put in place is that this is a team responsibility and so this is becoming part of our culture.'[2]

The design-build process as required by the Concurrent Product Definition (CPD) methodology can only successfully operate on the principle of partnership between Functions like Engineering, Manufacturing, Procurement and other departments as specific cases demand. It therefore necessitates a multi-functional character of teams. Design-Build Teams (DBTs) as owners of the design-build process are, thus, a direct, logical extension of the multi-functional approach. There are no DBTs which are not of a multi-functional nature. But there can be multi-functional teams without being DBT at the same time. Specific deliverables are at the focus of all of the DBTs' activities, and not the exclusive view of a particular discipline or Function.

A DBT is in charge to design, develop and build individual aircraft components to time, cost and quality, and can be described as 'a multi-disciplinary group of people who are collectively responsible for delivering a defined product or process. It is composed of people who plan, execute and implement life-cycle decisions for the system ... It includes empowered representatives (stakeholders) from all of the functional areas involved with the product – all who have a stake in the success of the program, such as Design, Manufacturing, ... and Logistics personnel, and, especially, the customer.'[3]

'One meeting [of a Boeing 777 design-build team] ... gave an idea of how these [multi-functional] things worked. ... There were twenty people present, representing the following 'organizations' – some of them other departments in Boeing, others from outside companies: Customer Service, Weights, Alenia (the Italian company who would manufacture the flap), Design to Cost,

1   Smith, P.G. and Reinertsen, D.G. (1998), *Developing Products in Half the Time* 2nd Edition, (New York: John Wiley & Sons), pp. 242–3.
2   Sabbagh, K. (1996), *21ˢᵗ Century Jet. The Making of the Boeing 777* (London: MacMillan Publishers), p. 82.
3   Anon. (1996), DoD Guide to Integrated Product and Process Development, (Washington, D.C.: Department of Defense, Office of the Under Secretary of Defense (Acquisition and Technology)), Version 1.0, 20301-3000, 5 February 1996.

Structures, Factory, Manufacturing Engineering, Tool Engineering, Materiel, Aerodynamics, and Quality Control.'[4]

In previous times, communication was minimal across the functional borders. Engineering did the design without the involvement of Manufacturing and tossed the drawings over to the latter once they were completed. There was poor communication of Engineering with units external to the Engineering Function. In a multi-functional DBT environment the *'timely manner'* requirement would ideally be fulfilled if the DBTs could be created in such a way that the communication streams across the functional borders *within* each DBT could be optimised,[5] while at the same time minimising them *beyond* a DBT's boundaries.

Minimisation of the need for communication across DBT boundaries can be achieved by early interfaces' freeze so that for a long time teams can work more or less independently from each other and simultaneously.[6] Among others, this requires that, for example, the product breakdown as well as the integrated schedule plan are structured in a way which facilitates early freeze of interfaces between modules of the product structure. If in addition teams are organised around the sub-structures or sub-systems of the product – with robust interfaces between them – instead of being organised by Function, much faster development cycles can be achieved. Therefore, when a business decides on an organisation structure it should take the degree of interdependence between teams into account.

For the fulfilment of the *'right level'* requirement DBTs need to be provided with the authority, the resources and all other levers necessary to make the vast majority of project decisions themselves. In other words, one needs to ensure that teams are sufficiently staffed and that the subsidiarity principle can be applied.

'Upper Management must provide a team setting in which the team takes full responsibility for the project. There are two parts to this issue. One is to weaken the linkages between the team and the remainder of the organisation so that the team can in fact move with some freedom. The other is to create a motivational structure where the team must indeed complete the project on schedule. ... The best way of helping the team take responsibility for the schedule is to place it in an environment where the external reasons for schedule slippage have been removed. By providing the required resources while establishing a separate identity for the team, upper Management encourages it to apply its own resources rather than rely on others who do not have a clear stake in the project's outcome.'[7]

'Decision making should be driven to the lowest possible level commensurate with risk. Resources should be allocated to levels consistent with risk assessment authority, responsibility and the ability of people. The team should be given the authority, responsibility, and resources to manage its product and its risk commensurate with the team's capabilities. The authority of team members needs to be defined and understood by the individual team members. The team should accept responsibility and be held accountable for the results of its efforts.

4   Sabbagh, K. (1996), *21st Century Jet. The Making of the Boeing 777* (London: MacMillan Publishers), p. 69.
5   It also needs to be optimised across the organisational hierarchies. This will be explained below.
6   This does by no means exclude later refinement of interfaces. However, the early interface freeze must ensure that the later refinement has only one direction: less space needed instead of more, less weight instead of more, fewer harnesses instead of more, and so on.
7   Smith, P.G. and Reinertsen, D.G. (1998), *Developing Products in Half the Time* 2nd Edition, (New York: John Wiley & Sons), p. 267.

Management practices within the teams and their organisations must be team-oriented rather than structurally-, functionally- or individually-oriented.'[8]

As far as subsidiarity is concerned, project decisions need to be made by the people working on the project daily. Otherwise, the project may be delayed every time such a decision is required. CPD needs subsidiarity. However, to ensure that the subsidiarity principle can be applied on the level of the DBTs is a subject beyond CPD. It leads us to the more general and larger problem of how to embed multi-functional teams within an existing company organisation. As will be shown in Chapter 6, embedding these teams leads directly into the creation of matrix organisations.

Availability of sufficient and skilled resources is another classical problem to projects. The classical solution to overcome resource bottlenecks is to outsource work, either as off-site sub-contracts or by bringing sub-contractor resources onto the site and into the teams.

As on-site sub-contractors do enjoy the same company-paid training and gaining of experience compared to the permanent resources, the company is at risk to eventually lose this investment when sub-contractor employees leave the site. However, this is a risk the project should accept as it is small compared to the gains in cycle time reduction. In addition, the company has the chance to monitor the sub-contractor's staff performance in much more detail and at arm's length compared to, for example, going through interviews. Offering high-performers a permanent role is then less risky to the company compared to other recruitment schemes. And communicating to on-site sub-contractor staff the possibility of eventual permanent employment with the company may also have a motivational effect after all.

In summary, if we want to explore the benefits of CPD – such as better integration, better quality and maturity at Entry into Service (EIS), much reduced rework, a significantly faster development cycle, reduced costs – we need to:

- create multi-functional DBTs;
- ensure that at start of the Development sub-phase – that is, at the time when thousands of individuals are to be allocated to DBTs – the development project is at a maturity level commensurate with the possibility to freeze interfaces;
- freeze interfaces so that communication across DBT boundaries can be minimised;
- ensure best possible communication within each of the DBTs;
- adequately staff the teams in terms of skills, quantity and quality; as well as to
- apply the subsidiarity principle wherever possible – by setting clear rules about what teams can decide on their own and what not (delegation of authority) – so that teams can take the vast majority of decisions themselves.

---

8   Anon. (1996), *DoD Guide to Integrated Product and Process Development*, (Washington, D.C.: Department of Defense, Office of the Under Secretary of Defense (Acquisition and Technology)), Version 1.0, 20301-3000, 5 February 1996.

THE GENERIC MULTI-FUNCTIONAL DESIGN-BUILD TEAM

Introduction

What are generic characteristics of multi-functional DBTs? It has been stated that representatives from different Functions should become members of such a team and that the teams should be capable of working as independently as possible. Apart from a Project Manager acting as a project and team leader,[9] it should be considered to include representatives seconded from the following Functions:

- Engineering (loads specialists, stressmen, designers, weights and masses specialists, integrators of engineering solutions);
- Systems Engineering (covering architecture design, CPD process, interface management, integration of Digital Mock-Up (DMU), Requirements Management, Verification & Validation);[10]
- Manufacturing (factory designers, jigs & tools' designers, logistics people, operations planners, specialists for concessions, work query notes and interfaces with other manufacturing sites);
- Procurement (purchasing and contract specialists, finance people);
- Customer Services (specialists for aircraft accessibility, maintainability, serviceability, reliability as well as product support);
- Information Technologies (to ensure support of all Information Technology (IT) tools and infrastructures needed);
- Materials (for example to ensure timely qualification of new materials);
- Quality (mainly process quality specialists);
- Project Management (schedule and resource planners, interdependencies managers, cost controllers, cost estimators, risk managers, change controllers and configuration managers, documentation managers, accommodation & facilities managers); as well as
- Human Resources (representatives dealing with the myriad of personal issues of individuals seconded into teams).

Figure 5.1 depicts a multi-functional team with only the major functions being represented.

---

9 Project managers are defined in this book as individuals which are familiar with the application of the Project Management methodology to projects. The terminology 'Project Manager' was first introduced by Gaddis, P.O. (1959), 'The Project Manager', *Harvard Business Review*, May-June 1959, pp. 89–9. Please note, that not all Project Managers are project leaders, but every project leader should also be a Project Manager.

10 In some companies, verification and validation are under the responsibility of the Quality Function.

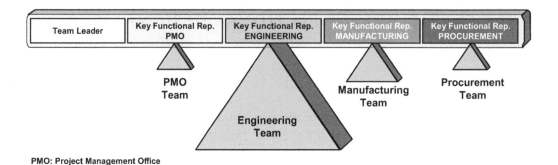

PMO: Project Management Office

Figure 5.1   The basic structure of a multi-functional Design-Build Team (major Functions only)

However, exactly which Functions should be represented in a DBT depends on circumstances such as:

- Is the multi-functional team the right organisational choice in the first place? There may be applications where multi-functionality does not add value, for example in aerodynamics and for loads calculation on aircraft level (both are pure Engineering domains), construction of new Manufacturing plants and facilities (pure Manufacturing domain) and so on.[11]
- Is the component to be developed a 'Make' or a 'Buy' part? If it is 'Make', Manufacturing should be represented, if it is 'Buy', Manufacturing and Procurement.[12]
- Are there enough skilled resources available? Is there sufficient workload for one individual representative in a team? Does the available budget allow to employ more resources? If not, key functional representatives could 'serve' more than one team as long as individual workload allows one to do so. In this case, those teams should be located close to each other and any key functional representative should not be a direct member of either team but should organisationally be located on a higher level.

The typical overall size of a generic team should be between 30–70 members, largely depending on the complexity of and the 'Make/Buy' philosophy related to the component to be design-build as well as on the possibility to have frozen interfaces.[13] However, the number of representatives of each Function in a team will be quite different. Engineering and Systems Engineering together usually represent the vast majority of team members, see again Figure 5.1. Thus, there must be some kind of hierarchical reporting structure

---

11 Multi-functional DBTs are usually not required for what is called non-specific design activities. These activities cannot be attributed to specific components but are applicable on aircraft level or major component assembly level. The subject of non-specific design activities is not further investigated in this book in any detail as in comparison with specific design activities fewer people are involved.

12 Even in the case of 'Buy', Manufacturing needs to be represented as procured items will still be assembled by Manufacturing at some stage.

13 It is important to note that multi-cultural teams should be smaller in size compared to national teams. This is because the complexity of building up confidence and relationships among the team members increases if different cultural backgrounds are involved.

within the Engineering community of a multi-functional DBT. For the representation of other Functions, there may be only a few or perhaps only one representative.

Project Leader

There must be someone who sets the day-to-day priorities in the multi-functional team and acts as a team leader: this can only be the teams' project leaders because they have the skills (or are supposed to have them) to do exactly that. While the key functional representatives manage the detailed work, the project leader concentrates on managing the key events of the project. Notwithstanding this overall share of responsibilities, the project leader must control the project in close cooperation with the contributing key functional representatives.

Managing a multi-functional team is one of the most difficult jobs one can imagine. It is strewn with many hurdles and stumbling blocks. The chances for a successful management will, however, rise if the project leader combines the following skills:

a) The project leader must be capable of coordinating and integrating the often differing functional views of the different team members. The multi-functional approach requires the ability to facilitate a discussion process towards a compromise agreed by all functions. The lack of integrative skills such as holistic thinking, balancing strength and weaknesses of team members, cross-cultural abilities and so on is a common cause of project failure. This integrative responsibility does, however, not replace responsibilities of the key functional representatives. It rather supplements them.

b) The project leader should be highly skilled in the classical management areas of organisation, planning, directing and controlling. However, these skills are somewhat conflicting with the skills required for (A).

'Planning and organising are important tasks of the project manager and tend to be done well by a person who enjoys process. Although a good plan is more often than not the result of a team effort, a project manager who likes to plan may have a tendency to take on the entire plan. The project manager who enjoys planning will see organising as just another part of a good plan, which indeed it is. Such a person may tend toward introversion and requires considerable order and discipline.

When we move on to the tasks of directing and monitoring, however, we see a requirement for another type of perspective. These tasks involve interactions with people. Directing requires that people be given assignments and that they be guided through these assignments through monitoring and feedback. These "people" interactions are often best accomplished by an extroverted type of person who likes to discuss situations with people and may not enjoy the paperwork associated with planning and reporting. Through this simple discussion, we note that the project manager is called upon to do many [conflicting] things...'[14]

c) This requires project leaders of a balanced personality. They should be able to shift gears whenever required as well as focus on issues coming across their desk

---

14 Eisner, H. (2002), *Essentials of Project and Systems Engineering Management*, 2nd Edition, (New York: John Wiley & Sons), p. 124.

in a certain, prioritised way, for example according to their importance and/or urgency. Typically a balanced personality can be found among generalists.

d) There is a better chance for successful multi-functional facilitation in aircraft development, if the project leader has some knowledge in aircraft (aerospace) project development and has some knowledge covering the basic understanding of how the aircraft company operates.

e) It is vital that the project leader has previously created a network of contacts within the company. In fact, personal friendships and alliances can become an important source of influence at times where obstacles need to be removed.

f) Most importantly, a project leader should be capable of exerting leadership. This will be further elaborated below.

Obviously, personal skills of the team leader are only one side of the coin. The other is that the organisation and company must provide the right levers. An adequate level of authority to develop the needed amount of reward and penalisation power must be provided to the project leader. For example, it must be ensured that the key functional representatives of each team are direct reports to the team leader, as discussed earlier. Also, team leaders should ultimately report to someone neutral in the company, someone who is not representing Engineering, Manufacturing or Procurement. Otherwise there is the danger that, for example, Engineering would dominate the decision process within the teams. The Manufacturing and Procurement team members would find it difficult to understand why they would report to someone from Engineering too. It is therefore recommended to have at least an Executive Vice President for Programmes in the company, who reports directly to the CEO and to whom all team leaders ultimately report. Once the formal corporate authority has been delegated to project leaders, it is entirely up to them to earn the additional respect needed to run the team.[15]

The project leader's role is similar to the role of a pilot at the controls of an aircraft. Project leaders continuously monitor the progress of the project through the evaluation system, watch for indications of present or future difficulty, and communicate to the appropriate functional specialists any need to change plans, schedules, budgets, and performance to reach the project objectives. The 'pilots' must also monitor these change signals to be sure they have been received, understood, acted upon, and that they do in fact produce the desired result.

In summary, the ability to create and lead a multi-functional DBT to project success lies at the heart of a project leader's roles and responsibilities.

---

15 With more and more team leaders working as Project Managers and 'as recognition grows of the nature and importance of Project Manager assignments, the need becomes more apparent for more formalized career development in Project Management. Because projects begin and end, Project Management assignments are less secure than functional assignments. Frequently, the best qualified people cannot be attracted to projects, because they must leave the known security of a functional department for an unknown future when the project ends. In an increasing number of organizations, the Project Management Function has been established as a part of the fairly permanent organizational structure. This provides a base on which to build continuity of project assignments, long-term security for project-oriented employees, and more effective career development programs.' From: Archibald, R.D. (1992), *Managing High-Technology Programs and Projects*, 2nd Edition, (New York: John Wiley & Sons), p. 89.

Project Management Office

Often, team leaders acting as project leaders are supported by a Project Management Office (PMO), which provides them with project status information as well as suggestions for future steps to be taken.

'A definite trend is observed over recent years to set up such an office in a number of industries. This reflects the coming of age of the disciplines of Project Management, which is taking its place as an area of functional expertise, along with Engineering, Marketing, Manufacturing, and so on.'[16, 17]

Compared with a cockpit crew of (older) aircraft, the PMO is the navigator and flight engineer (status control), whereby the pilot is the project leader. The PMO may be in charge of any of the following:

- creation and maintenance of architectures (Work Breakdown Structure (WBS), Organisation Breakdown Structure (OBS), Cost Breakdown Structure (CBS)) in close liaison with Systems Engineering in charge of the Product Breakdown Structure (PBS);
- schedule and resources planning;
- interdependencies control;
- work authorisation and control;
- work progress control;
- cost control;
- contract administration;
- documentation management;
- risks and opportunities management;
- accommodation and facilities planning (remember that the dynamics of multi-functional team building make this a challenging task);
- cost estimation (for example, for providing adequate data for 'Make/Buy' or technical configuration change decisions, to judge risk repercussions and so on);
- change control and Configuration Management;[18] as well as
- project monitoring and reporting.

The PMO is a service organisation providing services to the team leader and key functional representatives, providing them, for example, with aggregated and integrated scheduling information rather than controlling the internal schedules of each task. Each multi-functional team should at least have one member of the PMO.[19]

---

16 Archibald, R.D. (1992), *Managing High-Technology Programs and Projects*, 2nd Edition, (New York: John Wiley & Sons), p. 48.

17 For more information on the subject of PMO see for example: Johnson, J. and Horsey, D.C. (2001), 'The IT War Room', *The IT Software Journal*, Software MagCom June 2001, http://www.softwaremag.com/L.cfm?doc=archive/2001jun/WarRoom.html, accessed 10 October 2009.

18 The latter has originated as an Engineering discipline. However, in an integrated, multi-functional environment technical change repercussions also need to be analysed from a schedule, resources, cost, manufacturing and procurement point of view, among others. It is therefore better to manage it under a more neutral umbrella than Engineering.

19 In this case there really are two representatives of the company's Project Management Function in the team: the project controller belonging to the PMO as well as the team leader.

If in DBTs the project controllers are the only representatives of the Project Management Function besides the team leaders, they should at least act as the 'keeper' of the schedule plan and should control the costs.[20] In addition, within the limits of their workload, they may take on any other of the above mentioned roles. However, with so many multi-functional DBTs necessary to develop a commercial aircraft, this may not always be possible because of lack of adequate resources. In this case, it may be wise to only staff teams adequately on higher organisational levels to be able to manage all roles.

In order for project controllers to feed significant and realistic status and forecast data back to the team, they need to be fed regularly themselves by all key functional representatives with the latest updates. In:

- anticipating problems early enough by in-depth analyses of schedule and project cost data; and by
- suggesting ways of avoiding or correcting problems and issues

Project controllers can really be of high value. To make this iterative cycle of feeding and providing project data within a team a reality it is of significant importance that the data is pulled by the team leader in the first place. This is where some Project Management experience of – and the appreciation of – the added value of the Project Management methodology by the team leader will be very beneficial for the overall project success of the team. If team leaders do not themselves work with the plan they own, project controllers will be of little value.

Key Engineering Representative

The other very important role within the multi-functional team is the one of the leading Engineering representative. This stems from the fact that there is a need for people management of the large Engineering communities within the DBTs. After all, they represent the majority of the team members (as was outlined above). The key Engineering representatives in the teams need to give directions about how to design. They integrate Engineering solutions taking the constraints of the other Functions into account.

> 'The fact that both the P[roject] M[anager] and the ...[key Engineering representative] have, to some extent, overlapping responsibilities, suggests that it is critically important that these two people work together productively and efficiently. Friction between these key players will seriously jeopardise project success.'[21]

> 'The P[roject] M[anager] and the ...[key Engineering representative] must be able to work harmoniously together and both be dedicated to the success of the project.'[22]

An important member of the Engineering community led by the key Engineering representative within the DBT is the integrator for the DMU. Concurrent Engineering

---

20 But it must be stressed that both, schedule and costs should be owned by the entire team.
21 Eisner, H. (2002), *Essentials of Project and Systems Engineering Management*, 2nd Edition, (New York: John Wiley & Sons), p. 21.
22 Eisner, H. (2002), *Essentials of Project and Systems Engineering Management*, 2nd Edition, (New York: John Wiley & Sons), p. 130.

places the DMU at the heart of the technical quality of the design process. But the DMU is quite complex and requires proper management. It is therefore necessary to establish around the DMU the appropriate technical control processes. Otherwise all benefits of Concurrent Engineering are lost and there is the risk of more manual work or even loss of a baseline. Therefore, for each multi-functional team there should be one or several DMU integrator(s) whose job would mainly be:

- to ensure the validity of the data composing the DMU, especially those related to configuration;
- to perform problem analyses; as well as
- to ensure the correctness of the DMU (especially after integration phases) by checking that, for example, there are no 'holes', no erratic data, no clashes.

Hence, DMU integrator(s) contribute to the quality of the design process.

However, with different teams working concurrently for different airframe zones as well as different systems and their installations, all using the *same* DMU, there is also a role for a referee taking decisions in case of conflict. This referee should be an experienced person with lots of understanding for the problems associated with accommodating design solutions for a great number of different aircraft systems and associated equipment in a rather limited volume of zone-space. However, such a zonal DMU referee role is not necessarily to be found on lowest hierarchical team level, but rather on mid-level only, depending on the size of the zone.

Key Manufacturing Representative

New product developments are initially often seen by Manufacturing as a problem rather than an opportunity.

> 'After all, they [Manufacturing people] get measured on the basis of monthly shipments and gross margins. New products do nothing immediate to help either of these objectives. Instead, they tie up valuable people and equipment which could be used to do profitable business. ... There is simply very little short-term incentive for most Manufacturing organisations to support new products.'[23]

It is nevertheless important to overcome these objections. Senior Management in the company must ensure that Manufacturing seconds a sufficient amount of skilled people into the multi-functional teams. This will cause some problems to the running production and can only be done to an extent, which neither causes disruptions to nor compromises product quality of the Series Production.

What is needed within the DBTs is operational manufacturing knowledge to influence the product design and development process to achieve production time, cost and quality objectives (and therefore the Business Case). In other words: Manufacturing representatives in the teams have to ensure various 'Design to X' (DtX, with X=C for 'Cost', X=L for 'Lean', X=M for 'Manufacturability', and X=Q for 'Quality'). They also simulate, plan and deliver all supply, logistics and production processes as well as all jigs & tools required to achieve DtX.

---

23 Smith, P.G. and Reinertsen, D.G. (1998), *Developing Products in Half the Time*, 2nd Edition, (New York: John Wiley & Sons), p. 249.

By definition, Manufacturing Engineering is the 'glue' between Engineering and Manufacturing. Manufacturing will therefore usually identify Manufacturing Engineering representatives for secondment into multi-functional teams to deliver DtX, jigs & tools and relevant processes, see Figure 5.2.[24] This is fine as long as the latter can really represent the needs of Manufacturing Operations. However, experience shows that this is not always the case as one more filter in the multi-functional team communication between engineers and operational manufacturing is introduced.

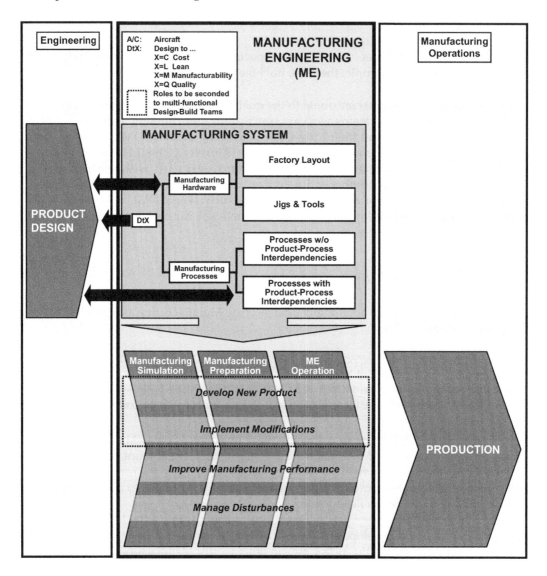

Figure 5.2   Typical Manufacturing Engineering roles

---

24 Note from Figure 5.2, that it is not necessary to second all Manufacturing Engineering (ME) tasks to multi-functional DBTs. This applies for example to the classical ME operations tasks such as generation of work orders.

A better way of connecting Manufacturing to the new product is therefore to select champions in charge of all Manufacturing issues related to the development. They in effect act as Project Managers for product manufacture. These champions would be members of the Manufacturing Function, but would also report to the project leaders (of the highest-level multi-functional teams). They would be in charge for all aspects of Manufacturing such as:

- planning and construction of new facilities and buildings;
- specification, design, procurement and deployment of jigs & tools;
- specification, procurement and installation of new machines;
- setting up of the manufacturing organisation for the new product;
- recruitment of white and blue collars in sufficient quantities;
- training of any of those;
- planning of Assembly Stage (AS) operations;
- logistics;
- operation of the production processes;
- transportation;
- and so on.

Thus, all Manufacturing staff associated with the new product – including Manufacturing Engineering staff – would be members of the Manufacturing Function and also part of the programme organisation, and would report to the champion. This link would ensure that valuable Manufacturing experience is fed into the design when it is needed most and that Manufacturing has full visibility on the product evolution in order to follow with facilities, building, jigs & tools and transportation means.

What needs to be ensured is that Manufacturing Engineers always keep in touch with the actual needs of the shop floor people, processes, methods and tools, that is, with Manufacturing Operations. Otherwise, their contribution may after all not result in ease of assembly and the aspect of manufacturability may not get the required attention in the teams. Within the DBTs, the 'build' aspect would possibly not be investigated as much as it should.

Key Procurement Representative

The key Procurement representative in a multi-functional team ensures that:

- 'Make/Buy' decisions are well prepared from a 'Buy' perspective (based on subjects like supplier price, quality, technology, reliability, financial strength, among others);
- contractual issues are sorted out as soon as possible and in line with the company's procurement strategies;
- financials with regard to Recurring Costs (RC), upfront payments (if any) and supplier claims resulting, for example, from technical changes are under control; as well as that
- suppliers are constantly monitored to ensure agreed levels of performance (for example with regard to promised parts' delivery dates and quality).

Procurement needs to establish strong communication links with Engineering (such as for technical specifications and associated changes), PMO (such as for financial and schedule data) as well as Manufacturing (such as for parts' deliveries and quality), which more than justifies the presence of a key Procurement representative in a multi-functional team.

LEADING MULTI-FUNCTIONAL DESIGN-BUILD TEAMS

Leadership in a multi-functional environment is all about generating teamwork and efficiency within a team as well as in the interaction with other teams.[25] [26] An efficient team:

- has clear, challenging and realistic objectives and standards, which are understood and accepted by all the members;
- performs a distinct task while focusing on its objectives and working energetically towards their achievement;
- has working procedures and plans understood and accepted by all members;
- strikes a balance between task success and meeting the needs of the individuals in the team and creating productive relationships between the members;
- has a mix of people, which enables the team to be effective. It is a mix not only in terms of the abilities needed to perform the task but in terms of the ability of the members to perform necessary team roles. For example, it will be a mix of people who can get things done, people who can produce ideas, people who can motivate and support, and people who can organise;
- has a flexible leadership, which is appropriate to the ability and willingness of the team to work on the task before them. For example, if the team members are low on ability and willingness, the leadership may need to be active and directive but if the team members are highly able and willing to perform the task there may be little need for leadership activity at all;
- has members who are committed to the team's objectives and to each other;
- has members who support and respect one another; who communicate freely and honestly with one another; who resolve conflicts; who listen to one another, who build on each other's ideas; who have established reciprocal confidence; and who can work on the basis of consensus and who recognise the interdependencies within the team;

---

25 'Teamwork is a term used to describe effective collaboration between members of a team. It is also often used to describe the effective way people work together although they may not be members of a specific team. For example people say that they want a "teamwork attitude" throughout an organisation, meaning that whenever people, for example from different departments, deal with one another, they collaborate and co-operate effectively together to fulfil the common purpose of the organisation.' From: Anon. (1998), *Total Quality Management and Empowerment* (Portsmouth: University of Portsmouth, Portsmouth Business School, Centre for Project and Quality Management), Unit 1d, Session 4, p. 3.

26 However, the aspect of efficiency of teamwork is typical for the view of American and European managers (like the author), who consider teamwork as a 'bringing together of people to solve problems and to decide upon the most effective solution'. It is worth noting that there are also other views about the aspect of teamwork. Asians, for example, consider teamwork as a means for 'team members to work together without friction, which solve problems in a nice, family-like way.' Quotations from: Hoffmann, H.-E., Schoper, Y.-G. and Fitzsimons, C.J. (2004), *Internationales Projektmanagement* (München: Beck-Wirtschaftsberater im Deutschen Taschenbuch Verlag), p. 121, free translation by the author.

- learns from experience by reviewing its performance, that is, its successes and failures, both in terms of task achievement and in terms of relationships and individual satisfaction within the team;[27]
- has members with similar levels of qualification and willingness to learn.[28]

It must be understood that creating and maintaining high levels of effective teamwork within a multi-functional environment represents a huge challenge for the leaders and the company overall. As a result of the complexities associated with multi-functional team organisations, a different management style is required compared to a purely line management one with direct reporting lines into one manager: there is less chance to be able to lead people through instructions and orders. Instead, people need to be led through conviction and motivation. This is in particular true for leadership exhibited by the leaders of multi-functional teams as well as by the key Engineering representative in the teams, who *de facto* is the leader of the largest functional community. Compared to management through authority and instructions, leadership is also the better approach in international projects with *multi-cultural* teams, as the latter require even higher motivation levels among team members to compensate for the higher levels of frustration resulting from cultural conflicts.

Multi-functional, multi-cultural team leadership means to encourage change through the development of visions, to agree upon objectives and to implement actions *together* with the involved people. It aims at motivating team members to act in a self-responsible way and to shape teams commensurate with the project progress.

The authority of leaders working in an integrated, multi-functional environment stems primarily from their personal abilities to earn such authority. More than in most other management positions they must elicit performance from others – who often are not even under their direct control – by relying on their displayed values and interpersonal skills rather than on formal authority.

> 'Through negotiation, personality, persuasive ability, competence, reciprocal favours, and the like, … [a team leader] may have a great deal of indirect power …, even if he lacks direct … power.'[29]

Functional bosses, if entrusted with the leadership of a multi-functional team, often experience difficulties related to the changeover from doing a specific type of work to managing the efforts of others. As R.D. Archibald rightly points out, 'it is difficult to convey a good understanding of the...role [of the leader of a multi-functional team] to functional managers whose experience has been wholly within traditional, functional organizations', where bosses prevail.[30] The former functional boss needs to be able to give up his focus on functional issues and needs to become a generalist.

---

27 Anon. (1998), *Total Quality Management and Empowerment* (Portsmouth: University of Portsmouth, Portsmouth Business School, Centre for Project and Quality Management), Unit 1d, Session 4, pp. 4–5.
28 Note that similar academic titles or degrees do not necessarily indicate similar levels of qualification, particularly not in multi-cultural teams. It is important to look more carefully.
29 Archibald, R.D. (1992), *Managing High-Technology Programs and Projects*, 2nd Edition, (New York: John Wiley & Sons), p. 81.
30 Archibald, R.D. (1992), *Managing High-Technology Programs and Projects*, 2nd Edition, (New York: John Wiley & Sons), p. 53.

A famous quotation, attributed to Gordon Selfridge, the founder of the department store Selfridges in London, says what is most important about the difference between a boss and a leader:

- 'the boss drives his men – the leader coaches them;
- the boss depends on authority – the leader on good will;
- the boss inspires fear – the leader inspires enthusiasm;
- the boss says "I" – the leader "we";
- the boss fixes the blame for the breakdown – the leader fixes the breakdown;
- the boss says "go" – the leader says "let's go".'

Warren Bennis, one of the foremost authorities on organisational development, leadership and change in the US, distinguishes between leaders and managers:[31]

- 'the manager administers; the leader innovates;
- the manager is a copy; the leader is an original;
- the manager maintains; the leader develops;
- the manager focuses on systems and structure; the leader focuses on people;
- the manager relies on control; the leader inspires trust;
- the manager has a short-range view; the leader has a long-range perspective;
- the manager asks how and when; the leader asks what and why;
- the manager has his eye on the bottom line; the leader has his eye on the horizon;
- the manager accepts the status quo; the leader challenges it;
- the manager is the classic good soldier; the leader is his own person;
- the manager does things right; the leader does the right thing.'

Leaders of multi-functional teams are even more exposed than bosses or managers of uni-functional teams. They are under intense scrutiny and are watched by more than one Function all the time. Fortunately, being more exposed also offers the opportunity for faster communication via the snowball effect: if a leader recognisably demonstrates positive leadership behaviours than people will not only know more quickly about this but larger numbers from different Functions will be willing to follow him. Positive leadership behaviours have a multiplier effect, which can make life of a leader much easier while having a successful team at the same time.

How can leaders of multi-functional DBTs show positive examples of respected behaviours and be a role model for the other members of their teams? There is a wealth of literature available on the subject of leadership and there is no need here to repeat in all details what is written elsewhere. However, for the interested reader Appendix A2 provides an overview of some classical behaviours for leaders of multi-functional DBTs.

---

31 Bennis, W.G. (2000), *Managing the Dream: Reflections on Leadership and Change* (Cambridge, MA: Perseus), p. 5.

LEADING TEAMS THROUGH THEIR TEAM PHASES

Introduction

Groups of people, much like individuals, go through predictable and noticeable stages of development. They mature in terms of task progress and in terms of interpersonal relations. At each phase of development certain problems can occur. For example, when a team is first formed, its members do not know how to work together. They have not yet developed their own skills in group decision making and they do not fully understand their own responsibilities.

One of the team leaders' most important functions is therefore to develop the project teams and to help them move through the various stages of their evolution. Team leaders do not only run teams on a day-to-day basis, expecting them to miraculously grow on their own. They rather build them pro-actively by engaging with the teams in a range of activities, which the teams deliberately choose and implement to develop specific aspects of team performance.

Phases of Team Development

The Miller Consulting Group provides an excellent description of the various development phases which a multi-functional team goes through.[32, 33] During the Affiliation phase, team members are getting to know and accept each other. It is the phase during which the team becomes familiar with individual behaviours of team members. Members learn which behaviours are acceptable and unacceptable to the team. The 'basics' are learned in this stage: communication, listening, problem solving and decision making. It is during this phase that leaders need to clarify objectives, roles and responsibilities, and describe the Project Plan.

Team members also learn to understand the cultural differences of team members and the importance of having different languages in the team.

> French team members tend to put a lot on emphasis on sovereign and cultivated appearance, while in Germany detailed professional knowledge stays in the foreground. For the French culture, the latter is generously ignored. In Germany, language is regarded as a means to sort out interests. French or English put a lot of emphasis on good expressions.

Cultural conflicts in a team could result in schedule problems as the sorting out of conflicts and the associated lower levels of motivation consume time. An important means during the Affiliation phase to avoid unnecessary increases in leadtime is therefore to make teams responsible for the integration of new team members and the personal, regular check by the team leader to find out about the well-being of new team members.

---

32 Miller, L.M. and Howard, J. (1991), *Managing Quality Through Teams. A Workbook for Team Leaders & Members* (The Miller Consulting Group).

33 Other team development models distinguish between the phases 'Storming', 'Norming', 'Forming' and 'Performing'.

Team members then go through an Adolescence phase, in which they become more comfortable with each other. Once team members have succeeded in getting to know each other sufficiently well, they start learning how to communicate with each other. The downside of more communication is that disagreement and conflict may start to arise as members begin to trust and express opinions. Team members may therefore start to exhibit rebellion, questioning and impatience. Typical unproductive behaviours during the Adolescence phase are attacking the person rather than the problem, as well as defending and blaming other team members. But team members also start feeling some 'team spirit' and begin to feel empowered.

The Productivity phase occurs after the team has learned to work effectively as a team. The team can now solve problems by utilising problem-solving processes based on the skills, competencies and talents of the members and the leader. Typical productive behaviours are valuing team members' opinions, seeking opinions of others, and acting as a single unit. Typical unproductive behaviours are complacency (as the team becomes too sure of itself) and forgetting that problem solving involves analysing all facts and data.

The Maturity phase is the ultimate phase of an effective team's development where it responds automatically to challenges. The team has a clarity of purpose and a sense of unity, and is functioning like a finely-tuned machine. Team members know each other, solve problems and make decisions very comfortably and without fear. Team behaviour carries over into everyday work. The investment and involvement of the members is directed towards the team. A high level of energy is evident.

However, there comes a time where individual team members wish to move on in their careers. Typically this occurs about three years after having joined the team and it is usually the best team members who want to leave first. Team members do not feel challenged any more as work in the multi-functional team becomes a routine. However, there is still a lot of aircraft development work to be done, even after three years. So leaders are faced with a conflict: on the one hand they would like to keep team members in their teams to ensure continuation of the effective teamwork. On the other, if individual team members become demotivated because they are not allowed to move on in their career, the efficiency of the teamwork will drop anyway. Leaders therefore should demonstrate support to individuals attempting to move on while at the same time trying to find a suitable replacement and while keeping control of the transition phase. The challenge for the leader during this transition phase is to ensure high levels of efficiency while managing people transfers at the same time.

Finally, when the project comes to an end, the multi-functional teams have to be dissolved. During this Ramp-Down phase leaders have to take personal care of finding new roles for individuals who did not manage to find a suitable new role themselves. This usually turns out to be very difficult, in particular if there is no immediate new aircraft development project following. Obviously, the wider company organisation needs to support the leader in this task. If the Ramp-Down phase is not managed well and former team members get frustrated about it, the next aircraft development project will find it difficult to identify sufficient people willing to agree upon their secondment into DBTs.

The Kick-Off Workshop

It is a good idea to hold a team kick-off workshop at the beginning of the Development sub-phase, that is, during the Affiliation phase of the team's own evolution cycle. This should be organised by the team leader and should last for a few days.[34] There is usually a lot of resistance to spend that much time on something like a kick-off workshop. Since team leaders are made personally accountable for achieving objectives, they often fear losing unrecoverable time if they were to organise such a workshop. However, forming a strong team with a common vision early in the Development sub-phase will certainly lead to cycle time reduction opportunities later.

There are a number of good reasons to hold a kick-off workshop. First of all it is of utmost importance that people get to know each other as team members. In a co-located, multi-functional environment people have very different backgrounds and have probably never worked together before. To establish an atmosphere of trust, reliance and respect for each other is therefore a major objective of the workshop. Note that the creation of reciprocal confidence is particularly important for international projects. During the workshop it should become clear that not everyone can be a specialist in each discipline and that it is the mixture of specialists and generalists in the same team which makes it strong: it needs team members who are pushing for better design and more advanced technology, others for keeping the schedule, again others to push for cost reductions.

In addition, teams do not only require members who are specialists in their respective fields, but also require members who are socially competent. Finally, to accept a role in the project means to accept a team role. Typical team roles include: the perfectionist, the optimist, the boss, the observer, the team player, the artist and so on. A successful team reinforces and combines the strengths of individual team members while reducing their weaknesses down to a level where they do not adversely impact the team's performance.

At the workshop, team members should discuss and formally share the objectives and clarify the roles and responsibilities within the team.[35] A team is in fact a group of individuals sharing the same objectives. Creating a team therefore means foremost that the common objective should be defined, communicated to and known by the team members. The list of common objectives is the list of the overall time, cost and quality (TCQ) objectives of the development project broken down to the level of that team: each team has its own schedule, cost and quality targets. However, different cultural backgrounds among team members could make it difficult to achieve consensus.

[For example] to identify the cultural differences in quality it is helpful to define in team workshops how individual team members regard process and product quality. In discussing the differences the team can then try to come to a common understanding of what quality means to the team.[36] [However,] members of cultures with low Uncertainty Avoidance regard the fulfilment of quality standards with low priority compared to members of cultures with a high Uncertainty Avoidance. The latter need room for creativity and spontaneous solutions.[37]

---

34 Workshops for international projects require about 50 per cent more time than for national projects.
35 In fact, usually a whole series of bi-lateral discussions before, during and after the workshop is required to gain acceptance of the project objectives and of the distribution of roles and responsibilities from all sides. However, the workshop is the ideal forum to sort out differing views.
36 Hoffmann, H.-E., Schoper, Y.-G. and Fitzsimons, C.J. (2004), *Internationales Projektmanagement* (München: Beck-Wirtschaftsberater im Deutschen Taschenbuch Verlag), p. 301, free translation by the author.
37 Hoffmann, H.-E., Schoper, Y.-G. and Fitzsimons, C.J. (2004), *Internationales Projektmanagement* (München: Beck-Wirtschaftsberater im Deutschen Taschenbuch Verlag), p. 302, free translation by the author.

Team members should identify the technical and social skills needed to achieve the objectives of the team, evaluate the team's present mix of skills and select the skill areas which are most in need of development. Once a common understanding about the objectives has been achieved, the team would then try to agree upon roles and responsibilities of individual team members. As R.D. Archibald writes: 'To be most effective, the members of a work group should actively participate in developing … [a responsibility matrix] chart to describe their roles and relationships. Such development resolves differences and improves communications so that the organisation works more effectively. The responsibility matrix is useful for analysing and portraying any organization, but it is particularly effective in relating project responsibilities to the existing organisation.'[38] Clarifying roles and responsibilities should include a common discussion about the role of the team leader as opposed to the role of other team members. The purpose is to identify common views of the team leaders' versus team members' roles and to negotiate workable compromises on those areas where there is a difference of opinion on roles.

The team should also define its role within the existing wider company organisation, emphasising in particular its relationship to other groups with which it will work. Team members could, for example, be asked:

- to state what they see as the team's contribution to the organisation;
- to identify three or four activities which are unique to the team; as well as
- to identify three or four activities which the team should share with others.

The team members then share and discuss their views and try to reach consensus on key shared activities as well as on each of the team's contribution items. These statements can then be used as a 'role charter' to guide the team and can be communicated to other teams in the organisation.

There is more to do at the workshop though. For example, the starting point of all work, that is, the specifications or requirements given to the team, should be discussed. If it is not known yet, the actions to close this knowledge gap as well as their corresponding dates of completion should be defined. Discussing and knowing about the requirements received by the team does in particular help younger engineers. They still lack the experience of having designed similar components for different aircraft programmes.

All members should understand the risk potential of the project to be achieved by their team. Identifying major risks to the team – and consequently the probability of project success – should therefore also be an activity to be performed at the workshop. In particular, high-risk activities should be analysed within the team and it should be discussed whether the team has the skills to handle them.

Another important element of the kick-off workshop agenda is to raise awareness among all team members of what is likely to happen over the course of the coming years while the team is working together to develop the aircraft. This is also a way of training on the methods and processes used in the company by the different disciplines. Many engineers do, for example, not know about the methods and processes of Manufacturing and vice versa (the same applies to Procurement, Quality, Project Management and so on). Yet everything is linked together and it is important to explain the whole picture. This should be done by someone senior with a lot of experience in how the company designs,

---

38 Archibald, R.D. (1992), *Managing High-Technology Programs and Projects*, 2nd Edition, (New York: John Wiley & Sons), p. 61.

develops, produces and maintains aircraft. Also, explanatory training material should be provided to team members to give them the possibility of revisiting the training contents at a later stage.

The team should then be tasked to convert the overall aircraft development process into a list of specific key activities it needs to do in order to achieve the objectives. This can best be done by the creation of a network plan or PERT-chart.[39] It should define the key activities, their start and end dates as well as their logical interlinkages and key milestones, taking into consideration the overall estimated leadtimes available to do the job. The completed chart will be imperfect and will certainly need much refinement during the weeks following the workshop. But it sets the scene and it will generate a common and agreed view of how to achieve the objectives. In Chapter 13 we will come back to the subject of holding a planning workshop.

During the initial workshop one is advised to also perform a cultural team analysis in order to find out about the cultural risks and opportunities which are associated with the composition of the team. It is most important to recognise how big the differences are between the cultures represented in an international team, because this is an indicator for potential conflicts. The analysis could for example be based on Hofstede's cultural dimensions. It can be used to establish a ranking of every team member along the cultural dimensions. While, for example, discussing each other's position on the Power Distance scale, team members will learn to understand why there are differences and how to deal with them. Visualising the results of the analysis will yield higher transparency of the cultural differences within the team. Finally, the team should work together and establish the 'war room' described in Chapter 6. Creating a 'war room' together is a first and simple team-building exercise.

Now, when would be the right time for the kick-off workshop to take place? Clearly, it is difficult to find the right timing within the Affiliation phase. This is because of the ramp-up of resources, which takes place during that phase:

- at the beginning of the ramp-up there are not enough team members around to justify a workshop;
- towards the end of the ramp-up major team objectives should already have been achieved as by then the team is supposed to work at full speed;
- arranging the workshop in the middle of the ramp-up means excluding half the team members from important workshop results.

A pragmatic solution to this problem would be to hold the workshop around the time when a third of the ramp-up has been achieved. Subsequently those team members who participated in the kick-off workshop would train and inform later newcomers about its result, as and when they arrive. For workshops where it is foreseen to make use of outside trainers or facilitators one has to bear in mind that there might not be sufficient trainers or facilitators available. This is because in commercial aircraft development there are many multi-functional teams seeking to hold workshops at around the same time.

During the later phases of the team's own evolution it may be useful to arrange team diagnosis meetings to identify team strengths and weaknesses, using, for example, a questionnaire covering aspects like leadership, commitment, team climate, achievements and so on. Then team members can share and discuss their individual views in order to

---

39 PERT: Program Evaluation and Review Technique. This technique was invented by DuPont in 1957–1959.

reach consensus about the direction in which the team needs to develop. The team may then also want to set specific team-building objectives and develops and maintains an individual plan for team-building and operation.

As a general rule for all the aforementioned team-building events and their agendas, ample discussion time needs to be provided. It must be ensured there is enough time for all members to express themselves and for the team to consider their reactions. If conflicts occur during the workshop it is best to deal with them openly and in a straightforward manner. A facilitator should in this case help the team to thoroughly discuss the conflict situation.

If with all these workshops and meetings people still feel a necessity to get to know each other better, then additional events not related to work, for example, barbeques, bowling evenings, common days out and so on, are good ways to satisfy this need. Finally, as was said earlier, events should be held to reward teams for outstanding achievements as a visible sign of recognition by the higher-level Management, if and when appropriate.

Intercultural Team-Building

Intercultural team building is particularly difficult. This is for a whole variety of reasons, among which are:

- values and beliefs can differ significantly so that decisions taken on the grounds of values and beliefs will often come as a surprise to others;
- sources of motivation can be quite different for people with different cultural backgrounds. For example, to gain and have power is generally more important in France, while higher income is a stronger motivator in Germany, the USA and UK;
- as a result of adhering to their religious beliefs, team members may periodically be less able to fully contribute to the achievement of the team's objectives;
- there may be different interpretations of same or seemingly similar words. The same terminologies often mean different things to different people.

At Airbus, a multi-national company, discussions about terminologies happen frequently. Here is one anecdote, which underlines this:

A French Airbus employee once wrote a procedure about how to attest the conformity of a physical component with its 3D manufacturing data. As the French word 'attester' means the same as the English word 'to attest', he concluded that this must be accordingly for the French word 'attestation'. As a result, he called the document 'Attestation of Conformity', a document, which was applicable to all Airbus sites, including the ones in the UK. There, however, no one understood what the title meant, as in English the words 'certification' or 'testimony' are used for what the French colleague wanted to express. Only after some clarifying discussions the word 'attestation' became a meaning for the British colleagues, which thereafter adopted this new terminology.

There is no easy answer of how to improve the effectiveness of a multi-cultural team. But awareness of cultural differences and behaviours like listening, learning, respecting others and being open minded for change facilitate the working of a culturally diverse multi-functional team. In fact, as Hofstede and Hofstede point out, 'the principle of surviving in a multi-cultural world is that one does not need to think, feel, and act in the same way in order to agree on practical issues and to cooperate.'[40] For intercultural encounters to be successful, partners must be safe to believe in their own values. 'A sense of identity provides the feeling of security from which one can encounter other cultures with an open mind.'[41]

RECRUITING LEADERS

The most important decision to be taken at the beginning of a new development is the careful selection of the leaders of the multi-functional teams on all levels of the organisational hierarchy. This is so important because 'a strong leader will be able to overcome many shortcomings and imperfect decisions, but a mediocre one will be stymied even by small obstacles. The leader is often the buffer insulating the team from inappropriate management practices that run rampant in the rest of the company.'[42]

What should be the professional competence of good leaders for multi-functional, multi-cultural teams? Should they be engineers coming from the Engineering department because they understand how engineers think and work, know the aircraft development processes inside out and can judge technological risks well? Or should leaders have a Systems Engineering background capable of structuring projects well, providing to them an architecture which everyone involved in the project can use? Or should leaders be coming from Manufacturing with a mindset of delivering on time? Or from Procurement in case much of the development work is outsourced? Or from almost anywhere, other industries perhaps, as long as they exhibit strong Project Management skills?

T. Flouris and D. Lock, the authors of the book *Aviation Project Management*, write:

'Although there are many projects in commerce and industry where the Project Manager can be successful even though they have not been training in the core technical discipline of the business, aviation projects do demand that technical skill. It is important that aviation Project Managers have either technical expertise or at least a high degree of appreciation of the industry. Aviation as a field is a complex aggregation of technical knowledge and skill sets. Further, most of its projects and operations are subject to scrutiny and approval from national or international regulating authorities. This is not an environment for the beginner, and the Project Manager who comes in as a complete outsider would have significant difficulty in coming to terms with the industry-specific jargon and the regulatory background.'[43]

However, while the search for the ideal team leader should never come to a halt, it is clear that individuals with good leadership behaviours are a scarce resource in any company. Commercial aircraft companies are no exception to this. The debate about skills and competencies necessary for team leaders in a Project Management role is therefore somewhat academic. It is very difficult indeed, if not impossible, to find sufficient project leaders (that is, more than a hundred for large aircraft development projects) with the required skills. Thus, with so many multi-functional teams required for

40 Hofstede, G. and Hofstede, G.J. (2005), *Cultures and Organizations: Software of the Mind*, 2nd Edition, (New York: McGraw-Hill), p. 366.
41 Hofstede, G. and Hofstede, G.J. (2005), *Cultures and Organizations: Software of the Mind*, 2nd Edition, (New York: McGraw-Hill), p. 365.
42 Smith, P.G. and Reinertsen, D.G. (1998), *Developing Products in Half the Time* 2nd Edition, (New York: John Wiley & Sons), p. 120.
43 Flouris, T. and Lock, D. (2008), *Aviation Project Management* (Aldershot: Ashgate), p. 89.

aircraft development, by far not every team leader will come close to the ideal leadership behaviours described above.

It must therefore be strongly underlined that the lack of skilled leaders is the single biggest threat to the whole concept of multi-functional teams supporting concurrent ways of working. This would therefore also represent a threat to the type of complex projects discussed here. In many ways, one can only relentlessly try to find suitable candidates who could eventually grow into leadership excellence – regardless of their functional background. This requires significant efforts for candidate selection and training. If skills are not abundant and – for other reasons – cannot be recruited from outside, strategic competence development is the only way to eventually gain the requested skills.

> 'Successful project leaders must therefore be both inspired people managers and skilled problem solvers. They must bring the right individuals and information together to develop the best solutions for problems, coach sub-teams to perform at a higher level, and have a working knowledge of all areas of their projects, including the technical side, marketing, operations, and supply chain management. Above all, senior executives must trust these leaders to make sound, fact-based decisions about the direction of their projects without always seeking input from above.
>
> Do such people exist? The good news is that they do, but only the very best project leaders can now perform at this level. In our experience, the potential of the best leaders is stifled by the current inflexible approach. Giving them enough flexibility and authority to make decisions will unleash their potential and raise the performance of their teams a few notches. Meanwhile, most project managers will have to upgrade some combination of their leadership ability, their problem-solving skills, or their cross-functional expertise – hardly surprising, since many of them are engineers promoted to management without training or even, in some cases, natural aptitude. Organisations now have the task of creating processes to spot project leaders with strong potential and to develop their skills.'[44]

However, for leaders on higher levels of the matrix organisation there should at least be one criteria they should comply with: They should already have been with the company for some time. This guarantees that they have a good understanding of the company's culture and processes. Also, and perhaps even more important, they have an established network of contacts within the company, which often is a crucial enabler to ensure that obstacles are being removed and things are being pushed forward.

If internally not enough leaders for the DBTs can be found or developed into that role, recruitment from external sources is the only alternative. However, recruiting candidates from outside may not lead to optimal team leadership as these people lack the network of contacts within the company or some technical knowledge of commercial aircraft development or both. And even if the new recruits are technically competent they will need time to come down on their individual learning curve – in terms of learning about the project itself as well as the corporate culture. Team leaders recruited from external sources should therefore initially not be installed on higher levels of the organisation. If team leaders remain as an external workforce, then another disadvantage is the usually very short notice, with which they can leave the project. This could create a sudden leadership gap, which usually requires unacceptable long leadtimes to get closed.

---

44 Holman, R., Kaas, H.-W. and Keeling, D. (2003), 'The future of product development', *The McKinsey Quarterly* No. 3, p. 37.

Once a leader has been appointed he or she should be kept. Changing the leader during mid-course will often cause disruptions. Team members have to adapt to a new management style, new priorities, as well as new ways of working associated with the replacement. Appointing the best team leaders is therefore a necessary request for the beginning of the Development sub-phase. It is worth spending significant effort on this.

MAIN CONCLUSIONS FROM CHAPTER 5

- The aggregate effect of myriads of delayed or poor decisions during a commercial aircraft development results in major unwanted consequences for aircraft quality, maturity, development cycle time and cost. Much care must therefore be taken to ensure that the right decisions can be taken in a timely manner and at the right level. This is an organisational challenge.
- The 'right decision' requirement is fulfilled by the multi-functional approach. It ensures that all stakeholders are not only concurrently participating in the design and development process but are encouraged to actively influence the latter.
- Design-Build Teams as owners of the design-build process are a direct, logical extension of the multi-functional approach. Specific deliverables are at the focus of all of the Design-Build Teams' activities, and not the exclusive view of a particular discipline or Function.
- In a multi-functional Design-Build Team environment the *'timely manner'* requirement is fulfilled if teams are created in such a way that the communication streams across the functional borders *within* each team are optimised, while at the same time minimising them *beyond* any team's boundaries.
- Members of multi-functional teams may include representatives from Engineering, Systems Engineering, Manufacturing, Procurement, Project Management, Customer Services, Materials, Quality, Human Resources.
- Leadership in a multi-functional environment is all about generating teamwork and efficiency within a team as well as in the interaction with other teams.
- Teams go through predictable and noticeable stages of development. They mature in terms of task progress and in terms of interpersonal relations. At each phase of development certain problems can occur.
- One of a team leader's most important functions is therefore to develop the project team and to help it move through the various stages of its evolution.
- There are a number of good reasons to hold a team kick-off workshop at the beginning of the Development sub-phase. It should be organised by the team leader and should last for a few days.
- Intercultural team building is particularly difficult for a whole variety of reasons. Unfortunately there is no easy answer of how to improve the effectiveness of a multi-cultural team. But awareness of cultural differences, certain behaviours as well as being open minded for change facilitates the working of a culturally diverse multi-functional team.
- It must be strongly underlined that the lack of skilled leaders is the single biggest threat to the whole concept of multi-functional Design-Build Teams supporting concurrent ways of working. This would therefore also represent a major threat to the objectives' achievement of the type of complex projects discussed here.

6

The Need for Co-Location and Organisational Balance

THE NEED FOR PHYSICAL CO-LOCATION

Do members of a multi-functional Design-Build Team (DBT) have to sit together to ensure the best possible intra-team and cross-functional communication? Do they need to be *physically* co-located, that is, sharing the same office space? As Boeing's Alan Mulally writes:

> 'One of my most favourite expressions is that the biggest problem with communication is the illusion that it has occurred. We think when we express ourselves that, because we generally understand what we think, the person that we're expressing it to understands in the same way. Well, in my experience I've found that's very difficult. When you're creating something, you have to recognize that it's the interaction that will allow everybody to come to a fundamental understanding of what it's supposed to do, how it's going to be made. And I think we should always be striving to have an environment that allows those interactions to happen, and not have things be separate and sequential in the process.'[1]

Co-location stimulates communication by interaction. So, co-location is certainly beneficial for the multi-functional design-build process based on concurrent ways of working. However, some people, especially in high-tech industries, have suggested that electronic media have somewhat superseded the need for physical co-location. We may wish to call this virtual co-location.

It is clear that with the impressive advancements of Information Technology (IT) systems, which include Internet- and Intranet-based tools for communication, design and planning (such as e-mails, video conferencing, Computer Aided Design (CAD), Product Data Management (PDM)), the need for physical contact between people can be and has been reduced. This is, for example, because people can send messages quickly and simultaneously to more than one person, messages can be forwarded and stored, people can work on tools concurrently with full visibility on what others are doing, and so on. It is a lot more tedious to arrange (many) meetings, and very difficult to ensure participation of all stakeholders, to obtain the same communicative effect. No wonder that about one-quarter of the weekly hours of engineers or managers working in aircraft development is allocated to IT-based communication, see Figure 6.1.

---

1   Sabbagh, K. (1996), *21ˢᵗ Century Jet. The Making of the Boeing 777* (London: MacMillan Publishers), pp. 24–5.

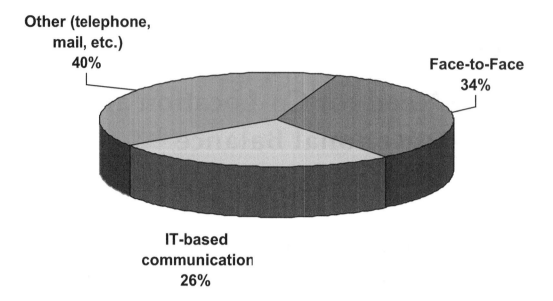

Figure 6.1   Typical time allocation of engineers and managers in aircraft product development by communication mode

Based on data provided by and in: Chase, J.P. (2001), *Value Creation in the Product Development Process* (Boston: Massachusetts Institute of Technology, Department of Aeronautics and Astronautics, Master of Science Thesis), p. 91.

However, communication between individuals is more than exchanging information which can be expressed in digital words. The German researcher Friedemann Schulz von Thun developed the so-called Four-Ears model, see Figure 6.2, which is widely accepted in communications research. According to this model, any communication between two individuals, that is, between a sender and a receiver, always contains four different messages:

1.   the message about the actual subject;
2.   a self-revealing message about the sender (for example: What accent is spoken? Therefore, where is the sender from? Is the sender awake or sleepy?);
3.   information about the relationship between sender and receiver (for example: Do the selected words, intonation and body-language show respect?); as well as
4.   an appeal: almost all communications are intended to provoke actions by the receiver (for example: to provoke admiration for the sender).

Any interpersonal conversation covers all four messages. But only half of what is conveyed in a conversation comes from the spoken words themselves. A surprisingly large 50 per cent (that is, the other half) is based on non-verbal communication. In projects like aircraft development where communication needs to be very intense indeed, it is wise to always be aware of the significant influence of non-verbal communication as well as of the three other messages beyond the actual subject. This is even more true for international projects where every exchange of information must be judged on the basis of its individual cultural origin. The correct understanding of the non-verbal messages becomes crucial and can only be grasped by carefully observing signs of the whole body,

Any communication contains messages about:

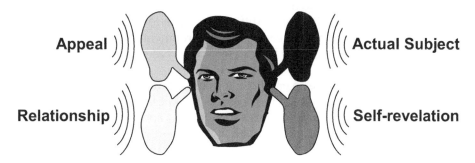

Appeal   Actual Subject

Relationship   Self-revelation

Figure 6.2   The Four-Ears model according to Friedemann Schulz von Thun

including voice and values. Clearly, computerised communication ignores all of this (at least for the time being). As expressed by T.J. Allen and G.W. Henn: it is 'bandwidth limited'.[2]
   There are other significant drawbacks of virtual co-location:

- Any written message solidifies the position of its writer and makes assumptions about how a reader of it will respond. This makes open communication difficult, if not impossible. In contrast, in direct verbal communication a position is progressively shaped by the reactions of the person(s) with whom one is conversing.
- Because written communication lacks the immediate feedback of an audience saying 'I agree' or 'Can you please explain' it invariably provides both, too much and too little information for its audience.
- The arrival time of a response to a written communication is not predictable. And what can one assume if there is no response at all?
- IT-based communication offers no opportunity for people on the project to experience the important interrelationships between information on the one hand and human behaviour on the other.
- Many team members will be computer literate, but many others will not. In particular, older persons are struggling with the latest IT developments. In fact, many of them are lost with these new technologies.
- Most people do need a 'social kit' at work. 'Speaking' to the computer all day long is not a substitute. In addition, important information external to a person's focus of the day may remain completely unnoticed without the chat at the coffee machine.
- Virtual co-location does not ensure 100 per cent dedication of team members to the programme. The whole philosophy of Concurrent Engineering is based on the assumption that cycle times will be much reduced if individuals are fully dedicated to the programme development. One of the biggest sources of delay is the sharing of time of team members among too many projects. Trying to keep focus on one project is hampered by conflicts of priority between projects. For an

aircraft development to be on time, as many team members as possible should therefore be fully dedicated to the project. Contrary to virtual co-location, physical co-location forces people to move away from their previous locations into an environment where it will be very difficult for them to work on different projects at the same time. In a physical co-location environment, full-time team members have nowhere to hide and they have to work on whatever problem is encountered during the project.

- Imagine a situation where an engineer in a co-located team goes to his colleague asking: 'Do you have an idea of the probable thickness of the rib?' He receives an answer and can continue working. Then he asks another colleague over the telephone for another type of information. And so on. Such is the daily work. Even the most modern virtual communication cannot compensate for the human mind. And the effort it would take to enter every bit of information received through walking around, telephone calls and so on into an IT system would be gigantic and would in the end only slow down progress.

Virtual co-location can therefore not replace physical co-location as information cannot be managed in a holistic manner. It also creates some problems if team members are not 100 per cent dedicated to the project. But it certainly is adding value – if and when properly used – on top of the advantages of physical co-location. The latter remains the principal means of communication of multi-functional DBTs. In fact, only physical co-location offers the potential for a well-integrated development. Getting people to talk to each other is the only truly effective way of transferring technical knowledge and thereby advancing the development process.

Once *co-located* multi-functional DBTs are installed, they will be the prime enabler for implementing the Concurrent Product Definition (CPD) process. Only then, the *right decision*, the *timely manner* and the *right level* requirement introduced in Chapter 5 can be fulfilled for the full exploitation of the capabilities offered by the CPD philosophy. No wonder that K. Cusick found out during her company survey – trying to pin-down some lessons learned from the application of integrated product development processes – that all surveyed companies stated co-location as critical for project success.[3]

In the early 1990s, IT tools and infrastructure were mature enough to make CPD a reality. Not surprisingly, this is also the time when Boeing and Airbus started to create co-located DBTs. In the DBTs that Boeing used to develop the 777, Engineering and Manufacturing (and other) representatives were grouped together for each of the 250 sub-systems.[4] Airbus introduced co-located teams for the first time in its French branch (formerly Aérospatiale) when developing the A340-600 and extended this principle to all other branches for the development of the A380.[5]

3   Cusick, K., *A collection of Integrated Product Development Lessons Learned* (La Mirada, CA: SECAT LLC), ftp://ftp.cs.kuleuven.be/pub/Ada-Belgium/ase/ase02_01/bookcase/se_sh/cmms/systems_engineering/sei_se_cmm_papers/ipd_ll_lkd.pdf, accessed 10 October 2009, p. 4.
4   Sabbagh, K. (1996), *21st Century Jet. The Making of the Boeing 777* (London: MacMillan Publishers), p. 67. http://www.boeing.com/news/releases/1995/news.release.950614-a.html, accessed 10 October 2009, indicates that there were 238 Design-Build Teams. So the number 250 mentioned by Sabbagh may be somewhat exaggerated.
5   See: Whitney, D.E., *Design-Build Teams at Aérospatiale*, http://esd.mit.edu/esd_books/whitney/pdfs/aerospatiale.pdf, accessed 10 October 2009.

'… Each of the many parties involved in the 777 must be made to meet other species of engineer face to face to exchange ideas about matters of mutual interest. The time was past when designers could "throw stuff over the wall" and wash their hands of it. This realization was what led Boeing senior managers to develop two linked ideas, one called Working Together and the other, which they had come across in Japan, called design-build teams.'[6]

'Working Together was to be the name of the first 777; it was to be painted on banners that went up around the factories, and on posters, baseball caps, badges and T-shirts. And it was to be repeated as a mantra in speeches and discussions between Boeing and its customers and contractors. But there was a lot riding on it. In 1992 Boeing believed in an almost theological way that this new way of working – whatever it was – could actually make a difference to the testable and costable quality of the plane they were making, an outcome that would not become demonstrable for another year if not two. But observing it in action in the new type of meetings that were happening every day in every 777 department revealed a very different atmosphere from that conveyed in a story from earlier Boeing days.'[7]

We know now that co-location is an important principle to make the integrated and concurrent approach a reality. But to what extent should co-location be applied?

Co-location first of all means that Functions identify suitable candidates to be seconded into multi-functional DBTs, sharing the same office space with other members of the same team. However, what also needs to be considered is the co-location of entire teams. Ideally, teams which share interdependencies should also be co-located as close as possible. This is because sooner or later every interface – even if frozen – requires communication in one way or the other. As distance plays a fundamental role in personal communication, teams which require more communication should be located closer to each other. For example, to help integration of systems into the Airbus A380 sections, the team 'Fuel System' was co-located with the team in charge of the wing and 'Avionics and Cockpit' was co-located with the 'Nose' team. Teams with no or only few interdependencies can be allowed to be further apart. To this end, it is important to firstly analyse the interdependencies between teams carefully. Only then it should be decided, which teams to locate where in a building.

However, in practice there are many limitations to this ideal picture. Available buildings are often not designed to the needs of co-location, neither in terms of size nor in terms of open office environment. Teams, which should be co-located because of their complex interdependencies, in reality often are not. This may be a direct consequence of workshare agreements for example.

For example, the multi-functional DBTs created at Airbus for the development of the A380 were located at Bremen (Germany), Filton (UK), Getafe (Spain), Hamburg (Germany), and Toulouse (France). Among others, Bremen hosted the teams in charge of the aircraft's flaps design-build because this is where historically the centre of competence for these components had emerged. The most complex interdependencies of the Bremen teams were, however, with the teams in charge of the wing's trailing edge located at Filton, again for reasons of available competencies. For the A380 development this remained unchanged, but it meant that co-location of these teams became impossible.

---

6   Sabbagh, K. (1996), *21st Century Jet. The Making of the Boeing 777* (London: MacMillan Publishers), pp. 59–60.
7   Sabbagh, K. (1996), *21st Century Jet. The Making of the Boeing 777* (London: MacMillan Publishers), p. 61.

Sub-contracting of DBT activities is another source of concern with regard to the co-location principles. Sub-contracting cannot always be done in a way which is compatible with the desired early freeze of interfaces. Instead, 'Buy' decisions are taken on the basis of a multitude of criteria, among which are cost reduction, risk mitigation, resource limitations, offset arrangements, market access and access to technologies.

Where this poses a problem for successful application of the CPD process, full co-location can perhaps be achieved temporarily during the early design phases. Functional representatives from the suppliers could be invited to join individual DBTs at the aircraft-developing company for a limited period of time until the common understanding of the design principles and interfaces is mature enough. Boeing, for example, invited hundreds of Japanese engineers to Seattle to join the DBTs during the early stages of the 787 development. However, for the Detailed Design phase, they returned to Japan to form the nucleus of integrated teams there.

> 'Teaming across multiple tiers of the supply chain early in the design process ... [fosters] innovation in product architecture, resulting in significant quality improvements, 40–60% cost avoidance, and 25% reduction in cycle time.'[8]

Similar to suppliers, aircraft manufacturers also involve airline customers in the design of a new airplane.

> 'An airline member of one team that was designing an electronics bay pointed out that the light was positioned directly overhead. That seemed like a logical place for a light, to our engineers. But the airline rep[resentative] explained that when a maintenance person is actually working in the bay, his head and shoulders block most of the light, making it very difficult to see. So, we changed the design, and put two lights on the sides of the bay. A small thing perhaps, but that kind of valuable customer insight was reflected in more than 1,000 design modifications to the 777.'[9]

Once a project is finished towards the end of the Development sub-phase, many hundred or thousand of developers need to go back to a home. This home can only be the functional home. Unless team members continue to work for new programmes there is a huge migration of staff to be managed back to the home Functions.

It must be stated that the overall dynamics of co-location, that is:

- creation of teams;
- dynamic sizing of teams (resulting from the aircraft company's own resources ramp-up but also from the invite to suppliers, customers and other stakeholders to join teams temporarily);
- co-location of teams;
- dissolution of teams; and finally
- return of team members to their home Functions;

---

8   Anon. (1999), *LAI: Systems Offering Best Lifecycle Value* (Cambridge, MA: Lean Aerospace Initiative).
9   Condit, P.M. (1996), *Performance, Process, and Value: Commercial Aircraft Design in the 21st Century*, (Los Angeles: World Aviation Congress and Exposition), 22 October 1996.

requires adequate accommodation and facilities management. This is a complex undertaking to be fulfilled by a role within the project itself. In a moment we will see which Function should be in charge of it.

A STIMULATING ENVIRONMENT FOR MULTI-FUNCTIONAL DESIGN-BUILD TEAMS

What is the right physical environment for a co-located team? Before trying to answer this question, it is worth looking at some of the interdependencies between communication on the one hand and architecture of buildings on the other. Team members working together on a project are obviously 'spatial' in the sense that they see each other not only in time but also in space. They meet others and discuss ideas with them in a space. And they are aware of the work of others primarily by seeing them. Physical space can be an important tool for generating awareness.

Architects organise the spaces in which people live, work and move by designing buildings which suit the needs of those people. But the architecture of a building also plays a role in *how* they live, work and move in those spaces. This is because the structure of space can initiate and influence social behaviour – the *intensity* of communication, for example. It can enable people to interact in real time, without barriers. Keeping this in mind, it is not a surprise that T.J. Allen and G.W. Henn, the innovators in analysing the interdependencies between architecture and communication during product developments, claim that 'if [by means of adequate architecture] you maximise the *potential* that people in an organisation can and will communicate ..., you will vastly increase the likelihood of knowledge transfer, inspiration, and hence innovation.'[10]

> 'Awareness is at the very center of the innovation process. A lack of awareness underlies much of the poor communication and inefficiency that we see in product development organisations. Engineers and other technical staff in large organisations are often unaware of the talent and knowledge housed within their own organisations. Awarenesss can be built through study, observations, and communication. The latter includes the exchange of thoughts, messages or information through speech, signals, writing, or behaviour. In the context of today's ... [multi-functional] product development organisations, it is communication that builds awareness.'[11]

Thus, managers should not only look at organisational structures to organise people into departments, programmes, multi-functional teams and so on, they should also look at how to configure the building space to encourage the very communication that spurs the development process. There really are two tools: organisational structure *and* physical space, which can – and must – be configured properly.

In order to find the most adequate architecture for a building to accommodate multi-functional teams, it is necessary to analyse the development processes which are expected to take place in the building. Who needs to talk to whom according to these processes, and when? Where do people need a calm environment to better concentrate themselves,

---

10 Allen, T.J. and Henn, G.W. (2007), *The Organization and Architecture of Innovation. Managing the Flow of Technology* (Burlington, MA: Elsevier), p. 2.
11 Allen, T.J. and Henn, G.W. (2007), *The Organization and Architecture of Innovation. Managing the Flow of Technology* (Burlington, MA: Elsevier), p. 85.

where do they need to communicate intensively? Where must there be open office spaces, where can there be walls? These are the sort of questions which need to be addressed.

With the start of the Development sub-phase, not only the growing teams need more space, but that space now also represents the 'battle zone' typical for a matrix organisation. It is now that real-time communication becomes so vital for project success. A perfect building architecture – designed to meet the needs of the development processes – becomes an invaluable tool for organising the teams, to promote intra-team coordination and to ensure communication between teams.

Ideally, all multi-functional teams should be co-located in a process-tailored office environment with as many team members as possible working on the same floor of the same building. Obviously, there are limits to the size of a building, so housing teams on different floors will inevitably become a necessity. But floors introduce separations between teams. In fact, 'vertical separation always has a more severe effect than an equivalent amount of horizontal separation'.[12]

> 'In most buildings, each floor is isolated. When we exit the elevator on a given floor of a building, we quickly forget about the existence of the other floors. There is a tendency for our mental image of the building to be limited to a single floor – the floor on which we happen to be.'[13]

Communication demands that teams need to be constantly reminded that there are people working on other floors for the same project too.

Fortunately architects have invented building designs where the effect of floor separation is reduced significantly. Figure 6.3 provides an excellent example for this.[14] It enables people to see across the other floors as well as their own, reminding them of the existence of those other floors, and of the people housed there. The architecture thereby helps to overcome the typical isolation of one floor from another. In fact, T.J. Allen and G.W. Henn claim it increases the probability of communication by an order of magnitude.[15]

Besides fully equipped working desks for employees, the office should contain sufficient meeting rooms. They are needed for all kinds of larger meetings but also for one-to-one discussions, for private telephone calls and as temporary offices for visitors. Rooms should be equipped with:

- flipcharts to provide explanations and to capture ideas and actions;
- wall display panels;
- PCs and screens;
- links to the company's intranet and e-mail system to allow for working as a mobile office;
- video conferencing and telephone communication tools for easier communication with other teams far away;
- workstations to dive into 3D digital models of the aircraft;

---

12 Allen, T.J. and Henn, G.W. (2007), *The Organization and Architecture of Innovation. Managing the Flow of Technology* (Burlington, MA: Elsevier), p. 71.
13 Allen, T.J. and Henn, G.W. (2007), *The Organization and Architecture of Innovation. Managing the Flow of Technology* (Burlington, MA: Elsevier), p. 73.
14 The creation of an atrium in the centre of a building is another example of reducing floor separation.
15 Allen, T.J. and Henn, G.W. (2007), *The Organization and Architecture of Innovation. Managing the Flow of Technology* (Burlington, MA: Elsevier), p. 75.

Figure 6.3   Example of a building architecture in support of co-located multi-functional Design-Build Teams

Source: Allen, T.J. and Henn, G.W. (2007), *The Organization and Architecture of Innovation. Managing the Flow of Technology* (Burlington, MA: Elsevier), p. 121, with permisson.

- windows so that others can see all 'brains' at work;[16] as well as with
- blinds to the windows, for example for one-to-one discussions or for situations where privacy is required.

The size of the project will dictate how elaborate the facility should be.

Clearly, physical space in the building should not be allocated departmentally but teams should be seated according to product areas or along process streams. This results in office arrangement with neighbouring people who often do not belong to the same department.

There should also be a 'war room'. This usually is a meeting room decorated with all the key information necessary for progress control, such as S-curves, list of actions, pictures of hardware, schedule charts, diagrams explaining functionalities and so on.

'Winston Churchill said, "This is the room from which I will direct the war." The prime minister was referring to the secure room deep in the ground near 10 Downing Street, and Whitehall, London. The bunker was created because of the relentless enemy bombings. During the height of World War II it operated 24 hours a day, seven days a week. It became the center of all major military activity in the United Kingdom. In this room, Churchill met with advisors, cabinet members, and military intelligence officers and made some of the most serious decisions of World War II.'[17]

But a clear building architecture can do even more to support the different types of communication, of which – according to T.J.Allen – there are three:[18]

- Communication for coordination. 'The right hand has to know what the left hand is doing'. This type of communication exists in nearly all organizations.
- Communication for information. This type of communication ensures that one keeps up-to-date. It increases in importance with the rate, at which knowledge is changing.
- Communication for inspiration. This type of communication is active in creating knowledge. In an organisation that relies on creative solutions to problems, communication for inspiration is absolutely critical.

The need for communication for coordination and information is usually well represented by organisational structures. But it is not suited to ensure communication for inspiration. This is where the structuring of space can help and allow for the uncertainty in interaction that lead to creativity. As the unintended, impromptu encounters are the ones that often produce the most creative ideas, it would be highly desirable if buildings

---

16 The transparency indicated by windows, both to the outside and inside is a metaphor for an orientation in different directions and is meant to support the idea that one has the freedom to make one's own decisions – that is, within the realm for which one is responsible – but also to influence the decisions of others seen through the glass walls. In terms of promoting interaction among seniors management in silo organisations, this ability goes beyond being merely symbolic to being part and parcel of transforming the firm and its decision-making processes. From: Allen, T.J. and Henn, G.W. (2007), *The Organization and Architecture of Innovation. Managing the Flow of Technology* (Burlington, MA: Elsevier), p. 17.

17 Johnson, J. and Horsey, D.C. (2001), 'The IT War Room', *The IT Software Journal*, Software MagCom June 2001, http://www.softwaremag.com/L.cfm?doc=archive/2001jun/WarRoom.html, accessed 10 October 2009.

18 Allen, T.J. (1986), 'Organizational Structure, Information Technology and R&D Productivity', *IEEE Transactions in Engineering Management* Vol. 33, No. 4, pp. 212–17.

are also designed in a way as to encourage 'people browsing', that is, open movement of people on and between floors.

The office should therefore also house areas where snacks and drinks can be consumed, and which serve as sources of serendipity communication. The aspect of serendipity communication is not to be underestimated. While team members usually communicate with other members of the same team as well as with members of other teams, with which their own team shares interfaces, serendipity communication ensures that latest news, problems and unexpected discoveries from other teams also reach individuals by chance. A coffee shop area ensures a good balance between planned and unplanned chains of communication.

> 'Another important element in stimulation communication is visual contact in real time. Qualitative observations lead us to conclude that people need to be prompted occasionally and reminded of the existence of potential technical communication partners. This holds true for all three types of communication… [For the communication for coordination], visual contact might remind an engineer that he needs to tell the person he sees about a design change. In terms of communication for information, visual contact could be a reminder that a certain person is the 'resident expert' to go with a given question. It is in the realm of communication for inspiration … that visual contact is probably the most important. If people do not see one another, they will not have the opportunity to interact and create the knowledge.'[19]

It is certainly worth analysing how architects can support the product development process by designing buildings, which represent suitable platforms for the co-location of DBTs – and whether the investment in such a building would pay off.

FINDING THE RIGHT ORGANISATIONAL BALANCE

DBTs working for the same project are difficult to manage and organise. This lies in the very nature of arranging multi-functionality: for the long duration of aircraft development projects, people need to be seconded from different Functions such as Engineering, Manufacturing, Procurement, Quality and so on, to the project in order to establish individual DBTs under the leadership of project leaders, see Figure 6.4. This results in new personal situations not only for:

- the seconded individuals; but also for
- the functional department heads; and even for
- the project leaders.

In fact, it generates fears among these individuals, which need to be addressed.

Not surprisingly, therefore, full-time secondment of functional representatives into DBTs led by project leaders is a delicate subject. One should, for example, not expect that seconded individuals cut all their ties with their home Functions. As they fear to be in a disadvantageous position with regard to their career and wages evolution compared to their peer colleagues, who keep staying in their home Function, many will try to resist secondment.

---

19 Allen, T.J. and Henn, G.W. (2007), *The Organization and Architecture of Innovation. Managing the Flow of Technology* (Burlington, MA: Elsevier), p. 73.

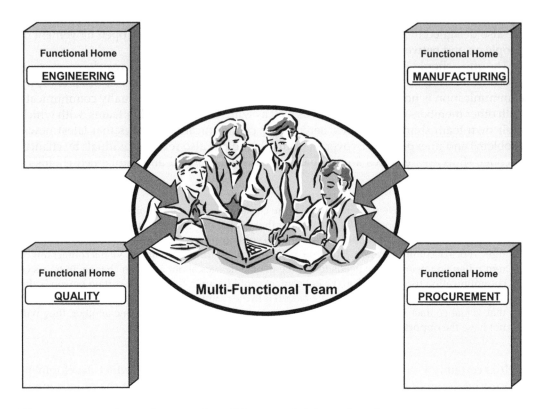

Figure 6.4   Secondment of staff into multi-functional teams

As R.D. Archibald writes: 'Projects are subject to sudden shifts in priority or even to cancellation, and full-time members of a project office are thus exposed to potentially serious threats to their job security; this often causes a reluctance on the part of some people to accept a project assignment.'[20] And: 'After observing what can happen to the people assigned to a project that is abruptly cancelled, or fails, or even when a project is successfully completed and the team members are rewarded with demotions or termination of their employment, people will not be motivated to follow in their footsteps.'[21]

In addition, it is felt important to keep the link to the functional homes in order to remain up to speed with latest developments in technology, methods, processes and tools, and in order to secure full attention of the functional top-level Management.

Functional department heads fear a perceived loss of power due to the sudden release of many of their men or women to a project. If the aircraft development is a major undertaking of the entire company, than indeed many or most individuals of a Function will be seconded into multi-functional teams. The remaining Function may become a much shrunk organisation.

In addition, functional bosses insist that the functional know-how must be kept and must be further developed under their control and within the Functions. As J.P. Womack and D.T. Jones, the lean experts known worldwide, stress it: 'Functions are where the

---

20 Archibald, R.D. (1992), *Managing High-Technology Programs and Projects*, 2nd Edition, (New York: John Wiley & Sons), p. 49.
21 Archibald, R.D. (1992), *Managing High-Technology Programs and Projects*, 2nd Edition, (New York: John Wiley & Sons), p. 132.

learning is collected, systematized and deployed. Functions, therefore, need a secure place in any organisation.'[22] Otherwise there could be a real danger of know-how being diffused into the wider organisation with the likely consequence that it will be lost for the company.[23]

'While at Boeing, I learned that one of the design teams ... had developed a design for a pressurized door, which was to be made out of a casting. They had done the design and received bids from the supplier and demonstrated that it would be cheaper than the conventional design. However, when it was reviewed by the senior functional management, it was abandoned. The Boeing company has a rule, "thou shall not make pressurised doors out of castings". This rule comes from having more experience and having built more aircraft than any other company in the world. They have learned that the risks of a cast part failing for this application are not worth the savings in cost. This is an example of the corporate knowledge that resides within the function.'[24]

Functional bosses also often feel that project leaders are interfering with their territory, again creating fears about loss of power and acknowledgment for being competent. The lack of understanding and of acceptance of the sharing of responsibilities required for effective CPD processes generates this fear of losing power and prestige. In cultures emphasising on power, control, and hierarchical position, this fear can be very strong.

Finally, project leaders acting as team leaders also have fears: they fear that they are not empowered enough to take decisions, that is, that they have little authority compared to the functional departments.

How can all these fears be reduced? Only by a well-balanced power sharing! First of all, the introduction of matrix organisations – with projects along one axis and Functions along the other – has proven to be beneficial for the management of aircraft development projects. The matrix organisation is a structure which allows the establishment of an organisational balance between projects and Functions. It tries to maximise the strengths and minimise the weaknesses of other organisational structures, such as uni-functional structures. Because of its net benefits, matrix organisations can be found in aerospace companies around the world.

'The matrix organisation evolved because project teams, while making intense focus and coordination possible, could not meet the challenge of keeping technical staff in close contact with new developments within their specialities. It traces its origins back to the late 1950s, when T. Wilson of the Boeing Company tried to accomplish both with a new organisational form for a major aerospace development program. The organisation he devised later came to be known as "the matrix".'[25]

---

22 Womack, J.P. and Jones, D.T. (1994), 'From Lean Production to the Lean Enterprise', *Harvard Business Review* March–April.
23 When a large portion of the technical staff of an organisation falls behind in knowledge, the organisation itself falls behind. Thus, we find that too widespread use of project team structure can lead to an erosion of a company's knowledge base. From: Allen, T.J. and Henn, G.W. (2007), *The Organization and Architecture of Innovation. Managing the Flow of Technology* (Burlington, MA: Elsevier), p. 36.
24 Hernandez, C.M. (1995), *Challenges and Benefits to the Implementation of Integrated Product Teams on Large Military Procurements* (Boston: Massachusetts Institute of Technology, MSc thesis), p. 95.
25 Allen, T.J. and Henn, G.W. (2007), *The Organization and Architecture of Innovation. Managing the Flow of Technology* (Burlington, MA: Elsevier), p. 37.

'The major benefits of the matrix organization are the balancing of objectives, the coordination across functional lines, and the visibility of the project objectives through the project…manager's office. The major disadvantage is that the person in the middle is working for two [or more] bosses.'[26]

It must be understood that team leaders acting as project leaders in a multi-functional matrix environment cannot be made *exclusively* accountable for the success of the project. This is already not possible because of the accountability question raised in Chapter 1: In a long duration project, such as a commercial aircraft development, new project leaders are joining and others are leaving it significantly more often than in short duration projects. New project leaders will be reluctant to accept the failures of their predecessors, so that in most cases there is a break in the continued accountability.

In addition, one has to admit that the failure of a commercial aircraft project would most probably jeopardise the entire company. Thus, there is always a shared responsibility among all project *and* functional stakeholders involved. Also from this viewpoint, functions must be given a chance to influence the progress of the development project. However, functional heads must accept that their influence on the project is based on:

- a clear understanding of their role compared to the role of the project leader; as well as
- assured lines of communication between seconded staff and functional homes.

As was said earlier, project leaders should have control over project-related matters. Functional managers within the team should be in charge of controlling functional expertise.

'For effective management by projects, both functional and project managers … must learn how their responsibilities are properly shared on projects. In oversimplified terms, the project manager is responsible for *what* (project scope) and *when* (project schedule), while the functional managers/leaders are responsible for *who* does the work and *how* the work is performed. *How much* (the project budget) is the responsibility of the project manager, but is usually based on the functional estimates.'[27]

As project leaders need to be empowered to do their job, key functional representatives in the team should operationally report to the team leader. Operational reporting includes everything related to the project's schedule, budget and scope, that is, all time, cost and quality aspects. However, in order to also implement functional expertise, functional staff within the multi-functional teams need assured lines of communication to their home functions. If key team members have already one direct *reporting* line to their team leader, what, then, is the other communication line to their home Function all about?

This line of communication is about being enabled – resource-, method-, process-, integration- and tool-wise – to do the required functional job. It is about *how* to do the job. Let us call this line the line of professional leadership: the functional homes exert functional expertise within the multi-functional teams via lines of professional leadership

---

26 Archibald, R.D. (1992), *Managing High-Technology Programs and Projects*, 2nd Edition, (New York: John Wiley & Sons), p. 45.
27 Archibald, R.D. (1992), *Managing High-Technology Programs and Projects*, 2nd Edition, (New York: John Wiley & Sons), p. 129.

between themselves and their functional representatives in the teams, see Figure 6.5. The professional leadership line must have sufficient strength in order for the functional representatives to be able to do their job properly. This can imply – and often does – to contradict views and opinions of the other functional representatives in the same team. In particular when contradicting any views expressed by the team leader the backing of the functional home may be needed.

The existence of two lines of communication and reporting may cause problems. In particular for international projects it is worth noting that people from collectivistic cultures with high Power Distance – that is, with an understanding for clear deep hierarchies – as well as high Uncertainty Avoidance (using the model by Hofstede and Hofstede introduced in Chapter 4), such as Spain or Greece, are often becoming insecure as a result of having to serve two bosses. For them, a professional position demands their personal loyalty to their (functional) boss. They will therefore find it difficult to work in multi-functional projects. As R.D. Archibald writes: 'In traditional corporate or national cultures where people have learned that every person should have only one boss, this attitude produces …[a] barrier to effective project management.'[28] The recognised deficiency of any matrix organisation (serving two bosses) has here its most severe consequence. Only careful leadership by the team leaders can minimise it.

The best way to ensure a strong line of professional leadership – and to reduce the hesitations of team members in accepting one reporting line as well as one line of professional leadership – is to ensure that both lines are given equal power whenever annual performance reviews for individual team members take place. Here is a suggestion of how this can work: Team leaders should not hold their annual reviews with their direct reports on a strictly bi-lateral basis, but should also invite those functional bosses to whom the direct reports have lines of professional leadership. During the review, the team leader discusses the performance of the individual against the set of previously agreed objectives and agrees with him or her the objectives for the next period. The functional boss discusses with him or her any issues with regard to the personal career, professional growth and development within the company. Then, the team leader and the functional representative together but alone, that is, without the direct report, discuss the annual salary increase and potential bonus for the team member, with the functional boss having the last word. The latter is necessary to ensure that by the time team members return to their functional homes they do not encounter huge disparities with regard to their wages. If it is performed in a professional way, the annual review becomes the 'oil in the gear' of the whole organisational machine of multi-functionality. Team members who have experienced annual reviews in this way, will be motivated to work in a multi-functional environment.

However, in international teams special care needs to be taken that the evaluating of performance results during the annual review is not biased by the team leader's or functional boss' cultural background. It is important to put in question one's own cultural background to find out how possible stereotypes or even prejudices might influence the evaluation of collaborators. In addition, for those team members who belong to a collectivistic culture with a strong context reference,[29] a discussion to evaluate performance

---

28 Archibald, R.D. (1992), *Managing High-Technology Programs and Projects*, 2nd Edition, (New York: John Wiley & Sons), p. 130.
29 The context of a verbal message comprises all the non-verbal messages delivered in the same situation. In every culture the context is important but different culture put different emphasis on the context. 'Context reference' can be defined as the degree to which cultures put emphasis to the context of a verbal message: the stronger the emphasis, the stronger the context reference.

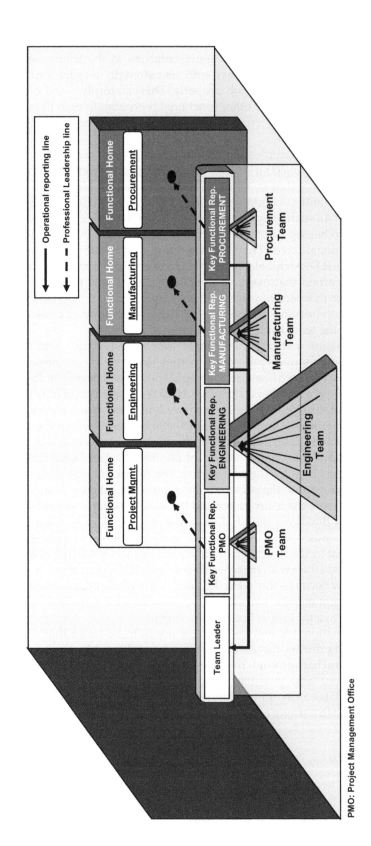

PMO: Project Management Office

Figure 6.5   The line of professional leadership

results will be unacceptable as affected team members could lose face. Societies with such a culture (such as Japan) have more subtle ways to provide feedback, for example by an intermediary.

In order to further reduce the fears of the functional department heads, some companies also use the 'trick' to distinguish between reporting lines and headcount. Seconded individuals remain on the headcount of their home Function. Consequently, the grading of the positions of the bosses in the Functions does not change as the number of headcounts remains unchanged.[30] But in order to also reduce the fear of the project leader, the operative reporting lines of seconded key functional team members are now directed towards the leaders of the multi-functional team, who themselves are usually on a different (for example Programmes or Project Management) headcount.

The complexity of headcount and reporting line issues, however, is only relevant for the key functional representatives of a multi-functional team. Double lines of reporting or any other type of formal communication do not need to be applied to the bulk of team members: they in any case remain on functional headcount and have one line of reporting to the key functional representative only, see again Figure 6.5. Organisationally, they essentially work in a similar environment as before in their home Function. Only the location has changed.

However, even if all fears were made obsolete, in practice the balance within a team between operational needs and functional expertise will always be difficult for the team leader to manage.

An example: a component could be newly designed with some weight benefits but is also available as a standard part without the weight benefits. Manufacturing and Procurement prefer the use of the standard part as it is also used on other aircraft programmes. The company would be able to achieve higher cost savings per unit of standard part. However, Engineering favours the new design because this would help to achieve weight targets. A team leader must be capable to resolve this conflict.

A Project Management role, such as the leadership of a DBT, is usually perceived to mean that the team leader has full responsibility of the team's project. The above example, however, shows that decisions which are deemed beneficial for one programme may have unfavourable repercussions for the company as a whole. If team leaders' decisions are overruled because of higher company interests, their accountabilities can no longer be exclusive. In fact, if anything can be said about authority for a project run by a multi-functional DBT embedded in a complex matrix organisation, it is that a project leader's authority is limited. This may come as a suprise but reflects reality.

Because limited authority is always a serious subject for project leaders, it has to be negotiated between them and their bosses. If this issue cannot be resolved, significant stress can arise in the relationship between the two parties. This may spill over to other team members too. Each team leader and key functional representative should receive guidelines from higher Management explaining the generic circumstances, under which escalation is requested. Essentially, such guidelines describe the shared responsibilities across all Functions or in other words: the amount of delegated authority. If, however,

---

30 Grading usually depends heavily on staff size.

according to these guidelines decisions can be taken within the team, the team leader should act and must be authorised to have the final word.

Finding the right balance between Project Management influence and functional influences has a major impact on the project's time, cost and quality objectives. Questions of allocation of responsibilities and reporting lines, of ways of working, of matrix-organisation versus line management organisation, to name a few, have therefore to be answered early – that is, before staff ramp-up at the beginning of the Development sub-phase – in order to avoid chaos during the aircraft development. The details of implementing multi-functional teams within an existing company organisation will be different from one company to another: aircraft have been and will continue to be developed within all kinds of organisational structures. However, whatever the detailed layout of the organisation, it should be communicated and explained to everyone who should know about it.

THE ORGANISATIONAL SIZE PROBLEM

Unfortunately, one multi-functional DBT is not enough to develop an aircraft. Boeing, for example, needed 250 teams in total with around 10,000 engineers to develop the 777.[31, 32] At Airbus, 110 teams with around 5,000 engineers were needed to develop the A380.[33] The average team size was therefore around 40–45 members per team.

> 'Garnet Hizzey, an ornately named British engineer, helped devise Boeing's design-build teams [for the development of the 777]. … In 1990 Hizzey and his colleagues looked at the tasks involved in designing a plane, and started to plan the best size and number of DBTs. "At that time we anticipated maybe we would have 80 or 100 teams. I think if we'd known at the time we'd end up with 250 we would have probably dead-ended it right there."'[34]

> 'Steve Johnson was in charge of the whole trailing edge, and had ten DBTs reporting to him. … [He said:] "And it's not the supplier's company and it's not the Boeing company. It's the Inboard Flap company, or the Outboard Flap company. They have all of the people necessary to design the structure, design the tools, develop the manufacturing plan, write the contracts – everything is in that little company. So we treat them like that."'[35]

Such large numbers of teams not only reinforce the necessity to do everything possible to facilitate communication, but also underline the necessity to establish a Project Management hierarchy. This is the organisational size problem.

The rule of thumb for the number of direct reports leaders can handle operationally without compromising their full management attention yields a number of around ten.

---

31 Sabbagh, K. (1996), *21st Century Jet. The Making of the Boeing 777* (London: MacMillan Publishers), p. 83.
32 There were a lot more people involved in the development than for the 747 jumbo jet which required about 4,500 people. Joe Sutter, the 'father' of the 747, writes in his book: '…I had 4,500 people reporting to me, some 2,700 of whom were actually engineers. The rest were managers, clerical and technical support people, and so on.', in: Sutter, J. (2006), *747: Creating the World's First Jumbo Jet and other Adventures from a Life in Aviation* (New York: Smithsonian Books, HarperCollins Publishers), p. 143.
33 Thomas, J. (2001), Die A380 – Eine große technologische und wirtschaftliche Herausforderung für Europa (Berlin: presentation to German MEPs), 29 May 2001, p. 19.
34 Sabbagh, K. (1996), *21st Century Jet. The Making of the Boeing 777* (London: MacMillan Publishers), pp. 66–7.
35 Sabbagh, K. (1996), *21st Century Jet. The Making of the Boeing 777* (London: MacMillan Publishers), p. 68.

This rule equally applies to professional leadership. Could therefore a higher-level project leader lead ten lower-level multi-functional teams? No, not if the rule of thumb is to be respected! With the help of Figure 6.6 one can see why this is not possible: each of the Functions represented in the n lower level multi-functional teams would also need one functional representative at the higher level to report to. This is because the functional homes would otherwise be obliged to exert their professional leadership to hundreds of representatives individually due to the high number of DBTs.

If, next to the project-managing team leader, the four main disciplines, Engineering, Manufacturing, Procurement and Project Management (Office), were all represented in each lower-level multi-functional team, then one would need four key functional representatives in the higher-level team (this will require yet another reporting line as will be described below). This means, the 1 + 4 individuals create themselves a multi-functional team, albeit on a higher level and of much smaller size (that is, consisting of five team members only). Again, the functional representatives in the higher-level team are direct operational reports to this team's leader. Consequently, the higher-level team leader now has n + 4 direct reports. As a result, a higher-level project leader should only lead up to five or six lower-level teams (Figure 6.6 depicts four!).

If the total number of required DBTs is n, than the number of hierarchical levels to manage n teams should be close to log.n/log m, assuming that each higher-level project leader can lead m lower-level teams. With three hierarchy levels in the project organisation and each project leader leading six lower-level teams, about 220 teams could be managed. (Note that Figure 6.6 depicts two levels only for reasons of simplicity!) In practice, of course, criteria like the ones mentioned above (building constraints, geographical locations and so on) can affect this number.

If there are, for example, three levels of multi-functional teams, this does, however, not mean that teams are similar on all levels. They are only similar in so far as to represent teams of a multi-functional nature with representatives from different Functions being team members. But they differ in size. Only the teams on the lowest level are really DBTs with sufficient team members to design aircraft components ready for build. Higher-level teams are much smaller and may only consist of one representative from each required Function. In addition, they are not directly in charge of design and build, but of integration and management of the lower-level teams.

For the development of its 777, Boeing used three levels of organisational project hierarchy, see Figure 6.7

'The [Boeing 777] plane was divided up into large areas of responsibility such as wings, empennage, fuselage and so on, and then each of these units was broken down into subcomponents, each the responsibility of a DBT [design-build team]. The wing, for example, was divided into leading-edge and trailing-edge teams but, because they were dealing with such large and complex pieces, the role of these teams was more supervision than strict detailed design. Reporting to the trailing-edge team were … [lower level] DBTs, each responsible for a single piece of the trailing-edge. … The teams, each with between ten and twenty members, were named after a piece of the wing trailing edge. They were: Flap Supports, Inboard Flap, Outboard Flap, Outboard Fixed Wing, Flaperon, Aileron, Inboard Fixed Wing and Gear Support, Main Landing-Gear Doors, Spoilers, and Fairings.'[36]

---

36 Sabbagh, K. (1996), *21ˢᵗ Century Jet. The Making of the Boeing 777* (London: MacMillan Publishers), pp. 67–8.

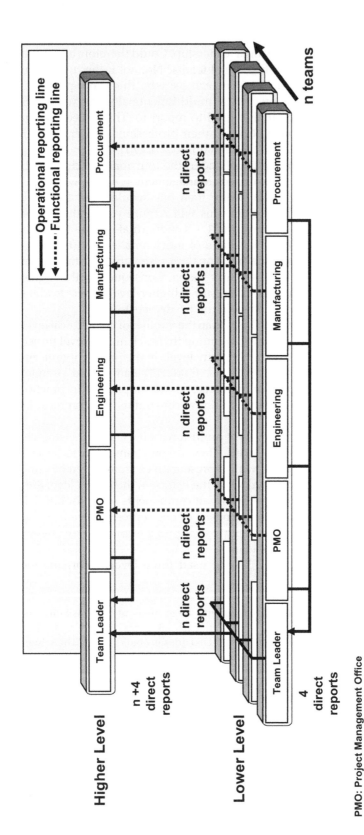

PMO: Project Management Office

Figure 6.6   The organisational size problem of multi-functional team hierarchies

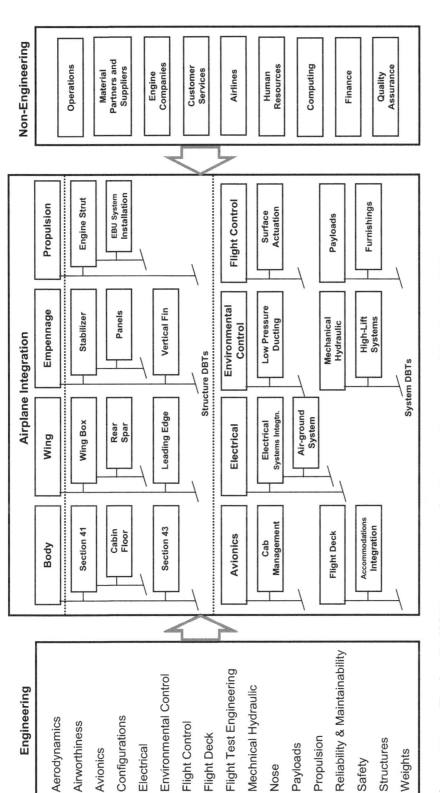

Figure 6.7   Design-Build Team structure for the development of the Boeing 777

Source: Breuhaus, R.S., Fowler, K.R. and Zanatta, J.J. (1996), 'Innovative Aspects of the Boeing 777 Development Program', ICAS 1996 Proceedings (International Council of the Aeronautical Sciences ICAS-96-0.4), p. LXXVII, with permission.

Airbus, too, used three levels of organisational project hierarchy for the development of the A380. It called each of its multi-functional teams on highest level ACMT (Aircraft Component Management Team), on medium level CMIT (Component Management and Integration Team) and on lowest level CDBT (Component Design-Build Team) to underline the different roles of these teams. The breakdown of the A380 into areas, for which ACMTs were made responsible, can be seen in Figure 6.8.

It has been said earlier that the lower-level team leaders should be direct reports to the higher-level team leaders. In principle, the same also applies to the lower-level key functional representatives in their relation to the functional representatives in the higher-level teams. The consequence of this would be that key functional representatives in lower-level teams could now face three different lines of reporting: to the team leader of the team to which they belong, to the key functional representative in the team on the next higher level as well as to their functional home.

As this may well get too complex, the problem may be solved in the following way (see Figure 6.9): On the one hand, key functional representatives of the multi-functional teams on lowest level only report to their team leaders as well as to the corresponding functional representative of the team on the next higher level. This principle also applies to slightly higher (but still low) team levels (if any).

Functional representatives on medium and high levels, on the other hand, also report to their respective team leaders but have a line of professional leadership with their home Function. They lack the reporting line to their next level up functional representative. Instead, decisions taken on higher levels are formally cascaded down to lower-level teams only via the line of operational reporting linking team leaders on the various levels.

As one realises, managing hundreds of DBTs leads to very complex organisations. Fortunately, concurrent ways of working do not put too much emphasis on detailing and outlining organisational issues, such as lines of reporting, but rather concentrate on best possible lines of communication.

SOME LESSONS LEARNED

The important point made in this chapter is that the concept of empowered multi-functional DBTs for aircraft development – required for the benefit of the CPD process, which in turn emerged thanks to advancements in IT – leads straightforwardly into a matrix organisation.

Implementing the CPD process requires excellent levels of communication. Admittedly, the implementation of the major enabler of good communication, that is, co-located multi-functional teams, makes the matrix organisation truly complex: as outlined above, one dimension of the matrix organisation is represented by the various Functions in a team, one by the hierarchical levels necessary to manage a given number of teams and one dimension by the professional leadership line of communication with the functional homes.

If not properly run, three dimensions and, thus, three major lines of communication and reporting can create new and unintended communication problems. It is already difficult enough to accurately illustrate the complex relationships within multi-functionally organised projects. One should therefore initially expect many problems when setting up this type of organisation for the first time. There will be lots of confusion and, thus, a partial unwillingness on behalf of the seconded staff to accept this way of working.

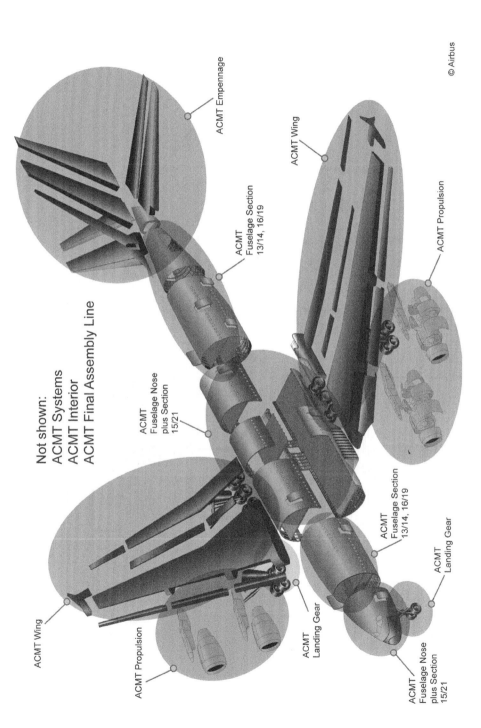

Not shown:
ACMT Systems
ACMT Interior
ACMT Final Assembly Line

ACMT Empennage

ACMT Wing

ACMT Propulsion

ACMT
Fuselage Section
13/14, 16/19

ACMT
Fuselage Nose
plus Section
15/21

ACMT
Fuselage Section
13/14, 16/19

ACMT
Landing Gear

ACMT
Landing Gear

ACMT
Fuselage Nose
plus Section
15/21

ACMT Propulsion

ACMT Wing

© Airbus

Figure 6.8   Responsibility scope for Airbus A380 multi-functional teams on highest organisational level (ACMT: Aircraft Component Management Team)

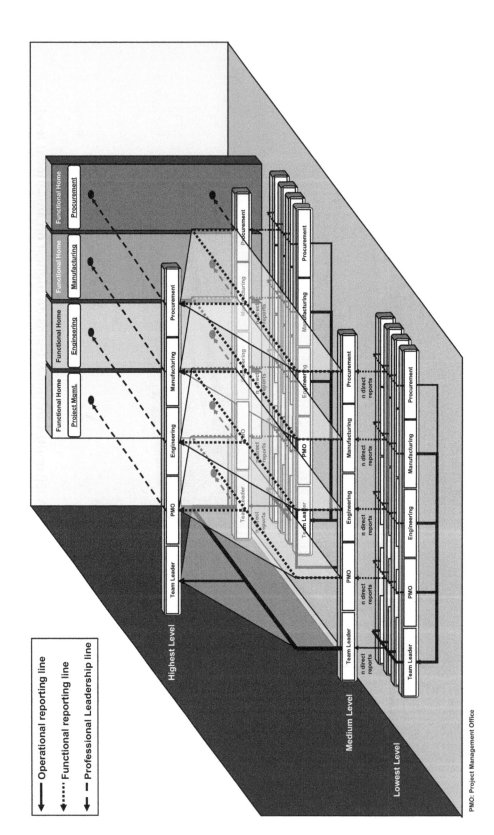

PMO: Project Management Office

Figure 6.9   How to avoid three lines of reporting in complex matrix organizations

Many newly appointed team members, who so far only were familiar with functional line management structures, will object to it.

Both, Boeing and Airbus used complex matrix organisations similar to the one described above to develop the 777 and the A380, respectively. As a result, a wealth of experience is available with this kind of organisation.[37] And indeed, people developing these aircraft as members of multi-functional DBTs have encountered problems which were experienced in very similar ways at both Boeing and Airbus. Typical problems resulted from the fact that:

- The multi-functional approach required team members to work with a diversity of perspectives from the onset. Many employees did not feel adequately prepared for such a change. They were rather used to work in a familiar environment with others who shared the mindset, training and language of a particular discipline.
- Some multi-functional teams worked better than others. Where they did not work well, team members represented more the interests of their functional homes because of an apparent lack of motivation or lack of understanding to use their functional skills in a collaborative manner.
- Moves across team assignments happened frequently for some team members, following the demand for special skills. This led to disruption of teams as they were originally set up and had learned to work together. Disruptions in team membership were viewed as contributors to some major design performance problems.
- Secondment of employees to multi-functional teams provided them with valuable developmental experience, but did not always give them the exposure they needed to advance vertically in their home organisation (that is, with a clear vision for the route to the top of their functional home). As a result, personal and company objectives often did not match.
- The evaluation of individual performance of members of multi-functional teams was often perceived to be unfair, in particular where links to the home Functions were weak.

It must be underlined that the initial introduction of multi-functional teams to an existing, functionally oriented organisation will lead to a cultural shock. It can only be turned into a sustainable and beneficial cultural change if leadership, care-taking of individual careers, staff training and communication are regarded by Management as fundamentally important to project success.

'The conflict in the "battle zone" [in the middle of the matrix organisation, i.e. within the multi-functional teams] is an absolutely necessary part of making the matrix work to advantage and realising the best outcome from this organisational structure. The optimal situation is the result of two forces within the matrix. One force should be working to get the product out into the market; the other force is holding back to guarantee product integrity. Conflict is, therefore, an inherent characteristic of a matrix organisation. … It is integral and intended to be there. It should not be eliminated but managed.'[38]

---

37 See for example: Brown, K.A., Ramanathan, K.V., Schmitt, T.G. and McKay, M. (1997), 'The Boeing Commercial Airplane Group: Design Process Evolution', *The European Case Clearing House Collection* Nr 397-037-1 (Bedford, England: Cranfield University).
38 Allen, T.J. and Henn, G.W. (2007), *The Organization and Architecture of Innovation. Managing the Flow of Technology* (Burlington, MA: Elsevier), p. 41.

'If priorities are not managed in a matrix organisation, the organisation will self-destruct.'[39]

What is needed to overcome these initial obstacles can be summarised as follows:

- highly skilled team leaders (project leaders);
- frequent, well-prepared communications and training (workshops, seminars, brochures, intranet information and so on) to all team members about how the multi-functional team structure should work;
- guidelines describing the roles and responsibilities, the levels of delegated authority, the signatory rules, the conditions as to when escalation should be used, as well as explanations of the agreed ways of working;
- written job descriptions for at least the key individuals of each team must be made available;
- multi-functional annual reviews with individual team members discussing the level of objectives achievement as well as career and growth developments;
- the clear demonstration that Senior Management of all Functions is in support of the concept of co-located multi-functional teams for aircraft development (for example by signing corresponding agreements, which are communicated and visible to people);
- a matrix organisation structure, which takes architectural aspects as well as interface aspects into account;
- a good and dynamic accommodation and facilities management; as well as
- a building and space environment in favour of the co-location principle.

However, on top of all of this, it is leadership skills which play a crucial role in the success of the design-build process. And it is leadership behaviours which must compensate for the organisational deficiencies associated with the complex matrix organisation outlined above to ensure project success.

Vance Packard, the co-founder of Hewlard Packard, once defined leadership as 'the art of getting others to want to do something that you are convinced should be done'.[40] Let us therefore now turn our attention to another important aspect of leadership not discussed so far: the timing for the setting of priorities to teams and team members during the Development sub-phase.

39 Allen, T.J. and Henn, G.W. (2007), *The Organization and Architecture of Innovation. Managing the Flow of Technology* (Burlington, MA: Elsevier), p. 38.
40 Packard, V. (1962), *The Pyramid Climbers* (New York: McGraw-Hill).

MAIN CONCLUSIONS FROM CHAPTER 6

- Getting people to talk to each other directly – rather than by electronic media – is the only truly effective way of transferring technical knowledge and thereby advancing the development process. Virtual co-location can therefore not replace physical co-location. The latter remains the principal means of communication for multi-functional Design-Build Teams.
- Distance plays a fundamental role in personal communication. Teams which require more communication should be located closer to each other. Where this poses a problem for successful application of the Concurrent Product Definition process, full co-location should at least be achieved temporarily during the early design phases.
- It is certainly worth considering designing and establishing a building architecture which supports the product development process and represents suitable platforms for the co-location of multi-functional design-build teams – rather then forcing people into an office space which happens to be on the market for rent but is otherwise not suitable to support the product development process.
- As one multi-functional Design-Build Team is not enough to develop a commercial aircraft, there is a need for an organisational hierarchy to manage all these teams. However, such a large organisation immediately raises many concerns, which need to be addressed with sensitivity and persuasion.
- It is leadership skills and behaviours which must compensate for the organisational deficiencies associated with the complex matrix organisation.

7

Getting Development Priorities Right

OBJECTIVES FOR DEVELOPMENT AND PRODUCTION

Having a certified, high-quality product in the market place at planned Entry into Service (EIS), fully satisfying or even exceeding customer expectations, combined with the requirement of staying within planned Non-Recurring Cost (NRC) targets, is *the one* major objective for the Development phase of any commercial aircraft. Safety, quality, time and cost are the paramount criteria for this phase. However, to lay the best possible foundations for delivering the aircraft programme's Business Case is *the other* major objective for this phase. Let us see why.

For the Series Production phase, continued customer satisfaction is, of course, the number one priority. But this phase is also decisive for the long-term well-being of the aircraft company. This is because sustainable profitability (and therefore the value of the programme to the company and its shareholders) is much more influenced by the Production phase than by the Development phase. The aircraft programme must therefore primarily achieve high revenues and lowest possible Recurring Costs (RC) during this phase.

To achieve high *revenues* is the accountability of the Sales department. However, as Sales departments tend to concentrate on numbers of aircraft sold rather than on Business Cases to be achieved, they need to be given lower limits for the price, below which the new aircraft should not be sold to customers. These limits are precisely set by the Business Case described in Chapter 4. In addition, it should be ensured that the aircraft programme's Management has a major role to play in the deal approval process. It does not need to be behind each and every sales contract, and in fact stays in the background as long as deals are within the allowable price limits, but it has a definite interest to ensure that overall revenues are met to deliver the Business Case.

As far as production *costs* are concerned, Management must also ensure that they are kept below the limits set by the Business Case. Otherwise, the overall planned profitability contribution by the programme for the company cannot materialise. How to achieve low RC? Surely by relentlessly striving for ever improving and leaner production processes once the aircraft is in Series Production. But even more important is an intelligent (that is, cost-minded) design, applied already during the Development phase to achieve lowest possible RC.

High revenues and low RC: Only a healthy, profitable company can continuously afford not to compromise on customer satisfaction. In other words, reaching for lowest possible RC is not a contradiction to achieving customer satisfaction. It is rather a prerequisite.

In conclusion: the Development phase is driven by objectives, which are genuine for this phase itself, but also by the objective to lay the foundations for achieving the programme's Business Case during the later Series Production phase.

SEQUENCING OF EFFORT PRIORITIES

Overall top-level time, cost and quality (TCQ) objectives are cascaded and converted as objectives to fully accountable team leaders. For the latter, they represent objectives which are all *simultaneously applicable*. However, it is unrealistic to deal with all objectives simultaneously and with the same level of management attention at any given point in time during the Development sub-phase. Even if it were realistic, it would not be very helpful as teams need clear instructions and clarity with regard to priorities, and should not be confused by making everything equally important. Focusing on one top priority at a time is easier to be communicated to and to be absorbed by the teams. There is less confusion about the current direction of the development path and also all management attention goes into the current focus. As a consequence, there will be a lot more ideas and results generated by the teams which contribute to the achievement of objectives.

However, does this mean that objectives should be set strictly sequentially? Certainly not. An airplane could never be built in this way. Instead, one should rather distinguish between an objectives-focused mindset on the one hand and the effort, which goes into objectives' achievement, on the other. While all objectives should be kept at all times in one's mind simultaneously, the effort which goes into achieving individual objectives should be of a quasi-sequential nature.

The fundamental challenges with this approach of quasi-sequential priority-setting are about timing and communication:

- When is the right time to switch from one priority to the next?
- How can this be communicated to ensure that the switch is understood and implemented in a coherent way by all affected teams?
- Is a coherent setting of priorities across all multi-functional teams at any given moment in time the correct approach or do different teams working for different components of the aircraft need different quasi-sequential phasings?

Challenging time, cost, quality targets for the development teams are not only to be cascaded vertically down the organisation hierarchy but targets also need to be phased in line with the development schedule.

It may be helpful to illustrate the problem by the following example: if the aircraft's wings are joined in the Final Assembly Line (FAL) to the centre box as a major component of the already assembled fuselage, then the centre box needs to be one of the first available major components. This is because it first has to be assembled as a sub-assembly, then possibly transported to another location to be mated with other parts of the fuselage, then transported again to the FAL where finally the wings can be attached to it. Assume instead, that the wings can be delivered to the FAL directly.

Given a launch date, which is common to both centre box and wings, there is typically less time available to develop the first centre box specimen compared to the first wing specimen. As a result, not all weight-saving possibilities may have been cashed for the centre box as there has just not been enough time available to do so. When the wing is joined to the centre box, the former may be more weight optimised in relative terms than the latter.

On the other hand, the wing team may have already spent more money on weight-saving activities compared to the centre box team, again in relative terms. As the first aircraft are usually used for flight tests and are not sold on to customers – or if so, under special conditions – the team developing the centre box could in theory make use of subsequent aircraft to further reduce the weight by changing the box' design.

However, the changed centre box design would probably lead to a new loads distribution across the joint. This in turn could easily result in design changes for the already more weight-optimised – and therefore more loads-change sensitive – wings. The consequence for the wing team would be relatively high additional costs and leadtimes due to change and rework.

Schedule planning must therefore not only be regarded as a discipline which helps achieving on-time delivery targets but is necessary to establish a sequential phasing of all kinds of targets in the first place. Teams should not decide for themselves how they want to sequence their priorities and when to switch from one priority to the next. Ideally, teams would be given clear guidelines from the top Programme Management-level for the phasing of their priorities, in line with the overall and current development schedule.

Progress of deliverables over time usually follows the shape of an S-curve if cumulated numbers are used, see Figure 7.1. This is true for both planned as well as actual performance data, respectively. As the name 'S-curve' suggests, the curve has a turning point, where the concave curvature changes into a convex one. Experience shows that it is a lot more difficult for the teams to get to the turning point than to achieve the bulk of the upper part of the S-curve. This is because the teams, the organisation, the methods and tools, and so on, everything necessary to achieve the deliverables, has at first to be geared up to start performing. Typically, therefore, actual performance is behind schedule for most of the lower part of the S-curve.

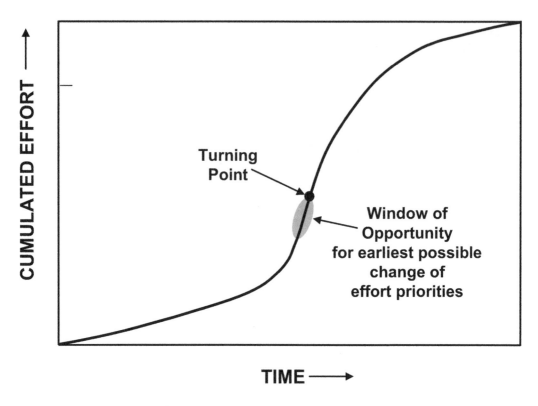

Figure 7.1   Finding the right timing for launching new activities targeted towards achieving objectives

However, once the system is up to speed and has started to deliver, actual performance increase is usually in line with plan (leading to a parallel actual S-curve compared to the baseline S-curve), sometimes even better than plan. Progress now seems to come almost automatically.[1] The switch from one priority to the next should therefore take place at the earliest opportunity once the actual performance (that is, the overall efforts dedicated to achieve an objective) has left the area with the largest concave curvature and is on good track to reach the turning point. The latter represents the latest date where the shift in priorities should take place. Figure 7.1 illustrates this.

But what should the quasi-sequencing of priorities now look like? In the following, a suggestion is provided for structural design work.

LOWEST POSSIBLE WEIGHT IS A MUST

In commercial aeronautics, the most basic requirement for superior aircraft performance is lowest possible weight and aerodynamic drag, while never compromising safety: an aircraft so heavy that it cannot take off the ground is not a viable product.[2] Nor is an

---

1   This is true except for the last 5 per cent or 10 per cent which are again difficult to achieve, because the focus of attention has diverted to new areas (that is, new S-curves).

2   As aircraft burn fuel to lift weight as well as to overcome drag, drag can also be expressed in terms of weight: the drag is equal to the amount of extra weight that would require the same amount of extra fuel to be burnt.

aircraft with so much drag that it cannot achieve its intended range. This is why weight and drag targets are fundamental for the aircraft design. The drag target – except for the relatively smaller parasitic drag – is already largely achieved at the beginning of the Development sub-phase as a result of the by then frozen outer configuration of the aircraft. The same cannot be said about the weight target: the weight target is known but in most cases not yet achieved at that point in time. Focusing the teams developing the aerostructure, the systems installation components (such as pipes, brackets and so on), the engines, the landing gear and other heavy components on weight reduction is therefore a first priority. As K. Sabbagh points out, this priority becomes a top priority at a time when about 25 per cent of the drawings are released.[3]

The 'hunt' for weight-saving ideas needs, for example, to be stimulated again and again during workshops, weight-saving meetings, brain-storming sessions and so on. All these events need to be supported by senior experts to ensure that really feasible weight-savings ideas emerge. 'Feasible' means that design solutions for the weight-saving ideas can realistically be envisaged in view of the funding and leadtimes available. Obviously, the design solutions are not implemented yet, that is, no weight savings have materialised. But at this point in time it is sufficient that judged weight-saving ideas bring the weight outlook close to the target weight. If multi-functional Design-Build Teams (DBTs) are properly driven, managed and facilitated through adequate leadership, one can be rather confident that all weight-saving ideas will eventually get implemented during the early stages of the Development sub-phase.

> As N. Forgeard, CEO of Airbus, outlines: 'We did set ambitious internal [weight] targets which are beyond what is necessary to deliver the performances guaranteed to our customers. This is a wise approach to make sure to fulfill our promises – and we do meet every performance guarantees – but also to have margins for future evolution of the aircraft. Always keep in mind that … [the A380] will still fly in 2050! As of today, we are within 1% of the targeted maximum take-off weight. This did not come straightforwardly. [The A380 programme manager] … and his team have been fighting repeated uphill battles for that and, at various points in time, we indeed have been above the target. We now have weighted several airframes and the most recent sections; we are happy and confident with the results. The issue of weight is over, period.'[4]

TIME IS PARAMOUNT

The aircraft industry is a cyclic one. In military aeronautics, one speaks of different fighter generations replacing each other as technology and demand evolve. In commercial aeronautics, the orders and delivery streams follow roughly a cyclical behaviour, see Figure 7.2. It is strongly correlated with the changes in growth of Gross Domestic Product (GDP). It is general economic growth (resulting in both more business travel and leisure trips), which drives demand for new commercial aircraft on top of replacements for older aircraft.

---

3   Sabbagh, K. (1996), *21st Century Jet. The Making of the Boeing 777* (London: MacMillan Publishers), p. 54.
4   Forgeard, N. (2005), *Airbus Annual Press Conference* (Paris), 12 January 2005.

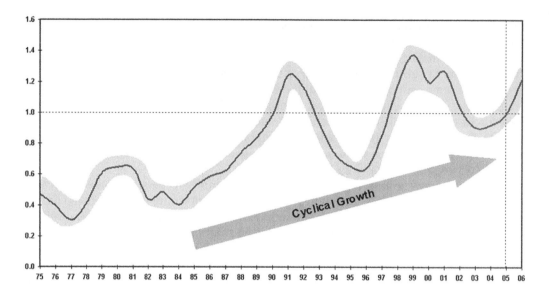

Figure 7.2   Deliveries of commercial aircraft with more than 100 seats (normalised to the year 2005)

To release a new product into the market during an economic downturn could easily result in a catastrophic situation for any aircraft-developing company: because of much reduced revenues, airlines are often not only cancelling earlier aircraft orders but have no financial means left to purchase the new product neither. Before launch of the development of a new aircraft, the Marketing department therefore has to come up with a robust analysis identifying the most favourable EIS date for the new aircraft programme. This usually is the time when large numbers of older generation aircraft have to be replaced, coupled with a vision about the next suitable upturn of the economy.

Depending on the targeted EIS date, there may not be much time left between this date and the date of the launch of the aircraft development. In addition, the competitors are not inactive and may be trying to hit the same EIS timing with a competing product.

Once a market opportunity has been identified, the aircraft manufacturer should therefore make every effort to hit the planned EIS as precisely as possible and not to be too late compared to the competition. As W. Spitz et al. point out, 'getting to market earlier [compared to the competitor] means that the company will have more opportunities to dominate a particular market segment before a competitor can react. If a company can lock in more customers, it has a better chance of both producing more units and smoothing the production run over the product's life-cycle and thereby realize its learning economies. By getting to market faster, the forecast for the product and the expected profitability of the program are more likely to be realized.'[5, 6]

---

5   Spitz, W., Golaszewski, R., Berardino, F. and Johnson, J. (2001), *Development Cycle Time Simulation for Civil Aircraft*, NASA/CR-2001-210658, pp. 1–3.
6   'In real life we see competitive advantages lasting as little as 6 months in the computer industry, to 1 to 3 years in the auto industry. Using those industries as the time scale, a competitive advantage in the airframe industry should be currently expected to last 2 to 4 years before it is rendered as standard.' From: Spitz, W., Golaszewski, R., Berardino, F. and Johnson, J. (2001), *Development Cycle Time Simulation for Civil Aircraft*, NASA/CR-2001-210658, pp. 2–9.

'Once one company has successfully entered [the marketplace], ... its competitors may face a more difficult time justifying a new program (with similar characteristics); the second or third entrant into a product category may face greater variability in the demand for their products, and therefore are less likely to realize learning economies. As a result, one can expect that the products offered in the market will be differentiated; once one competitor is first in the market, others will wait until new technologies can be integrated into their product offerings and thereby provide a significant benefit to leap frog the initial mover.'[7]

'The market clock measures the time it takes to respond to opportunities in the marketplace. It starts ticking when a customer opportunity appears and continues inexorably until the customer's need is filled. The market clock is unforgiving. It keeps on ticking whether we are working on the project or not, and with each passing minute we pay the cost of delay. The cost keeps accumulating steadily even when nobody is working on the project. When you view projects this way you will recognize that a week of delay in starting the project has the same economic cost as a week of delay at the end of the project. Thus, we should treat a week spent at the front end of a project with the same care that we would treat a week consumed at the very end. Such an attitude is sadly lacking in most companies. Instead, they have what we call a "burn rate" mindset. They worry about activities in proportion to the rate at which they consume, or "burn", money. Such a mindset assumes that there is no cost as long as no money is being spent. ... This mindset is a natural consequence of failing to understand the true economics of product development.'[8]

Finding ways to achieve a successful development within the limited leadtimes available is therefore absolutely essential in order not to miss the market opportunity. But there are other reasons to apply 'Time to Market' techniques: companies, for example, which have managed to shorten their development cycles have generally also experienced a reduction of their expenses in the order of 10 per cent to 30 per cent.[9] NASA claims that a 50 per cent reduction in cycle time would translate into a 25 per cent decrease in RC, a 61 per cent decrease in NRC as well as a 36 per cent increase in sales due to both, increased demand from lower price and market share theft from early market entry.[10]

But 'perhaps the best reason for using Time to Market as the centerpiece of a change program is that it tends to drive the basic changes in desirable directions. In order to get to market quickly, everything else must be working well, too: people from different departments must communicate effectively, non-values adding activities must be eradicated, customers must be involved intimately, and Management must be supportive.'[11]

7   Spitz, W., Golaszewski, R., Berardino, F. and Johnson, J. (2001), *Development Cycle Time Simulation for Civil Aircraft*, NASA/CR-2001-210658, pp. 4–10.
8   Smith, P.G. and Reinertsen, D.G. (1998), *Developing Products in Half the Time,* 2nd Edition, (New York: John Wiley & Sons), p. 53.
9   Smith, P.G. and Reinertsen, D.G. (1998), *Developing Products in Half the Time,* 2nd Edition, (New York: John Wiley & Sons), p. 13.
10 Spitz, W., Golaszewski, R., Berardino, F. and Johnson, J. (2001), *Development Cycle Time Simulation for Civil Aircraft*, NASA/CR-2001-210658, pp. 7–3.
11 Smith, P.G. and Reinertsen, D.G. (1998), *Developing Products in Half the Time,* 2nd Edition, (New York: John Wiley & Sons), 20.

However, rapid development is often perceived as to have too many disadvantages. For example, what 'one might expect to pay for accelerated development is in higher manufacturing costs for the launched product. The reasoning is that less time will be spent fine-tuning the design for production or working the kinks out of the production system. However, one ... [acceleration technique] is to get manufacturing people involved in the early design decisions that determine manufacturability. ... The net effect is shorter cycle time while also reducing manufacturing cost. The experience of many companies that have adopted these techniques bears out these simultaneous savings.'[12]

One could also be convinced that with shorter development cycles product quality or performance might be compromised. 'The answer here depends on how you view quality. ... If you view quality as satisfying customer requirements, then we do take advantage of this opportunity by understanding better what our customers need ... and providing them with only what they will value. ... Innovation in general requires time, so providing innovation that has little or no value to the customer can waste cycle time.'[13] The introduction of multi-functional teams considerably helps to achieve both, short cycle times as well as improved quality.

The pros and cons of cycle time reduction need to be carefully analysed. Penalties often have to be paid by the aircraft company to its customers if a contractually agreed EIS date cannot be met. This has to be added to the equation. Then a rational decision can be taken whether the benefits are worth the costs involved. Experience shows that in most cases they are.

There is quite a big difference between aircraft projects on the one side where governments are the customers and commercial ones on the other. The public hand is bound to the annual legislative budgetary process. Thus, annual cost overruns are more difficult to be accepted than schedule slippages. Governments keep a strong eye on not overrunning their annual budget allowances. As a consequence, public contractors have insisted on implementing management methods and tools, which concentrate on keeping the annual cost under control and still deliver a product fulfilling the specifications. This is basically achieved by carefully planning the actual Development sub-phase before the latter is launched.

However, commercial aircraft projects look for business opportunities in the worldwide market place. There usually is a limited time window for an opportunity to be cashed. Everything is therefore targeted to finish the development early enough to meet this window of opportunity. Failure to do so can blow the whole Business Case. Within certain limits, therefore, commercial aircraft projects need to focus more on schedule achievement than on annual budgets.

In addition, aircraft determined as development aircraft in support of the overall product validation process (for example by performing flight tests) do not have to be perfect. Only Series Production aircraft must be perfect. Development aircraft only have to be fit for purpose, thus reducing the effort (such as leadtimes) to design and build them.

---

12 Smith, P.G. and Reinertsen, D.G. (1998), *Developing Products in Half the Time,* 2nd Edition, (New York: John Wiley & Sons), 14.
13 Smith, P.G. and Reinertsen, D.G. (1998), *Developing Products in Half the Time,* 2nd Edition, (New York: John Wiley & Sons), 14.

In this book, many ways to achieve leadtime reductions are discussed: multi-functional teams, Concurrent Engineering, overlapping, early frozen and robust interfaces, to name but a few. But the starting point for achieving significant leadtime reductions is always careful planning, see Figure 7.3. Detailed project and schedule plans should therefore be established by the teams at the beginning of the Development sub-phase. This will be described in much more detail in Chapter 14.

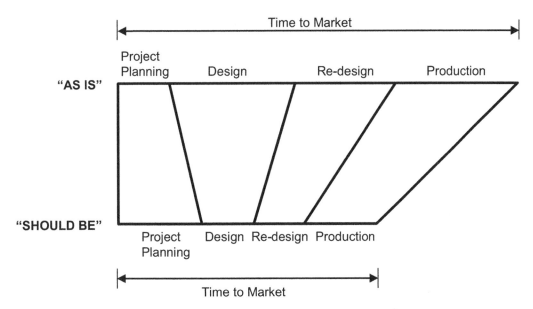

Figure 7.3   Reducing Time to Market

As far as the schedule plans are concerned, a bottom-up approach is usually applied. However, this approach rarely yields the short leadtimes required. Instead, the initial plans will usually show significantly longer leadtimes. Thus, the schedule plans have to be challenged in many dedicated leadtime reviews and workshops (effort!) until it can be demonstrated in the plan that the EIS date can be met. Team leaders subsequently have to ensure that planned milestones will be met. In summary, leadtime reduction efforts to meet the EIS date should be number two in the sequence of effort priorities.

CUSTOMERS ARE FOCUSED ON MATURITY

Time, cost and quality targets are laid down in the Business Case as objectives of similar importance. However, there is more at stake, such as the reputation of the aircraft-developing company, the motivation of the employees as well as the corporate culture. A company may lose customers even for its existing products, if the same customers are disappointed by the quality of the newly developed product which they also ordered. In commercial aeronautics, the number of customers is rather limited. Therefore, on top of financial losses, a development failure might also have significant consequences for other areas of the company's business. Aircraft customers have to operate the product for many years to come, sometimes for decades. To ensure that customers are entirely satisfied with the product not only binds them to the aircraft company but also contributes enormously to the well-being of the latter.

Another top priority for aircraft development therefore is about maturity at EIS: delivering what was promised to deliver. The aspect of maturity includes – but is not limited to – that:

- the aircraft is safe and fully certified;
- it has the performance such as speed, range and turn-around times as guaranteed;
- it does not cost the customer more to operate, maintain and refurbish than foreseen;
- the reliability of systems as well as the overall dispatch reliability of the aircraft meet or exceed the specifications; as well as that
- the aircraft's maintainability, supportability and accessibility are as required or better.

Improving the design to meet maturity objectives is a lengthy undertaking. However, its main window of opportunity is during the Develop Configuration and early Development sub-phases, with some fundamental decisions already taken during the Feasibility phase. After having reached the turning point of the Time to Market S-curve, maturity aspects should prevail against the other objectives until early stages of the Detailed Design period.

ACHIEVING THE BUSINESS CASE

As explained in Chapter 4, the aircraft Development sub-phase has been launched on the basis of a robust Business Case. It will be robust if:

- skills from Finance, Costing, Pricing, Sales and Market Forecasting, Engineering, Manufacturing as well as Planning, among others, have been brought together for its establishment;
- conservative assumptions on markets, sales, revenues, and cost levels have been made;
- a large set of scenario and sensitivity analyses has been performed; as well as if
- audits by various internal and external teams have taken place.

If robust, the Business Case represents a sound basis for reference, which can only be changed by approval of the Business Case stakeholders. Management entrusted with the aircraft development is committed to deliver the Business Case, that is, to achieve the appropriate Return on Investment (RoI) to all parties providing funding.

Delivering the Business Case depends almost entirely on the revenues received through aircraft sales and the costs associated with its production and support. Keeping or not keeping the development costs within budgets plays a relatively minor role in the achievement of the overall Business Case, see Chapter 4. From a Project Management perspective (with limited influence on the revenue side) one should therefore primarily concentrate on the reduction of RC.

The standard approach to RC reduction during development is the Design to Cost (DtC) philosophy. It follows an approach where alternative feasible design solutions with different product unit cost repercussions are elaborated in order to select the cheapest among them. However, while cost should never be forgotten as a design driver in the first place, in real life of commercial aircraft development there are some constraints, which limit the scope of DtC:

- most probably there are not enough resources (or NRC) budgets available in and for the DBTs to work in parallel on various design alternatives for the purpose of cost optimisation;
- if design solutions have already been optimised for weight, leadtime and maturity, there is often not much design opportunity left for cost reduction;
- after having spent much effort on weight and leadtime reduction as well as maturity optimisation, the end of the Detailed Design period is probably the earliest moment in time where one could dedicate a lot of effort to DtC activities. However, while typically 80 per cent or more of the Life-Cycle Costs (LCC) of an aircraft programme are already determined at the beginning of the Development sub-phase, this contrasts with opportunities for further cost reductions of 20 per cent at most.

As a result, it is probably reasonable to claim that in commercial aircraft development the DtC philosophy can only be applied at limited scale. Instead, RC savings still have to be realised later during Series Production phase by relentlessly striving for improved production processes.

However, Management does have another option: it can chose to dedicate considerable funds into research and development of technical design solutions for individual elements – such as low-cost brackets – as well as design principles – such as standardisation of brackets – which are mature and ready to pick at start of the Development sub-phase.[14] If these elements – specifically designed to ensure low production costs – are developed *prior* to having to take care about weight, leadtimes and maturity, and if Management requires them to be implemented without compromise, then there is a reasonable chance that associated RC objectives can be met.

DEVELOPMENT COSTS: PRIORITY DEPENDS ON FINANCIAL EXPOSURE

There is always one 'death to die'. Having said all of the above, it is usually difficult to keep the NRC of the aircraft development within the targeted limits. One could argue that as long as the Business Case is satisfied overall, achieving the NRC targets individually is of lesser importance.

However, NRC results in a significant cash outflow over a relatively short period of time, some years perhaps. Cash availability affected by and cost of financing required for the aircraft development will therefore impose absolute limits to the financial exposure of the company. Otherwise, the company could go bankrupt as a result of too much of a negative cash flow.

---

14 Brackets are a classical example for the trade-off between low production costs and lowest possible weight: on average standardised brackets are heavier compared to weight-optimised brackets but they offer significant savings in production, such as fewer part numbers, easier logistics, standardised assembly, and so on.

As long as these limits are not surpassed and running out of liquidity is not a threat, achieving the NRC objective should come last in the sequence of effort priorities. However, if the limits are tight – and they often are (that is, there is not enough own financing capacity available to cover sudden and unexpected downturns of the running business) – then limiting the financial exposure becomes the top priority, even before the weight priority. Planned development activities may well have to be delayed or even suppressed to save the company.

EFFORT PRIORITIES FOR AIRFRAMES, SYSTEMS AND ENGINES

All major objectives such as weight, time, maturity, cost and so on are of equal importance. However, the same should not be true for the timely sequence of prioritising activities required to achieve these objectives. There are different windows of opportunities which must be respected. The proposed sequence of prioritising activities related to the achievement of objectives is depicted in Figure 7.4, applicable for airframe design work.

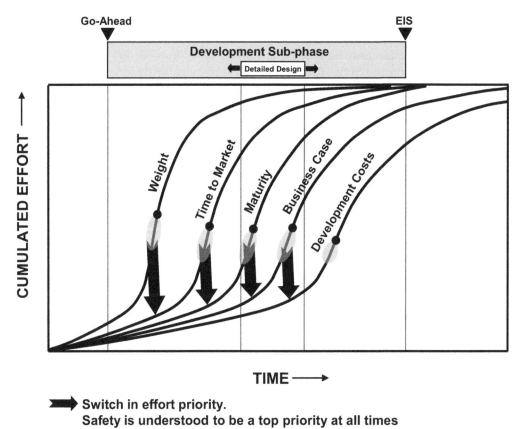

Figure 7.4   Proposed sequencing of efforts for airframe design work during commercial aircraft development to achieve equally important priorities. For aircraft systems' development, 'Weight' and 'Maturity' should be exchanged

For the development of the aircraft's on-board systems (for engines see below) the sequencing is generally proposed to be different: 'Maturity' and 'Weight' should be exchanged by each other. In fact, the systems' weight contribution to the total aircraft weight is smaller compared to the airframe weight. On the other hand, the maturity of the entire aircraft is predominantly determined by the maturity of its systems.

Yet again for the development of the very heavy engines the sequencing of the first three priorities should be 'Weight', 'Maturity' and 'Time to Market', provided the aircraft could fly with different, existing engines until the new engines are available. Most modern commercial aircraft are designed to be capable of flying with different engines. Replacing engines at a later stage is then comparatively easier compared to replacing on-board systems.

Note that some effort will after all always go simultaneously into all objectives at all times. The sequencing of priorities is therefore not strict but only quasi-sequential with considerable overlaps. As was said earlier, these proposed sequences of priorities should be analysed with regard to their applicability to individual components and their associated DBTs. Taking the interdependencies between teams into account, they should be optimised for the entire project as a whole. This may well mean that the activities' sequence for individual teams can differ. Whatever the outcome of this analysis, the results must be communicated to teams and must be transferred into personal objectives for team leaders and the key functional representatives in the DBTs.

Earlier in this chapter it has been said that top-level requirements need to be cascaded down to individual teams. In the next chapter it will be discussed how this can be done in the multi-functional matrix environment.

MAIN CONCLUSIONS FROM CHAPTER 7

- The Development phase is driven by objectives, which are genuine for this phase itself. But during this phase design teams must also lay the foundations for achieving the programme's Business Case during the later Series Production phase.
- Time, cost and quality objectives are cascaded and converted as objectives to fully accountable team leaders. They represent objectives which should all be *simultaneously* achieved. However, it is unrealistic to deal with all objectives simultaneously and with the same level of management attention at any given point in time during the Development sub-phase.
- Focusing instead on one top priority at a time is easier to be communicated to and to be absorbed by the teams.
- One should therefore rather distinguish between an objectives-focused mindset on the one hand and the effort, which goes into objectives achievement on the other. While all simultaneous objectives should be kept in one's mind at all times, the effort, which goes into achieving individual objectives, should be of a quasi-sequential nature.
- Teams should not decide for themselves how they want to sequence their priorities and when to switch from one priority to the next. Ideally, teams would be given clear guidelines from the top programme Management level for the phasing of their priorities. This phasing must not only be in line with the overall development schedule, but is in fact a driver for the schedule planning in the first place.

Managing Requirements and Risks

PART III

Managing Requirements and Risks

8

The Engineering of Requirements

WHAT ARE REQUIREMENTS?

In the context of product development, requirements represent an essential means of communication between an issuing and a receiving organisational party. The issuing party establishes requirements to describe its needs. The receiving party demonstrates its possibilities and capabilities by accepting requirements, see Figure 8.1. Typically, requirements are issued by an external customer to a company (or an organisational party within it), or by the latter to an external supplier. In larger projects, such as commercial aircraft development, requirements need to be managed by different organisational parties *within* the same company as well.

Issued requirements describe needs;
Received requirements – once agreed – describe capabilities.

Figure 8.1   Requirements represent an essential means of communication

However, a first draft of a set of requirements is often written not by the issuing party but by the receiving party. This is because the latter often has a lot more experience from previous projects (involving different issuing parties) than the former. An aircraft specification is an example for this, as will be discussed in a moment.

Requirements can be categorised by four different types of contents:[1]

- how the product should be like, describing forms and fits (that is, product requirements);[2]

---

1 Some sources, such as IEEE 1233 (1996), define as many as 25 different types of requirements. Other sources use only two categories: performance requirements and constraints.
2 Form – the physical properties of the product; fit – external interfaces/environments that the product has to interface with.

- what the product should be capable of doing, describing functions (that is, performance requirements);[3]
- how the product is to be designed, manufactured, assembled, tested and so on (design and production requirements); as well as
- how the product development and manufacture is to be managed (process requirements).

Process requirements describe all the specific integrative management processes to be applied for a given project. Design and production requirements are collected by the company's standard design and production handbooks. Whatever it needs to satisfy both process as well as design and production requirements can be regarded as belonging to the know-how asset of a company.

Only product and performance requirements really are an expression of the customer needs. Note that product and performance requirements need to cover not only requirements for the flying aircraft, but all aspects related to the wider aircraft system or even to the air transport system as depicted in Figure 1.1 too. This includes aspects such as airport characteristics, people-related requirements, requirements related to consumables, operational requirements, regulatory requirements and so on. Note also that all functionalities identified during intense functional analyses should be traceable to performance requirements.

> 'A basic concept is that every aircraft element must have at least one performance requirement. This concept is only logical because if it were not true, then elements would have no functions to perform and therefore no reason for existence.'[4]

From a Systems Engineering perspective, all requirements can be regarded as constraints except if they are performance requirements. Constraints cannot be traced to functionalities.

Requirements which depend on outcome of analyses (that is, solutions), which in turn have been initiated to fulfil a requirement, are called 'derived requirements'.[5] For example, the aircraft's Maximum Take-Off Weight (MTOW) is a derived requirement determined from extensive trade-offs during the early design phases. Many sub-system requirements are in fact derived. It is therefore fair to say that many requirements are just not known at the start of the development project. They are gradually established as the project evolves.

THE AIRCRAFT SPECIFICATION

The highest-level document, which contains product and performance requirements, is the aircraft specification. It is the conversion of the (external) customer needs into highest-level (internal) requirements to be used by the aircraft-developing company. In the aeronautics business, the aircraft specification is a contractually binding document. It is an expression

---

3   Function – the functionalities (including performance) that the product must provide.
4   Jackson, S. (1997), *Systems Engineering for Commercial Aircraft* (Aldershot: Ashgate Publishing), p. 36.
5   Another way to look at derived requirements is through the principle of requirements 'harmonics.' Requirements harmonics consist of requirements, which result in solutions, which, in turn, result in other requirements, which result in other solutions and so on.

for what the aircraft producer promises to deliver to the customer. In addition, it serves to help the developing teams to acquire a customer-oriented and common vision of the product. Customer-oriented thinking involves that the specification should not just be a list of product features but rather of benefits it offers to the customer.[6]

Historically, specifications were written by the Engineering department alone. This often caused tensions and arguments later in the design process as many non-engineering aspects were not covered. In the modern multi-functional project environment, it is important that all stakeholders, that is, Engineering, Manufacturing, Customer Support, Sales and Marketing, are given the opportunity to actively contribute to this document. Apart from this, allowing all necessary Functions to 'sign-up' the aircraft specification is also an important psychological step in ensuring that the development has a good start.

When compiling the specification, the company should, of course, also ensure that customer(s) are heavily involved. As an example, '777 customers had on-site representatives working side by side with Boeing designers to ensure that the new airplane filled their needs. United Airlines, All Nippon Airways, British Airways, and Japan Airlines had teams of two to four engineers on site who were actively involved in developing the 777.'[7] Also, Airbus invited a whole range of potential customers and around 50 potentially hosting airports to do the same for the development of its A380.[8] Involving the customers early, that is, during Develop Configuration phase, contributes significantly in reducing development cycles. However, involving the customer(s) is not only beneficial for the aircraft-developing company. It in fact also helps and supports the customer(s), which may have difficulties in expressing their 'user' requirements in a complete and consistent fashion, and in a technical language that can be understood by an aircraft developer.

However, even a specification established in such a multi-functional way – including the views of the customer – does not provide an ultimate guarantee for project success.

> 'The specification is only a pointer toward the customer. It is customers, not specifications, who will buy our products. Although the specification is a valuable tool for doing this pointing, it remains only a pointer. We cannot hide behind the specification if it proves to be wrong. Thus, customer involvement is an essential subject, both to construct an accurate specification, and to continually provide a sanity check on what we are developing.'[9]

This may be the reason why developers often regard the time and effort it takes to establish a robust specification as not worth it. However, this mindset would simply delay sorting out controversial issues until much later. Failing to invest in a good specification will most probably have adverse leadtime and cost effects to the project: changes, which need to be implemented later, consume relatively more time and cost than changes, which occurred earlier in the development process, see Figure 8.2.

---

6   There is an easy test to identify whether a requirement is a feature or a benefit: If a requirement in the specification does not allow to identify whether the customer would be paying more to have this requirement, then it is not a benefit.

7   http://www.boeing.com/commercial/777family/pf/pf_background.html, accessed 28 April 2004.

8   Thomas, J. (2001), *Die A380 – Eine große technologische und wirtschaftliche Herausforderung für Europa* (Berlin: presentation to German MEPs), 29 May 2001, p. 12.

9   Smith, P.G. and Reinertsen, D.G. (1998), *Developing Products in Half the Time*, 2nd Edition, (New York: John Wiley & Sons), pp. 92–3.

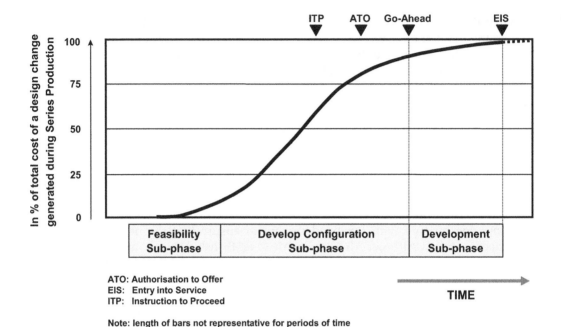

ATO: Authorisation to Offer
EIS: Entry into Service
ITP: Instruction to Proceed

Note: length of bars not representative for periods of time

Figure 8.2   Total cost of design changes generated at different stages
                during aircraft development (qualitative)

OVERVIEW ON HOW REQUIREMENTS ENGINEERING WORKS

Whatever the type of the individual requirement, the collection of a set of unique requirements must in the end define what is expected from a product or process. The importance of Requirements Engineering for project success stems from the fact that poor management of requirements unnecessarily consumes leadtime and cost, in particular where requirements are:

* over and above what is necessary to meet the needs of the issuing party leading to an over-specification of the product;
* omitted, leading to an under-specification of the product with resultant dissatisfaction of the issuing party;
* incorrect, leading to a misspecification, again with resultant dissatisfaction of the issuing party; as well as
* ambiguous, leading to misinterpretations, with the potential for additional dissatisfaction of the issuing party.

Failing to engineer requirements in a professional way at every organisational level only decreases the probability of project success. Poor or neglected requirements are responsible for around 60 per cent of all design errors. Even worse, unclear or incorrect requirements at the beginning of the Development sub-phase will certainly and directly lead to immature products.

'Although product developers always seem to have the opportunity to eliminate the largest category of error, they choose, or worse their managers decide, to rush headlong into constructing the wrong product. And thus they pay many times the price for the product than they would have if the requirements had been gathered and managed correctly at the very beginning.'[10]

An unambiguous understanding of the required objectives is therefore an essential prerequisite to achieve a high-quality product with highest possible maturity at Entry into Service (EIS).

In order to manage requirements well, Requirements Engineers facilitate the following process, see Figure 8.3:

- individual requirements as well as whole sets of requirements need to be agreed in terms of content and applicability between the issuing and receiving parties. This aspect of Requirements Engineering is called Requirements Elicitation;
- requirements need to be cascaded to all multi-functional Design-Build Teams (DBTs) as well as to all suppliers. This is called Requirements Cascading;
- when Engineering has matured its design concepts just prior to the Detailed Design period, it can then be verified whether the design solutions are compliant with the agreed requirements or not (Requirements Verification);
- finally, by the time test equipment is operational – either any of the test facilities such as systems' and engine testbeds or the entire aircraft during ground or flight testing – it can be validated whether the final product will be compliant with the product and performance requirements (Requirements Validation).

Since commercial aircraft development is an evolutionary process with many iterations and feedback loops, requirements will evolve over the duration of the project. Consequently, it does not suffice to simply capture the requirements once, but the above mentioned steps of the Requirements Engineering process must be repeatedly facilitated. In the following, these process steps are discussed in more detail.

REQUIREMENTS ELICITATION

Before facilitating the process leading to any agreement on a requirement, Requirements Engineers ensure that requirements have the right standard and quality to make them unambiguous. By starting to talk to both the issuing and the receiving party they kick-off the elicitation process.

First of all, Requirements Engineers challenge both the issuing and receiving parties to make sure that there is a common understanding whether a requirement is really mandatory or just 'nice to have'.

---

10 Eigner, M., Haesner, D. and Schmitt, R. (2002), 'Requirements Management and Traceability', *Precision Lifecycle Management*, p. 4, http://www.usb-muc.com/requirement_ management/Whitepaper_RMT_Web. pdf, accessed June 2004.

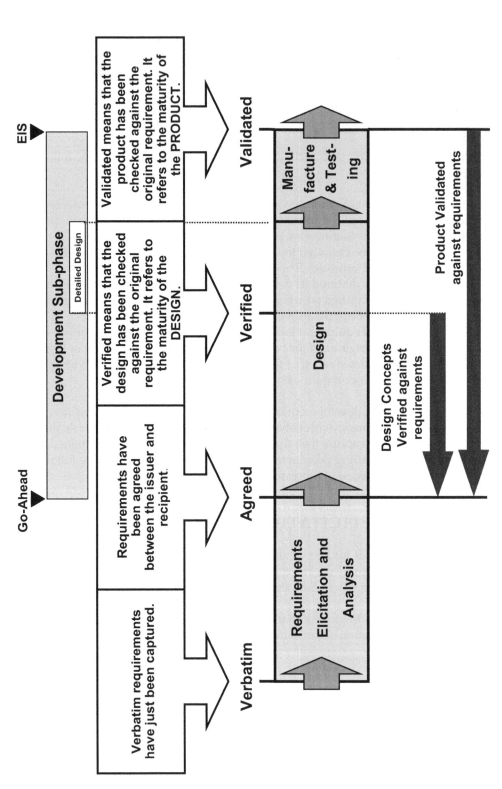

Figure 8.3   The Requirements Engineering process

If the issuing organisation is expecting full compliance with a requirement then the use of 'shall', 'will' and 'must' enforces this. International practice is to use the word 'shall' only. If the issuing organisation of a requirement can tolerate a certain amount of non-compliance with a requirement then international practice is to use the word 'should'. If the term 'should' is used in a requirement, it must reasonably be expected that the recipient of the requirement may not comply with it.

They also challenge whether it is in fact possible at all to fulfil the requirement within the given cost- and timeframe, or within other existing constraints. Challenging each requirement in this way can already avoid significant cost and leadtime problems at later stages.

Secondly, it will be checked whether all mandatory and principally achievable requirements are in fact verifiable because otherwise prove of compliance will be impossible. 'Verifiable' means that there are no options as to how well the recipient complies with the requirement. The latter can either be verified by full compliance or it cannot be verified at all. A good test of verifiability is to propose already at this stage – that is, during Requirements Elicitation – a verification method and to attach this information as an attribute to the actual requirement.

Thirdly, Requirements Engineers need to ensure that the requirement's text is as concise as possible, yet as comprehensive as possible.

To be concise and comprehensive at the same time, requirements statements should be:

- *Simple.*
  - *Each requirement should contain one statement only (if a requirement contains more than one it should be decomposed).*
- *Complete.*
  - *The requirement has no missing text, no 'to be defined' or 'to be confirmed', no reliance on headings or drawings or other text or other requirements;*
  - *There is an 'actor' identified – the 'item', there is an 'action' identified –the 'verb', there is an 'object' identified – the' noun'. If needed, there is a 'condition identified' of the action.*
    The [Subject/Noun] shall [verb] at [object/quantity] [adverb] [condition/object].
    Example: 'The Aircraft shall fly at 36,000 ft for 10 hours'.
- *Clear.*
  - *The requirement has only one possible meaning, features short sentences and paragraphs, uses correct grammar, uses explained terms, acronyms and definitions, and only defines a single 'need', that is, is atomic. Any person who should have a say on the requirement content should be able to understand it immediately. The interpretation of the requirement should not result in different understandings by different people. Thus, the requirement is unambiguous.*
- *Correct.*
  - *The requirement accurately states the need for the right component.*
- *Modifiable.*
  - *The requirement can be easily changed, due to minimal interdependencies, for example with other requirements.*
- *In the right format.*
  - *The requirement – if intended to be further processed and used by a database – needs to comply with a prescribed format.*
- *Independent from yet existing solutions or implementations.*
  - *The requirement should state what is to be developed and not how.*
- *Free from assumptions.*
  - *Requirement statements should be based on used scenarios confirmed by the customer or client.*

Fourthly, it needs to be checked whether the set of developed requirements is consistent. For example, no requirement should be in conflict with, or should duplicate, other requirements. At functional interfaces, it needs to be checked whether there are at least two corresponding functional requirements, that is at least one corresponding requirement from each side of the interface.[11]

For example, system B performs a function using a quantity provided by system A. Say A is a volt generator delivering 28 volts. The requirement for A could be: 'System A shall deliver electrical power at 28 volts.' Then, the requirement for system B might be: 'System B shall provide 20 cfm of air, using the 28-volt power supply of the system A.' Thus, there are two requirements at this functional interface. If a requirements analysis was performed and failed to identify the two requirements, then the set of requirements is not consistent.[12]

Finally, although the requirement's text is the explicit means to convey what is needed between parties, additional information may have to be added by the Requirements Engineers to assist in the management and communication of each requirement. This is provided in the form of attributes attached to every individual requirement. Examples for typical attributes include:

- a unique identifier (that is, an ID code);
- the name of the owner of the requirement;
- the rationale for the requirement;
- the maturity of the requirement (such as 'draft', 'verbatim', 'mature', 'released' and so on);
- the requirement's priority, expressing the importance of the requirement (for example, to the customer or to ensure safety);
- any assumptions required for the understanding of the requirement;
- the verification/validation means as described earlier; as well as
- the status whether the requirement is verified or validated, repectively, or not.

Further information provided by attributes may, for example, contain links to drawings, schedules, technical information, data tables and so on. All extra information should be rigorously referenced.[13]

Once Requirements Engineers have led the issuing and the receiving parties through the Elicitation process, discussing:

- the importance and priority of the requirement (mandatory versus 'nice-to-have' only);
- the formulation of the requirement according to the required text standards; as well as
- the establishment of the requirement's attributes;

---

11 As S. Jackson states: 'Functional interfaces are the most neglected type of interface. And, yet, they are the most important because they characterize the whole purpose, that is, the function of the interface. With the goal of "completeness" in mind, it is well to remember that there are at least two associated functions for every interface.' Jackson, S. (1997), *Systems Engineering for Commercial Aircraft* (Aldershot: Ashgate Publishing), p. 69.

12 Example based on an example provided in: Jackson, S. (1997), *Systems Engineering for Commercial Aircraft* (Aldershot: Ashgate Publishing), p. 70.

13 Note that this means that requirements themselves should not consist of graphics, tables, charts but only of plain text.

both parties will have come to an excellent understanding about each and every requirement. Misunderstandings and errors are by then virtually impossible and agreement on the requirement is merely a formality. Once requirements have been agreed, the Elicitation phase terminates. This phase should be completed around the start of the Development sub-phase (see Chapter 3) as the subsequent verification process will have to make use of frozen concepts to demonstrate that the design meets the requirements.

THE CASCADE OF REQUIREMENTS

In Chapter 5 the organisational size problem for the management of 100 or so multi-functional DBTs was described. As a consequence of this problem, an organisational hierarchy with a certain number of levels is required to manage this many teams. Assigning component design and build accountabilities to different teams implies that teams want to know what to design and build and how to work and operate. In other words: they need all the requirements which are relevant to them but not more and not less. Not only do they need to have all the requirements applicable to them but they also need all requirements captured in one concise document. This is important as all team members are than able to read the whole set of requirements in a time-efficient manner. They do not have to analyse *all* the aircraft-level requirements themselves and individually in order to gain knowledge about the requirements applicable only to them.

While process as well as design and production requirements usually do not depend on organisational constraints,[14] product and performance requirements do. Thus, in order to ensure that aircraft requirements in the specification agreed with the customers reaches the multi-functional teams, they need to be cascaded from the top to the bottom, from every higher level to every lower level.

The first step of the cascade is that the requirements contained in the aircraft specification are transferred into requirements for the next level down, whereby all the requirement standards outlined above should be respected. In aircraft development, the next level down requirements document for the aircraft itself is the Top Level Aircraft Requirements Document (TLARD). It is, for example, further cascaded down into the:

- Top Level Systems Requirements Document (TLSRD);
- Top Level Structural Requirements Document (TLStrRD); as well as into functional requirements documents such as the
- Top Level Maturity Requirements Document (TLMRD);
- Top Level Maintenance, Reliability, Supportability Requirements Document (TLMRSRD);
- Top Level Quality Requirements Document (TLQRD);

to name but the most important ones. The TLARD contains requirements for all high-level functionalities the aircraft has to fulfil as well as constraints the aircraft designers will have to respect.

Different systems on the aircraft will contribute to the fulfillment of the functionalities' requirements. In commercial aviation, systems are grouped according to Air Transport

---

14 A company's design and production handbooks usually apply to all teams of the company.

Association (ATA)-chapters, each covering specific functionalities, see Figure 8.4.[15] In the TLSRD, the functionality requirements contained in the TLARD are therefore re-grouped to generate more detailed functionality requirements on an ATA-chapter by ATA-chapter basis. The same applies to any constraints contained in the TLARD. Note that not only on-board systems but also the airframe is broken down into various ATA-chapters. However, product requirements contained in the TLARD related to the structural ATA-chapters are re-grouped and edited in the TLStrRD.

Once the TLSRD and the TLStrRD have been completed, specific requirements documents for individual ATA-chapters can be generated. This includes individual Systems Requirements Documents (SRD) as well as Structural Requirements Documents for different aircraft major assemblies (for example the wing (ATA 57), leading to the Wing Structural Requirements Document).

From here, the cascade of requirements should continue down to the development teams on lowest organisational level. This can be the aircraft company's DBTs or teams with vendors and suppliers.

One could imagine every top-level aircraft requirement to be attributed with a 'team applicability code' to be able to filter the complete set of requirements for each DBT. The applicability code could in this case be just another of every requirement's attributes. It would define for which aircraft component or which organisational party a specific requirement would be applicable to. Filtering requirements by the same applicability attribute would generate a set of requirements for the component or the organisational party described by the attribute. For example, filtering all requirements by the team applicability code allocated to a specific multi-functional DBT would yield a set of requirements for that team.[16]

However, the textual formulation of requirements to be used by lower-level multi-functional teams may well have to be different from the corresponding requirements' text in the aircraft specification. It is a key problem for the cascading of requirements that the wording and content of requirements has to be adapted to the specifics of the different organisational levels. Simple filtering will therefore not suffice and a lot more involvement of the Requirements Engineering team than a simple 'copy and paste' of higher-level requirements is required.

In practice, therefore, a first draft of the applicable requirements are to the largest extent written by the teams themselves, based on individual team members' experience from previous projects. For the Requirements Engineers it is now vital to link all the team-generated requirements to the higher-level aircraft requirements to ensure traceability and consistency. While scope completeness on the highest level needs to be ensured by direct agreement with the issuing organisation – in most cases the airline customer – scope completeness on lower levels can be ensured by applying traceability measures, which ultimately link each lower level requirement to a requirement on highest level.

This exercise will also determine whether a team has too many or too few requirements. If there are too many and no traceability can be established back to any of the higher-level aircraft requirements, then Requirements Engineering has to ensure that the issuing and

---

15 ATA is a system, standardised by the Air Transport Association of America, for categorising (in chapters) functionalities belonging to an aircraft.
16 As requirements on the one hand are about the product and its creation and on the other are exchanged between different organisational units, it would ease the understanding and management of requirements if the product breakdown and the organisation breakdown would overlap as much as possible. Ideally, therefore, the applicability codes should be homologue to the Product- and/or Organisation Breakdown Structures.

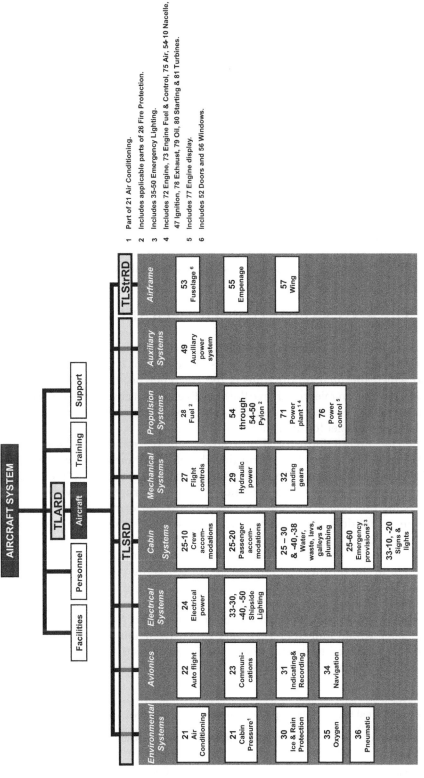

Figure 8.4 Top requirements documents and their ATA-chapter correlations for commercial aircraft development

TLARD: Top Level Aircraft Requirements Document
TLSRD: Top Level Systems Requirements Document
TLStrD: Top Level Structural Requirements Document

Based on: Jackson, S. (1997), *Systems Engineering for Commercial Aircraft* (Aldershot: Ashgate Publishing), with permission.

1 Part of 21 Air Conditioning.
2 Includes applicable parts of 26 Fire Protection.
3 Includes 35-50 Emergency Lighting.
4 Includes 72 Engine, 73 Engine Fuel & Control, 75 Air, 54·10 Nacelle, 47 Ignition, 78 Exhaust, 79 Oil, 80 Starting & 81 Turbines.
5 Includes 77 Engine display.
6 Includes 52 Doors and 56 Windows.

receiving parties discuss the problem and resolve it. If there are too few, then it needs to be analysed why aircraft-level requirements are not represented on team level. Additional requirements may have to be generated on team level.

The cascade described so far here was very much of a vertical, top-down nature. But some teams will also receive requirements from a horizontal direction. This is because Systems Engineers working on the basis of the TLSRD will generate requirements themselves, for example the System Installation Requirements Documents (SIRD) and Equipment Installation Requirements Documents (EIRD) as introduced in Chapter 3. They provide important information for teams in charge of structural design. It is these requirements which ensure that the design of pipes, connectors, brackets, fixtures and so on complies with the needs of the aircraft systems. Another example for horizontal requirements are requirements issued by the Final Assembly Line (FAL) such as Conditions of Supply (CoS), handling and transportation requirements and so on.[17]

For teams to have the full picture of what is requested, any horizontal requirements must also be included in the concise set of team-level requirements. Only then are they able by themselves, with the support of Requirements Engineering, to cascade requirements further down, such as to lower-level teams or to suppliers and sub-contractors.[18]

VERIFICATION AND VALIDATION OF REQUIREMENTS

As we have seen earlier, requirements represent a structured and formalised way of communication. However, what counts in the end – with good or bad communication – is the degree to which the final product meets the customer needs. Albeit important, communication is only one enabler to achieve the objective to meet customer expectations. It is not a guarantee for achievement of objectives. Therefore, it is necessary to *demonstrate* to customers what they will get for their money.

In the commercial aircraft industry this is an extremely complex process, called Requirements Validation, required to confirm that the physical, manufactured product meets the requirements. It comprises a whole series of tests necessary to validate the product and performance requirements. As S. Jackson notes: 'Testing assures that every aircraft element performs the function it was intended to perform, performs it to the expected level of performance, and does not perform functions it was not intended to perform.'[19] The customer is often directly involved in the testing. The testing of airworthiness-related requirements is also monitored and controlled by the regulatory authorities as it represents an essential element in the certification process. Tests are mainly performed on the test benches described in Chapter 3 and during other ground and flight tests.

However, it would be rather risky to wait until the aircraft is flying (or almost flying) to check its compliance with the product requirements for the first time. Any major deficiencies identified at that stage would indeed be very, very expensive to eliminate. Therefore, an intermediate step is introduced to check whether the *design* meets the requirements. This compliance check is performed before the start of the Detailed Design period and is called

---

17 FAL requirements could also be considered as another set of top-level aircraft requirements. However, they are generated later compared to those other top-level requirements, which need to be available before start of the Development sub-phase.

18 For the latter, a company usually has standardised templates to be used as technical product requirements (that is, a specification as seen from the supplier's perspective) represent an integral part of the purchase contract.

19 Jackson, S. (1997), *Systems Engineering for Commercial Aircraft* (Aldershot: Ashgate Publishing), p. 132.

Requirements Verification, see again Figure 8.3.[20] In comparison with the Requirements Validation process, non-compliances identified during the Verification process result in corrective actions, which are a lot less expensive. A similar verification shall be performed to check whether the design and production requirements and principles have been fully applied.

Traditionally, four different means of compliance are applied:

- analysis (for example by means of simulation or similarity);
- demonstration;
- examination or inspection; as well as
- testing (much alike as for the validation process).[21]

'Analysis is any kind of mathematical, computational, or logical task performed to verify a requirement, which cannot be verified in any other manner. In addition, analysis is used as an initial type verification to assure that the aircraft meets the requirements early in the development phase. The results of the analysis will be confirmed by actual flight tests later in the program. Analysis can also be used in conjunction with testing to extend the envelope of the test results. For example, if the aircraft is tested at certain points in the flight envelope, computer simulations can be used to predict the aircraft behaviour in regimes beyond and in between the actual flight data points. ...

Simulation is a type of analysis using computers. ... [For example,] the flight handling characteristics and aircraft performance characteristics can be simulated and used to verify performance requirements long before they are tested in flight. ...

Similarity is a frequently used method of verification. It is based on the assumption that another component, which has met the same performance requirements and operated in the same environment will meet its own performance requirements. In the aircraft industry this similarity is based largely on in-service data, that is, on past experience of another component in another aircraft. ...

Demonstrations are similar to tests but do not require any instrumentation of any sophistication. Typical demonstrations might include the functioning of an emergency alarm, for example. ...

Examinations are the easiest type of verification. They are simply a visual confirmation that a requirement has been met. There can be both drawing examinations, to determine that the required equipment has been included in the drawing, and hardware examinations, to confirm that a piece of equipment has been installed on the aircraft.'[22]

Verification of design concepts is performed at Critical Design Reviews (CDR). However, a CDR which lasts for one to a few days usually does not provide sufficient time to verify all requirements. CDRs therefore need to be well prepared during the weeks ahead, so that at the CDR itself only non-compliances are discussed. The CDR really is only the end point of the much longer Verification process.

---

20 According to the Quality Standard EN 9100:2001:
    'Verification shall be performed in accordance with planned arrangements to ensure that the design and development outputs have met the design and development input requirements'.
    'Design and development validation shall be performed in accordance with planned arrangements to ensure that the resulting product is capable of meeting the requirements for the specified application or intended use, where known.'
21 Verification tests and validation tests are often confused as the same type of tests – such as wind tunnel tests – can be both, a validation as well as a verification test. The difference is that the verification test is used to verify the design while the validation test is used to verify whether the final product meets product and performance requirements.
22 Jackson, S. (1997), *Systems Engineering for Commercial Aircraft* (Aldershot: Ashgate Publishing), pp. 132–3.

As there are very many Verification & Validation activities, a Verification & Validation Plan is required. It outlines the concepts, schedules, methods (such as simulation, test, calculation and so on), roles and responsibilities and the key stakeholders for the individual Verification & Validation processes. To be effective, the Verification & Validation activities need to be planned, monitored and drumbeated in exactly the same way as so many other activities during the Development sub-phase.

MANAGING REQUIREMENTS CHANGE

Specifications and requirements will most probably not be perfect in the beginning. Early issues of the corresponding requirements' documents are based on a somewhat limited knowledge about for example the customer needs. After all, it may simply be too early to understand all the requirements describing the aircraft's on-board systems, despite knowing that poor requirements will most probably lead to poor systems design at a later stage.

'The specification is by nature an imperfect representation of an imperfect product.'[23]

However, an imperfect specification is not so much a problem as it might seem at first glance. What is important, though, is that the essential cornerstone requirements are captured. Initial gaps can be filled thanks to the experience of skilled team members if they have the big picture. They know from previous programmes what needs to be done, even without having all requirements available. Fortunately, thus, knowledgeable DBTs represent a corrective factor wherever early requirements' freeze is impossible.

However, later issues of the many requirements documents should provide the complete set of requirements and must be subject to proper change control. The latter aspect is fundamentally important as really 'frozen' specifications or requirements will hardly ever exist. One would certainly have an easier life with frozen requirements, but trying to freeze requirements in complex projects has turned out to be foolhardy.

Ideally, requirements change control should include the capture of potential cost, schedule, performance and maturity repercussions for each requirement changed and for each team or supplier impacted by the change. However, there simply may not be enough time to get the feedback from each impacted team. In many cases, therefore, a rough estimation of the repercussions on aircraft level must suffice.

SUMMARY

Only by rigorously analysing what is required can the aircraft company or the multi-functional DBT understand the commitments that need to be made. Commitments have to be explicitly agreed and understood on both sides – the issuing and receiving parties – for them to be consistently met on time. Requirements Engineering is a discipline ensuring that a customer's needs can be expressly and demonstrably met through explicit, unambiguous, related and traceable statements. The latter define what is to be achieved

---

23 Smith, P.G. and Reinertsen, D.G. (1998), *Developing Products in Half the Time*, 2nd Edition, (New York: John Wiley & Sons), p. 103.

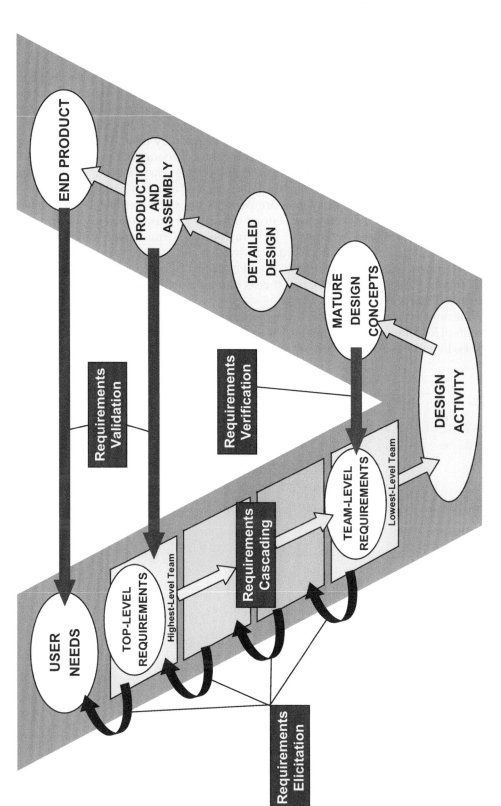

Figure 8.5   Verification & Validation as essential steps to a Requirements-based Engineering

at each stage of the aircraft development life-cycle and on each organisational level. An overview can be seen in Figure 8.5.

Unfortunately, many project leaders do not attribute much attention to the benefits of Requirements Engineering. This may be because requirements by their very nature limit the creativity of engineers. Or maybe it is because the repercussions of failing to do good Requirements Engineering only come to light later in the development process. Many team leaders may have moved on to another position by then. Or it may be their first aircraft project all together and they are lacking the experience of how important the earlier design phases are for achieving the project objectives. There is sufficient evidence that many projects failed because of unavailable, unclear or unknown requirements.[24] As H. Eisner claims: 'Poor requirements are perhaps the most mentioned issue when examining reasons why ... development efforts ultimately result in lack of performance as well as cost and schedule overruns.'[25] But professional Requirements Engineering is not only a risk mitigation activity to avoid negative cost and leadtime repercussions. It is an essential prerequisite to achieve excellence in maturity and quality. Project leaders on all levels should therefore make sure that the set of requirements given to them and their teams fulfils all the criteria outlined above.

It is time now to turn our attention to the management of risks, which – after analysing the benefits of Requirements Engineering – is another tool to support various Management disciplines, in particular Rapid Development.

---

24 See for example: Standish Group & Scientific America, Survey done in 1994 in the US of 365 managers/8380 projects.
25 Eisner, H. (2002), *Essentials of Project and Systems Engineering Management*, 2nd Edition, (New York: John Wiley & Sons), p. 225.

MAIN CONCLUSIONS FROM CHAPTER 8

- Requirements Engineering is an essential prerequisite to achieve excellence in maturity and quality.
- Lack of Requirements Engineering leads to adverse cost and leadtime repercussions.
- During the Elicitation process, Requirements Engineers facilitate the discussion between the requirements issuing and receiving parties in an attempt to come to an excellent understanding about each and every requirement.
- The cascade of requirements is a particular problem if large-scale organisations are required for product development. This is because the wording and content of requirements has to be adapted to the specifics of the different organisational levels. Simple 'copy and paste' of higher-level requirements is not sufficient.
- A first draft of the requirements' documents are therefore often written by the teams themselves, based on individual team members' experience from previous projects. The requirements contained therein must nevertheless pass completeness-, traceability- and consistency-checks with the higher-level aircraft requirements before becoming applicable to the teams.
- A dedicated set of Verification & Validation Plans is required for the individual Verification & Validation processes. To be effective, the Verification & Validation activities need to be planned, monitored and drumbeated in exactly the same way as so many other activities during the Development sub-phase.
- Freezing requirements in complex projects early is foolhardy as change is very often unavoidable. But an imperfect set of requirements is not so much a problem as long as essential cornerstone requirements are captured. Initial gaps can be filled thanks to the experience of skilled team members if they have the big picture. Knowledgeable Design-Build Teams therefore represent a corrective factor wherever early requirements' freeze is impossible.
- The many requirements documents must be subject to proper change control. However, when looking at the possible repercussions of a requirement's change, a rough estimation on aircraft level must suffice.

9

Managing Risks

WHY RISK MANAGEMENT?

It lies in the very nature of all projects that they are associated with risks. This is because projects comprise many uncertainties that have the potential to adversely impact upon project objectives some time in the future.[1, 2] Consequently, the management of projects must always deal with the risks associated with a project. In fact, the management of risks cannot be separated from effective Project Management, which deals with schedule, cost and performance objectives: Risk Management is an essential and integral part of Project Management. In other words, the famous project compromise-triangle of time, cost and quality objectives becomes a quadrangle by the addition of the risk aspect, and life-cycle value can best be achieved through a balance of all four aspects.

> 'To reduce cost at constant risk, performance must be reduced. To reduce risk at constant cost, performance must be reduced. To reduce cost at constant performance, higher risks must be accepted, and to reduce risk at constant performance, higher costs must be accepted.'[3]

Risk Management as a Management discipline provides the early warning system which each projects needs. But Risk Management is more than this: it tries as much as possible to reduce the potential impact of risks by applying a rigorous process. Omitting Risk Management can have significant adverse consequences in terms of cost, leadtime, quality and performance variances. While this statement is true for all projects, it becomes even more important for projects applying Rapid Development techniques such as concurrent ways of working and overlapping. As we saw in Chapter 4, this is because in Rapid Development decisions often have to be taken without knowing the full picture, thus accepting additional risks.

Risk Management is about anticipating and mitigating adverse impacts which will occur with a less than certain probability, see Figure 9.1. Risk Management therefore cannot be expected to deal with impacts which have already occurred, nor with impacts which will certainly occur in the future. It is therefore distinctively different from the management of issues (sometimes referred to as 'crisis management'), where at very short

---

1   Other project risk definitions include: 'An uncertainty which matters', 'A potential deviation from what was promised to the shareholders'.
2   Similarly, opportunities are uncertainties which could impact a project favourably. The principles of Risk Management also apply to the management of opportunities. In fact, one way of risk mitigation is the cashing of opportunities. The latter should therefore always be undertaken with the same rigour than the former.
3   Shishko, R. et al. (1995), *NASA Systems Engineering Handbook* (Pasadena: Jet Propulsion Laboratory), NASA SP-6105.

notice significant resources are made available to solve them, or real work-arounds, where the problems have to be solved by finding a by-pass because there is no other option. This distinction gets often ignored and the need for mitigation actions is therefore masked by all the effort associated with ongoing issue and crisis management. Prior to potentially becoming a real issue, Risk Management identifies risks and reduces their potential repercussions through the implementation of mitigation plans.

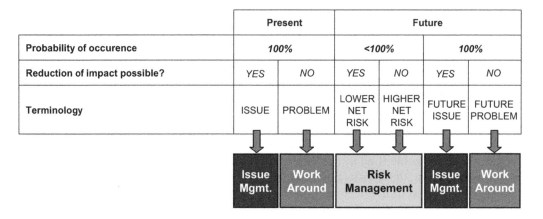

	Present		Future			
Probability of occurence	100%		<100%		100%	
Reduction of impact possible?	YES	NO	YES	NO	YES	NO
Terminology	ISSUE	PROBLEM	LOWER NET RISK	HIGHER NET RISK	FUTURE ISSUE	FUTURE PROBLEM
	Issue Mgmt.	Work Around	Risk Management		Issue Mgmt.	Work Around

Figure 9.1   Risks versus Issues, scope of Risk Management versus Issue Management

For the purpose of this book – that is, the definition of essentials for integrative management of commercial and complex product development projects – it is sufficient to get an understanding of what needs to be done by the multi-functional Design-Build Teams (DBTs) to control and mitigate risks. As in real life of commercial aircraft development projects (with their Project Management constraints described in Chapter 4) it is never possible to deal with *all* risks in a thorough way, a pragmatic Risk Management process would include the following four steps, see Figure 9.2:

- identification of risks;
- assessment and prioritisation of all identified risks;
- risk mitigation only for high-priority risks; as well as
- risk status monitoring and reporting.

Once a risk is identified and assessed as being important enough to launch a mitigation, Risk Management requires both some mitigation planning as well as the actual execution of reasonable plans.

How people deal with risks is based on their individual risk culture, which depends on three factors, see Figure 9.3:

- the project role of the risk identifying person;
- the personality and mentality of that person; as well as
- the person's cultural background and the cultural environment in which the person works.

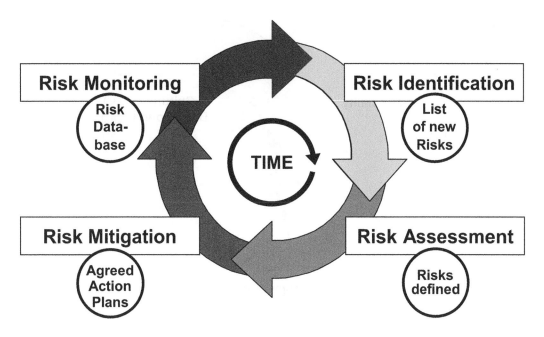

Figure 9.2   The four steps to Risk Management

A project leader, for example, has a different view about the risks of the project than, say, the Quality representative within the project team. Thus, the former may well deal with risks in a different way than the latter, depending on the project role. As far as the personality is concerned, a more optimistic person might deal with known risks differently compared to a more pessimistic one. Also, the cultural aspects should not be neglected. In strong collectivistic, masculine cultures (such as Spain) it is usually very difficult to gain acceptance for the application of Risk Management techniques in the first place. The opposite is true for individualistic, feminine cultures (such as Sweden). Note that the relative weighting of the three factors might change over the lifetime of a person and can be very different from one person to another.

In large projects like commercial aircraft development, where as a result of different involved risk cultures the complexity of Risk Management is certainly increased, it is strongly recommended to appoint dedicated risk managers. While risk managers should drive the Risk Management process through its four steps, they should be as independent from the development project as possible. This will encourage people to identify risks. Thanks to their independence, the risk managers will be able to list risks in an anonymous way and can, for example, also protect those team members who have identified risks associated with the work of their bosses.

RISK IDENTIFICATION

The initial step of the Risk Management process is about knowing what the project risks are. Risk Identification really is not only about identifying risks, but also about accepting, addressing, communicating and taking responsibility for risks. The word 'initial' does by no means imply that Risk Identification is only undertaken once, for example,

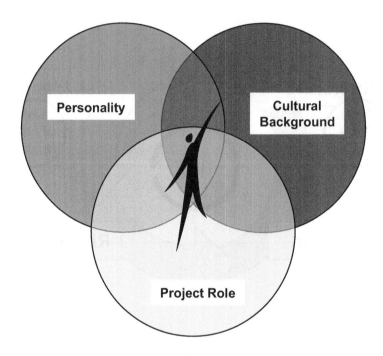

Figure 9.3   Factors of individual risk culture

at the beginning of the project. On the contrary, everyone should feel encouraged to raise and identify risks at any time. Encouragement should come from the leaders. Among others, leaders should encourage risk taking by accepting that errors can be made and by ensuring excellent communications in all directions.

> 'Product development is no more than a sequence of problems that arise and have to be solved, and if people are discouraged from talking about problems, nothing much will happen. In particular, if the messenger is shot from bringing bad news, the early warning system will vaporize.'[4]

> 'Freedom to fail is ... [an important] factor in keeping an early risk warning system vital. Communication on potential problem areas must be kept open, both horizontally and vertically.'[5]

In the multi-functional team environment described in Chapter 5, risks should be raised and discussed during team meetings dedicated to the management of risk. The additional benefit beyond simple Risk Identification is that a common view for what can go wrong emerges among team members. Such meetings not only raise risk awareness – a prerequisite to make Risk Management work – but will also lead to a favourable behaviour whereby a team assumes responsibility for proactive Risk Management. What team leaders and team members have to bear in mind, though, is that it will never be possible to identify all risks and that there will always be surprises. Risk Management is

---

4   Smith, P.G. and Reinertsen, D.G. (1998), *Developing Products in Half the Time*, 2nd Edition, (New York: John Wiley & Sons), p. 236.
5   Smith, P.G. and Reinertsen, D.G. (1998), *Developing Products in Half the Time*, 2nd Edition, (New York: John Wiley & Sons), p. 236.

not a science and the existence of surprises in not an argument at all against effective Risk Management.

There are many other ways in which risks can be identified and captured. Free brainstorming is one of them and the one which is most widely used. A dedicated risk workshop could be another. Whatever the chosen way, team leaders should be aware that cultural backgrounds also play a role when it comes to Risk Identification. For collectivistic cultures, identifying risks jointly will yield better results compared to those achieved through individual identification, while individualistic cultures or cultures with a small Power Distance prefer individual Risk Identification.

> 'Risk Management in international projects must be adapted to the multi-cultural circumstances. Only if this can be achieved can risk management really contribute to project success.'[6] 'The adequate form of risk communication in international projects depends on the one hand on the risk culture, on the other on the differing interests of individuals participating or being affected by the project. The question whom to inform about which risks and in what way becomes a political question then.'[7]

Risk identification can be stimulated if a hierarchy of proven risk categories is already available which may have developed during previous aircraft development projects. High-level categories could for example include:

- technical product risks;
- market risks; as well as
- process risks.

Technical product risks are risks to not achieving a satisfactory answer to the top-level aircraft requirements, that is, the specification. Market risks may arise if it is identified that the specification itself misses the target of satisfying customer needs. As the Development phase of a commercial aircraft lasts many years, the market risks can be very high. If a market risk materialises, this most certainly results in a significant increase in cycle time. Process risks are risks to not being able to meet time, cost and quality objectives resulting from lack or inadequacies of processes, methods and tools.

On a lower level, risk categories might include any of the following:

- management and organisation
    - accountabilities, level of authority, decision-making processes and so on;
- communication and teamwork
    - lack of teamwork and flexibility, low trust, fear culture, cultural barriers, errors and delay not reported, fear of discussing taboo subjects;
- priorities
    - lack of clarity on relative priorities of targets;
- process adherence
    - concern over lack of adherence to design processes;

---

6   Hoffmann, H.-E., Schoper, Y.-G. and Fitzsimons, C.J. (2004), *Internationales Projektmanagement* (München: Beck-Wirtschaftsberater im Deutschen Taschenbuch Verlag), p. 291, free translation by the author.
7   Hoffmann, H.-E., Schoper, Y.-G. and Fitzsimons, C.J. (2004), *Internationales Projektmanagement* (München: Beck-Wirtschaftsberater im Deutschen Taschenbuch Verlag), p. 273, free translation by the author.

- human resources
  - concern over lack of right people with experience in key areas;
- Configuration Management
  - concern over Configuration Management processes and authorities;
- Project Management
  - concern that effective Project Management support cannot be provided;
- Information Technology (IT)
  - concerns about fall back positions; lack of effective IT systems support;
- quality
  - concern over quality in design given other priorities;
- maturity
  - concern over downstream cost resulting from lack of design maturity;
- supply chain
  - exposure to weak suppliers, sole source suppliers;
- customisation
  - concern over preparedness for customisation requirements;
- certification
  - lack of 'in charge' to coordinate certification overall, specific certification concerns;
- external threats
  - exposure of transport-, manufacturing-, and test-facilities to strikes or natural disasters;
- finance
  - lack of financial visibility;
- customer
  - key customer order cancellations, global events;
- weight and performance
  - concern over technical challenges to meet contracted performance requirements;
- time and schedule
  - concern over meeting schedule targets; as well as
- industrial planning
  - concern over ramp-up capability given schedule requirements, transformation to serial production.

However, encouraging the identification of risks by looking at established risk categories should not limit the scope of Risk Identification. It must always be possible to get identified uncertainties accepted as risks – even if they do not fall into any of the known categories.

Once a risk is identified, it needs to be described in a way which allows people not familiar with the risk to understand what the risk is all about. Semantics should therefore be used in the description which is simple and clear and which distinguishes between a consequence ('The risk is that …') and a cause ('The risk is caused by …'). Risk mitigation will concentrate on the cause of the risk (see below). It is, thus, important to understand what the cause is, in particular for people who are not involved in the Risk Identification process but will be in charge of the mitigation plan implementation. The importance of selecting the right semantics for the purpose of better communication cannot be over-emphasised. In particular for multi-cultural project environments it is recommended that everyone uses carefully selected semantics in order to avoid misunderstandings which could arise as a result of the differences in cultural backgrounds.

Often, the same or almost the same risk is identified multiple times by different individuals in the organisation. Risk Management should then try to group risks around 'mother' risks to ease mitigation efforts, monitoring and reporting.

There are some problems when going through the Risk Identification process, depending on personality. Not all individuals working on the development project are capable or willing to identify risks. Others are not willing to describe risks. Optimists, for example, often overlook risks. In many individualistic cultures, such as the USA, risks are often not communicated until it becomes unavoidable. Busy managers or team leaders often believe they do not have the time to describe risks. They rather regard this activity as troublesome. Only training and coaching can improve risk awareness, leading to the comprehension that investing time in identifying and describing risks pays off easily.

RISK ASSESSMENT

If risks are continuously identified, the number of risks which emerge over time can be very high. In this case, it is necessary to prioritise risks (risk ranking) as teams will usually not be able to manage all of them. It is therefore necessary to assess risks against predefined criteria once their potential impact has been described, using the semantics rules outlined above.

Assessment criteria are usually established along the lines of severity and probability of occurrence. Severity is estimated by assessing all the potential impacts of a risk on schedule, cost, performance and quality, while taking their relative importance into account. The product of severity with probability of occurrence is called criticality. It is the aspect of criticality which should be in the focus of all Risk Management. This is because even if the severity increases with time – which it usually does – bringing down the probability of occurrence would still successfully mitigate the risk.[8]

When defining severity assessment criteria one also needs to bear in mind that the organisational size problem discussed in Chapter 6 leads to some kind of risk escalation rule. Note that this rule would need to cover the cumulated effects of risks too. Basically, Management on all levels needs to make up its mind above which severity thresholds it would like to have visibility on risks.

This typically leads to a debate about micro- versus macro-management of risks. Micro-managing risks could mean reducing the number of unpleasant surprises but bears the risk of high costs for the management of the risks. Macro-managing risks is cheaper but some essential risks may be overlooked. These could turn out to be very, very expensive indeed if they would materialise. Compared to the cost of business repercussions resulting from materialised risks, efforts which go into their prevention will most likely be significantly less expensive. Thus, micro-management of risks is a better approach than risk macro-management.

---

8   However, as the likelihood of occurrence is extremely difficult to estimate, reporting of risks may well have to be based on criticality *and* severity. Assessing the probability of occurrence usually depends on purely subjective judgements. Again, it should not be forgotten that Risk Management does not deliver precise indications about where one stands with each risk. It rather provides guidance to do judgements on what can go wrong which is better than to have no guidance at all.

However, not all risks can ever be mitigated. This would absorb too many resources which usually are not available. Mitigation actions are not for free. Thus, it is important to identify the right risk thresholds, above which micro-management should start. Appropriate impact thresholds could for example be defined according to the following principle:

- above an upper *severity* threshold (that is irrespective of a risk's probability of occurrence), where risks could become extremely damaging to the overall aircraft development project, all risks require close attention by top Management;
- below a lower *criticality* threshold it is considered that the mitigation of risks cannot be afforded due to lack of a dedicated financial budget or resources or both;
- in the range between lower and upper threshold multi-functional teams are implementing risk mitigation plans without the need to brief the top Management.

In particular thresholds for criteria related to cost need to be selected carefully. This is because a cost impact assessed by a lower-level multi-functional team may have significant repercussions on that team's overall financial outlook (and therefore on the achievement of the personal objectives of the team leader(s), too), but it may only have a minor effect on a project's financial outlook on aircraft level. Thresholds must ensure that top Management is not bothered with impacts which are negligible on its level. A similar approach is required for all other quantifiable risk impacts, such as impacts on product quality or maturity.

As Risk Management is intended to support Rapid Development, schedule impacts are obviously of particular importance for severity assessment too. But how are schedule impacts measured in a multi-functional team environment? For example, should a team describe a schedule impact only as a delay to the delivery milestone of its own deliverable or should it analyse and describe all the schedule repercussions until, say, aircraft Entry into Service (EIS)? It depends on the team. A team focusing on component deliverables should be capable of analysing potential schedule impacts because it knows the Critical Paths of the plan. Beyond the point of delivery of a component, though, it may not have an in-depth visibility on the project's Critical Path. Thus, for DBTs, schedule impacts can realistically only be expressed with regard to the delivery dates of 'their' components.

Teams on higher level, however, should use the information provided by the lower-level teams and propagate the impacts further through the project's schedule plan to identify the impact with regard to, say, the EIS. This requires that a communication structure is up and running which captures schedule repercussions at lower levels in order to feed them into a higher-level plan. One way of doing this is to use a fully integrated plan (see Chapter 13) but a regular reporting on risk Key Performance Indicators (KPIs) – representing schedule delays at hand-over points – may also be sufficient.

RISK MITIGATION

As teams cannot deal with all risks at the same time they will only address those which have been assessed of being of high criticality. Where risks have been assessed to justify management attention (on whatever level) a plan for their mitigation should be established. Wherever possible, risks should be mitigated through the application of appropriate and known business processes. But there will be many risks which request *dedicated* mitigation plans. The latter essentially consist of a rolling and regularly updated series of actions in an attempt to eliminate or reduce the net impact of risks over time.[9] Mitigation actions should be incorporated into the resourced project schedules and should be implemented as planned.

Among the risks to be mitigated there may be easy ones and not so easy ones. The natural reaction would be to get rid of the easy ones first as this demonstrates action and motivates all stakeholders. However, from a short cycle time perspective it is much more important to tackle all those risks first which lie on the Critical Path. And these ones are usually the more difficult ones to mitigate.

In any case, residual risks may remain even after all appropriate mitigation actions have been put in place. Regular risk reporting should ensure that there is visibility about the net impact situation of given risks, that is, success or failure of mitigation plans should be communicated in order to take further actions.

To ensure that mitigation plans are taken seriously, each risk should be 'owned' by a named individual. It is the accountability of the risk owner – who is not necessarily the person who has identified the risk – to define appropriate mitigation actions, ensure their implementation and report on progress. In addition, teams should budget the costs for risk mitigation as a provision. In fact, the programme's Business Case should have already included a provision for the perceived cost of potential actions required to mitigate risks away. Finally, a contingency fund should be held at appropriate level of the organisation to cover additional mitigation costs not covered by provisions as well as to deal with materialised repercussions where residual risks could not be mitigated away.

RISK MONITORING AND REPORTING

All multi-functional teams on all levels of the organisation should hold risk review meetings to ensure that the risks are accurately described given the latest schedule plan, assumptions and project circumstances, and that they are up to date. Mitigation plans should be reviewed to ensure that they remain appropriate and are on track, and to determine whether additional actions are required. Teams should also discuss changes to the schedule plan and/or business circumstances to enable identification of new risks. Reviews should be held on a regular basis and should be attended by the team leader and the key functional representatives. Results of the reviews with regard to:

- new risks identified;
- risk descriptions (cause, repercussion);
- risk assessment according to the predetermined severity/criticality criteria;
- risk ownership;
- existence and status on mitigation actions; as well as
- residual risks net of mitigation actions;

should be kept in a central database which can then serve as the source from which risk reports can be extracted.

---

9   Net impact is equal to the residual potential impact of a risk after a mitigation plan has been put to action.

A risk report should include all risks with potential repercussions above the agreed thresholds. In addition, other (that is, below-threshold) risks may well have to be included to bring them to the attention of higher-level Management, for example where:

- risk ownership is disputed;
- important mitigation actions outside of a team are not defined or agreed;
- the opportunity to mitigate the risk could be missed, for example where urgent contract signature is required; or where
- appropriate mitigation actions cannot be identified.

Reports should also include trend lines with regard to the net evolution of risks over time (number of new risks minus number of mitigated risks) in order to provide confidence to Management that project risks are coming more and more under control.

All data related to Risk Management could be managed using a source database. However, it would only be but one database of the many which should be integrated into a common architecture for the aircraft development project. This would remedy one of the major deficiencies of Risk Management as it is applied with many companies today: as a stand-alone methodology without connection to the other Project Management processes. As such it is often regarded as an administrative burden which is perceived as not to deliver any added value. However, before entering into the explanation of how Risk Management can be integrated into the full scope of Project Management processes, we still need to look at how costs are allocated to projects in order to get the full picture. By doing so, we are now entering the complex field of project architectures.

MAIN CONCLUSIONS FROM CHAPTER 9

- Risk Management must be an essential and integral part of Project Management. In other words, the famous project compromise-triangle of time, cost and quality objectives becomes a quadrangle by the addition of the risk aspect. Life-cycle value can best be achieved through a balance of all four aspects.
- Risk Management becomes even more important for projects applying Rapid Development techniques such as concurrent ways of working and overlapping.
- Risk Management is about looking ahead and is therefore distinctively different from the management of already existing issues and problems. This distinction often gets ignored and the need for mitigation actions is therefore masked by all the effort associated with crisis management.
- In international projects such as commercial aircraft development it is worth remembering that the way in which people deal with risks depends on their individual risk culture.
- Everyone should feel encouraged by their team leaders to raise and identify risks at any time. Risk taking is sometimes necessary to overcome project obstacles. Leaders should, thus, also encourage risk taking by communicating that some degree of errors is acceptable.
- In the multi-functional team environment risks should be discussed during dedicated team meetings. The additional benefit beyond simple Risk Identification is that a common view for what can go wrong emerges among team members.
- Risk Identification can also be stimulated if a hierarchy of proven risk categories is already available.
- Risk should be described using semantics, which allows people not familiar with the risk to understand what the risk is all about.
- Micro-management of risks is a better approach than risk macro-management.
- The monitoring of potential risk propagation through the project's schedule plan and across different teams is particularly difficult in large-scale project organisations. Special care must be taken on this.
- The programme's Business Case should not only include a contingency to cover issues and problems which will inevitably materialise sooner or later. It should also include a provision to cover all perceived costs associated with potential actions required to mitigate risks.

PART IV

Integrative Project Architectures

10
Allocating Costs to Projects

INTRODUCTION

With the huge financial risks involved in developing commercial aircraft, professional project cost control during the Development sub-phase is obviously crucial. Unfortunately, in many Engineering-dominated companies cost control is still practised as cost monitoring. The difference is that with *monitoring* there is no pro-active attitude involved to influence and shape the design and development process in a way which is favourable to keep costs under control, while with *control* there is. In addition, the job is often left to some cost 'controllers' only rather than being a task for everyone. What is needed is a mindset that costs are essential for project success. Team leaders in particular should demonstrate to everyone through genuine behaviours that the control of costs is vital, even if in the sequence of effort priorities outlined in Chapter 7 Non-Recurring Costs (NRCs) fall behind other priorities.

However, while everyone in the project organisation should be conscious of how to avoid extra cost, at least the members of the Project Management Office (PMO) in each of the multi-functional teams should have an in-depth knowledge of how cost control is exercised within the company. The latter is based on some essential processes and techniques, comprising any of the following:

- How are costs allocated to projects in a company? This question will be addressed in this chapter.
- What are the essential cost categories for project cost control? Again, this question will be addressed in this chapter.
- How do cost categories relate to the cost allocation for work packages used in projects? This will be described in Chapter 13.[1]
- How are budgets set on a periodical basis and how are actual costs monitored against them? This subject will be dealt with in Chapter 15.
- What contingency policy should be used to cope with risks or issues resulting, for example, from late changes? Again, this subject will be addressed in Chapter 15, taking into account what was said about Risk Management in Chapter 9.

---

1 Note that the comprehension of the integrated project architecture described in Chapter 11 is a prerequisite for the understanding of Chapter 13.

AN EXAMPLE FOR A COMPANY'S COST ALLOCATION STRUCTURE

Many classical cost accounting structures are based on cost elements, costs units and cost centres, see Figure 10.1. Cost elements are defined in a company's Chart of Accounts to define the type of cost it wants to control, monitor and report on. Cost elements typically include labour costs, material costs, depreciation as well as cost for services like electricity, water, telecommunication, taxes, fees and so on. The definitions for cost elements are mutually exclusive, that is, there are no overlaps. Each cost-relevant value item – with a value item being, say, $1,000 – can therefore be attributed by a single cost element, that is, it is either labour or material or depreciation cost, and so on. The so characterised costs are called Costs by Nature.

Cost units are created to determine what the costs have been generated for. They are units of the aircraft company's added value-generating business processes. A cost unit could, for example, be an individual work package or a set of tasks of a new development project or even an entire individual aircraft in Series Production. As such, cost units are the ideal link to project cost control. They allow for the monitoring of period results and form the basis for price calculations, among others. Costs allocated to cost units are called Costs by Destination.

A cost centre is a distinct area of the company. The purpose of its creation is to identify where costs have been generated. The demarcation line around a cost centre usually follows the borders of organisational units but costs centres can, for example, also be created to respect technological, economical, project or geographic constraints. All employees of the company belong to a cost centre. Thus, knowing their name (or personal ID) one can identify the cost centre they work for. Cost centres are classified as:

- primary cost centres, which are the only cost centres with direct deliverables to cost units; and
- secondary cost centres, which deliver services to primary cost centres. They include departments such as Human Resources, Accounting, Procurement but also site services such as power supply, restaurant services, site security and maintenance services, to name but a few.

Costs allocated to cost centres are called Cost Centre Costs.

Depending on how cost-relevant value items are allocated to cost centres and units, one differentiates between primary and secondary costs. Costs, which can be directly attributed to cost centres, cost units or Profit & Loss are called primary costs. They are directly booked by the various accounting departments within the company to those destinations. Secondary costs represent intra-company business relationships.

Within most primary cost centres (that is, type (I) in Figure 10.1), total planned costs are set in relation to a reference unit, for example, total planned manhours or machine hours, see Figure 10.1. Thus, the costs are associated with a workload measure and are therefore sometimes called workload costs. As a result, within a type (I) cost centre, there is proportionality of costs to the workload reference unit. This relation is called the hourly rate or charging rate. It often differs from one cost centre to another. For example, there may be different charging rates for aerodynamics research compared to structural design. A centre's charging rate for a period is calculated taking all periodical costs of that centre into

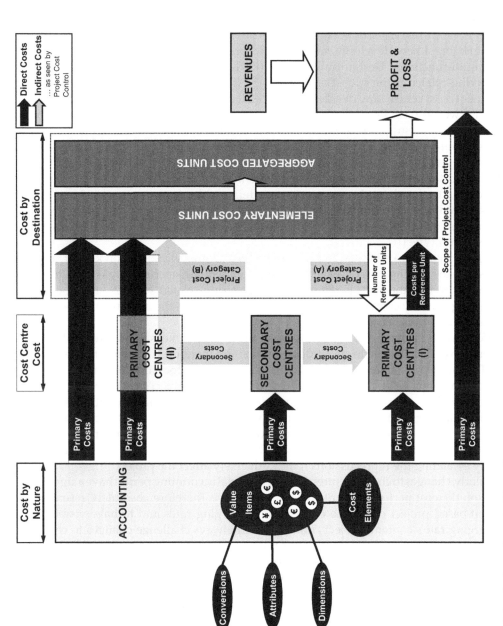

Figure 10.1　　Principles of cost allocation

account – including the allocations from secondary cost centres – divided by the (nominal) workload, expressed in number of reference units, of that centre during the same period.[2]

However, there also are primary cost centres, of which the costs are directly allocated to cost units without making use of any workload reference unit (that is, they essentially just 'transfer' primary and secondary costs to cost units), see Figure 10.1. This type of primary cost centres (type (II) in Figure 10.1) is often created for the costs generated by the staff managing a new development project (whereas, for example, all Engineering staff in the multi-functional Design-Build Teams (DBTs) would usually belong to the larger Engineering-related primary cost centres of type (I)).

A type (I) cost centre may decide to procure all or parts of the intended results, which otherwise would have been generated by the internal workforce as agreed with a cost unit, from external suppliers as *deliverables*. For this subcontracted part of the workload, there is no reference unit any more involved between the cost centre and the cost unit. Instead, the associated workload-related costs are in this case charged against the cost unit as primary costs. The same is true whenever a primary cost centre purchases non-workload-related goods and services as deliverables from a supplier. Here again, the associated costs are directly charged against cost units. A type (I) cost centre may therefore also act as a type (II) cost centre. In reality, thus, most cost centres are mixtures of type (I) and type (II).

Secondary costs are allocated to the primary cost centres via allocation keys. This way, it is possible to steer the flow of costs. For example, adequate steering would use allocation keys to ensure that the costs generated by Engineering-related services are allocated to Engineering primary cost centres only. Without this steering, the costs would possibly be allocated to all primary cost centres.

Unfortunately, a change in the definition of the allocation keys can make a substantial difference to costs allocated to specific cost centres and therefore ultimately to specific cost units. In other words: financial success of a specific project does also depend on whether the allocation keys used by the company to estimate the project's costs are the same as the ones used later to monitor them. With, for example, commercial aircraft development cycle times lasting for many years, this may well be a challenge. As it is the company's Management who defines the policy for how costs are allocated, there is not much a project leader can do to influence allocation keys – other than keeping track of their changes and possibly escalating the reprecussions if they adversely affect the project.

Similarly, changes to charging rates *during* a financial accounting period have a significant impact on the cost performance of a cost unit too. It is therefore absolutely essential for cost unit-based project control to expect stable charging rates and to have visibility on the charging rate's content. Project control should always challenge changes to charging rates. Also, project control should seek commitments from the cost centres for the stability of their charging rates. Only too often projects get caught by increase in charging rates, where the root cause for the increase lies outside the project itself. As commercial aircraft developments take many years it will be difficult to achieve a commitment for a constant charging rate over that same period. However, commitments should at least be provided on a periodical basis in line with the periodical budgetary process.

---

2   Sometimes costs such as the depreciation for buildings and equipment are included in charging rates. This is often the case if buildings and equipment are shared by more than one project. If, however, those buildings and equipment are for the exclusive use of a single project, then they should not be included in the charging rate. They should rather be treated as primary costs directly allocated to the project's cost unit(s).

A PROJECT'S COST CATEGORIES AND THEIR ATTRIBUTES

Cost controlling a project is somewhat less complex compared to the whole scope of a company's Finance department. For example, a project's cost controlling does not deal with accounting or balance sheets. Nevertheless it is complex enough to give headaches to many people.

The Business Case, on the basis of which the aircraft development project has been launched, provides the reference against which project control must demonstrate cost performance. The project's cost control should therefore be structured along the same categories which have been used to establish the Business Case (see Chapter 4), and the Business Case should not include objectives which later cannot be monitored.[3] Only then will the project be able to evaluate whether it has achieved the objectives of the Business Case or not. In essence, this means that much care has to be taken when setting up the cost content of the Business Case.

For reasons of practicality it can be assumed that costs fall into two cost categories only when entering the scope of project cost control. They are also depicted in Figure 10.1 and include:

a) reference units, such as hours, booked by individuals or machines against a certain activity or milestone, and multiplied by the applicable charging rates (that is, applying costs per reference unit); as well as

b) costs directly charged against cost units, such as for sub-contracted work packages or externally procured items such as services or aircraft components.

Please note that category (a) is only applied for the relation between type (I) cost centres and cost units, whereby (b) comprises costs allocated to cost units by type (II) cost centres as well as directly from Accounting. In order to capture the entire workload-related cost a cost unit has negotiated with a cost centre, it may not only have to consider cost category (a), but also cost category (b) if the cost centre has sub-contracted part of its workload, see Figure 10.2.

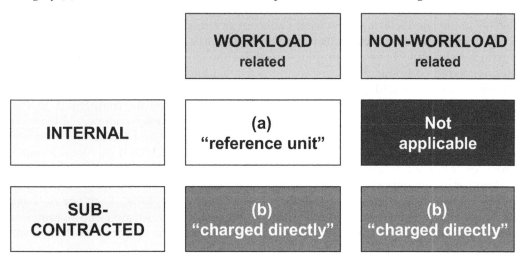

Figure 10.2   Cost categories and workload

---

3   It might, for example, be difficult to monitor the status of objectives achievement if different accounting rules are applied by the various companies collaborating in a development project.

Value Item Attributes for the Needs of the Company

Besides allocating cost elements to cost-relevant value items, one can imagine allocating other, additional attributes to value items. This usually becomes necessary in order to not only fulfill the controlling needs of the project but also those of the company. What indeed is often overlooked by team leaders is that project control also needs to take place within the existing company's financial framework of rules for accounting and operative planning, among other. As a result, cost item attributes need to be selected such that the needs of, for example, the company's regular quarterly or yearly financial closures as well as of its operative planning are secured.[4] Among others, such attributes therefore need to allow for differentiation between:

- capitalised versus non-capitalised value items;
- cash- versus Earnings before Interest and Taxes (EBIT)-relevant value items;
- value items relevant to different life-cycles (such as costs for Research and Technology (R&T), Development, Cost of Sales, and so on).

Assets of the company, such as machinery, jigs & tools and buildings, are value items to be capitalised. As any aircraft development needs new such items, they represent an increase in assets of the company. It is in fact the associated production or procurement costs of such assets, which need to be capitalised. And it is only the periodical depreciation costs for these asset values, which represent cost-relevant value items to be considered by a project's cost controlling.[5] Sometimes it is difficult to determine whether a value item needs to be capitalised or not. The Finance department of the company has to provide advice and needs to establish clear guidelines for project controllers to take appropriate decisions.

Cash is obviously important to be monitored on a company level, particularly in the aircraft business. This is because the cash outflow of an aircraft development is always high and absorbs a significant portion of the company's liquidity. When Boeing developed the 747 in the late 1960s it almost went bankrupt as a result of too high a cash exposure. Even a cash exposure which under normal circumstances would be associated with only medium risks could have catastrophic effects for the aircraft company. For example, if an unexpected economic downturn would occur, leading to suddenly reduced revenue streams, a company could get into deep trouble. Aircraft development is a risky undertaking and cash exposure monitoring is therefore an absolute must for the company. Also, the more cash is available, the easier it will be for a company to raise commercial loans to fund new aircraft programmes. Project cost planning, control and forecasting – clearly identifying which of the value items are cash-relevant and which ones are cost-relevant – provides important data required to perform overall cash monitoring on company level.

Another highly important aspect is the company's profitability, of course. Profitability can be expressed by a variety of Key Performance Indicators (KPIs). Each one of them has its strengths and weaknesses. One KPI widely used is EBIT. An aircraft programme's contribution to the company's EBIT would take into account the revenues resulting from

---

4   The forward look of a company's operative planning usually covers the next three to five years.
5   Capitalised value items do not appear on a company's statement of profit and loss. Only the periodical depreciation costs for the capitalised value items are captured on the cost side of the statement. This ensures that investments do not heavily affect the profitability of the company in any given period. It rather reduces company profitability in a controlled way to have sufficient funds available for future investments.

the sales of aircraft and related support services as well as the costs associated with the development, production and support of the aircraft. Costs would include depreciation of capitalised jigs & tools, buildings and so on as mentioned above.[6]

Cost-relevant value items for the life-cycle phases Research and Technology (R&T) as well as Development need to be separated from the Series Production costs (called Cost of Sales) if a company's annual closing is outlined according to the Cost of Sales methodology. This is more difficult than it might seem at first glance. The aircraft will continue to be developed (for example, retrofitted, upgraded and so on) for many years – even after Series Production has started – even if the associated development effort can be considered to be much smaller compared to the effort during the Development phase. Therefore, a company needs to decide at which point in time the Development phase of the project should formally end and the Series Production phase should commence. The expected date of Type Certification or the Entry into Service is usually a good choice for this financial transition point.[7] It should also decide about how to collect costs related to Development after the Series Production phase has started.

Value Item Conversions for Project Control

Project control cannot only work with attributes attached to value items but also needs value items to be represented in different conversions. Formulae are needed to convert a value item from one conversion to another. Necessary inputs for conversion formulae include, among others:

- exchange rates (such as to the US$, €, UK£, and so on); as well as
- the economic condition (Current Economic Condition (CEC), economic condition of year xxxx (EC xxxx).

The Finance department needs to provide project controllers with the necessary conversion rules for the application of currency and economic conditions.

Value Item Dimensions for Project Control

Finally, project control needs to take into account what can be described as value item dimensions such as:

- the timely period (month, quarter, year) or the date of validity (end of year, for example); as well as
- the target, commitment, budget, actual, forecast or outlook dimension[8]

---

6   The main difference between cash and EBIT-related costs from a project's cost control point of view lies in fact in the different ways in which capitalised items are treated: for a given period, the cash-related view takes into account the real periodical expenses associated with the project. The EBIT-view would only take the periodical depreciation costs for that period into account. The operative planning of a company is usually more cash-oriented while the periodical financial closure is more EBIT-oriented.

7   Knowledge of the agreed end of the Development phase is also important as target or forecast completion costs (that is, the whole costs until completion) make reference to this date.

8   Compared to forecasts, 'outlooks' in addition superimpose on the forecasts a judgment of the risks and opportunities associated with the current way of how the project operates. Such judgments should be made

of a value item. How to monitor actuals against previously allocated targets and budgets will be described in Chapter 15. Commitments are the costs for which the target- or budget-receiving units in the organisation have commited to do the work. If there is still a delta between the target and budget on the one hand, and the commitment on the other, then this signifies that the convergence process to be described in Chapter 15 has not yet come to a satisfactory closure.

Forecasts (such as for the end of the year) can, for example, be calculated by an extrapolation of the current expenditure's run-rate or by a bottom-up calculation using the detailed interdependencies of an integrated schedule plan. However, with hundreds of cost centres involved in a commercial aircraft development project it may be difficult to generate forecasts on a cost centre by cost centre basis. This is because the effort associated with its calculation is tremendous and therefore is not done often during a financial period. It may also not be known which cost centre is contributing which workload to the project during any given period. And even if this were known for the current period, it may not be known so well for subsequent periods. In order to make forecasts, cost controllers therefore group cost centres together and calculate a corresponding charging rate for this cost centre group. They also make a judgment on how this group charging rate may evolve over the next periods. Doing a forecast on this basis is much easier, albeit not perfectly accurate. It is also associated with risks as at the end of a given period it may turn out that the judgment on the charging rate was quite incorrect, which again can lead to unpleasant surprises for project leaders as costs are suddenly rising.

Besides the cost categories and attributes mentioned so far there are many other criteria by which it must be possible to aggregate project-related costs for reasons of good project cost control. Among others, they are linked to the product structure, the project organisation as well as to the implemented processes. However, these criteria are not genuine financial criteria but rather depend on product, process and resources architectures. We will now look at these architectures, keeping in mind that Cost Control might require specific views on the project.

---

by the team leaders together with their project controllers. Examples for risks could include:
- unexpected changes to company-internal charging rates; as well as
- higher than expected supplier claims resulting from technical changes.

MAIN CONCLUSIONS FROM CHAPTER 10

- From a project's cost control perspective, projects are composed of cost units to which costs are allocated – either directly or via cost centres.
- This is done using essentially two different cost categories:
  - reference units, such as hours, booked by individuals or machines against a certain activity or milestone, and multiplied by the applicable charging rates (that is, applying costs per reference unit); as well as
  - costs directly charged against cost units, such as for sub-contracted work packages or externally procured items such as services or aircraft components.
- Besides the cost categories (and some attributes), there are many other criteria by which it must be possible to aggregate project-related costs. However, these other criteria are not genuine financial criteria but rather depend on product, process and resources architectures.

11

Designing an Integrated Architecture

INTRODUCTION

Complex projects like the development of commercial aircraft require the management of an enormous amount of data and information. Experience over many decades has proven that the use of structured architectures, that is, systems of interlinked modules, provides significant support to the management of a project. This is because architectures allow for an *organised* collection and sharing of business and technical information related to a project and product.[1] As E. Rechtin expresses it: 'The essence of systems is relationships, interfaces, form, fit and function. The essence of architecting is structuring, simplification, compromise and balance.'[2] Architectures also allow for a graphical representation of project aspects, thus capturing the imagination of the people's mind and easing communication.

Two types of architectures are generally used to support the management of a project: Breakdown Architectures and Sequence Architectures. Breakdown Architectures are also called Breakdown Structures. With the latter, a single module is broken down into many other modules: lower-level modules are interlinked with a single top-level module via modules on one or more intermediate levels. Breakdown Structures represent hierarchies. This is why the graphical representation of Breakdown Structures essentially has the shape of a triangle (using two dimensions, see Figure 11.1) or a pyramid (using three dimensions). Classical Breakdown Structures include the:

- Product Breakdown Structure (PBS);
- Work Breakdown Structure (WBS);
- Organisation Breakdown Structure (OBS); as well as the
- Cost Breakdown Structure (CBS).

But there are a few other Breakdown Structures in use during an aircraft development project, such as the:

---

1 It must be noted, though, that the benefits of structured architectures are regarded differently in different cultures. In Germany, for example, it is common to initially establish a clear architecture, subsequently followed by working on the individual subjects. In cultures with strong context relations (such as France, Brazil) it is important to sort out the entire scope of subjects first while architecting a structure is of lesser importance.

2 Rechtin, E. (1991), *Systems Architecting* (Englewood Cliffs: Prentice Hall).

- Functionality Breakdown Structure (FBS);
- Systems Breakdown Structure (SBS);
- Infrastructure Breakdown Structure (IBS); as well as the
- Generic Process Architecture (GPA).

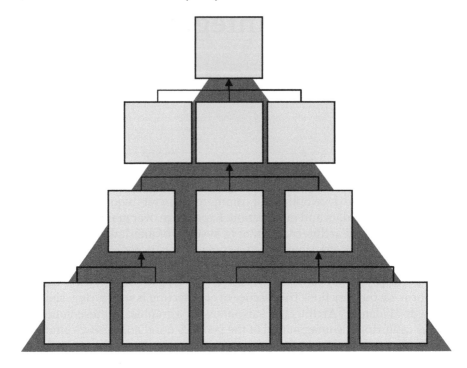

Figure 11.1   Shape of a classical Breakdown Structure

Sequence Architectures describe input → output relationships, see Figure 11.2. Compared to the static Breakdown Structures they are dynamic structures as the dimension of time plays a predominant role. Sequence Architectures can be regarded as horizontal architectures because time dependencies are traditionally depicted along horizontal axes. Classical Sequence Architectures include the:

- Build Process Architecture (BPA); as well as
- any schedule plan.

However, modules of both Breakdown Architectures and Sequence Architectures require attributes to identify incoming and outgoing links from and to other modules respectively. Let us call them Type I attributes. Essentially, they consist of names (or codes or the like) representing those other modules with which a module shares incoming and outgoing links. Type I attributes describe the structural build-up of an architecture.

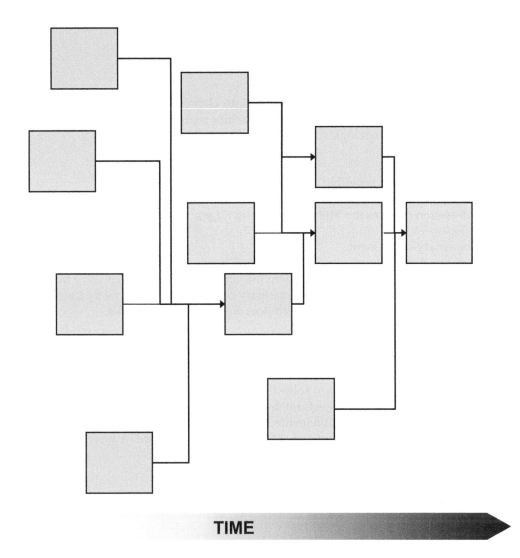

Figure 11.2   Shape of a classical Sequence Architecture

Note that for Breakdown Architectures each module but the one on the top has one outgoing link only with the module on the next higher level. In addition, modules on all levels except on the lowest level have more than one incoming link from lower level modules. As a consequence, there are, for example, no modules with one incoming *and* one outgoing link. This may well be the case for Sequence Architectures. For the latter, there may be one or more links coming in *and* coming out of modules, respectively: all combinations are possible.

CLASSICAL ARCHITECTURES

The Product Breakdown Structure (PBS)

The PBS decomposes the entire final aircraft into elementary components which then cannot be broken down any further. It is a structure representing the aircraft in a zonal view. A typical breakdown logic could be – from top to bottom:

- aircraft;
- major component assembly (such as the wing);
- section (such as the Fixed Trailing Edge);
- sub-section (such as the Mid Fixed Trailing Edge);
- ...
- elementary component.

There are many ways to gradually zoom into smaller and smaller zones of the aircraft until the aircraft is decomposed into elementary components. How exactly the aircraft is broken down therefore depends on the choices engineers wish to make.

The Functionality Breakdown Structure (FBS)

Designing an aircraft also means to solve the problem of finding the right architecture of functional modularity. The architectural design of functionalities does not look into how these functionalities are to be implemented in hardware, software or human modules. It rather concentrates on the functionalities that are to be performed. A typical breakdown logic could be – from top to bottom:

- aircraft functionalities;
- generation of lift;
- generation of lift at low speeds vs. high speeds;
- ...
- elementary functionality.

Note that for the benefit of development cycle time reduction, elementary functionalities should not go across the boundaries of component modules (that is, equipment) wherever possible. Thus, there clearly should be a strong link between PBS and the FBS.

> For example, a door could have two functionalities: to open/close and to lock/unlock it. This door may consist of four modules: the actual door, the lock and two hinges. Obviously the lock/unlock functionality would belong to the lock. But if the two hinges would be different in design the opening/closing functionality would go across component modules of different types with different development cycles. Faster development could be achieved, if the hinges were of the same type and could therefore be regarded as belonging to the same module. No surprise that by far the majority of all doors have two hinges of the same hinge type.

Note that the principle of elementary functionalities not crossing the boundaries of component modules implies that a component module can serve more than one functionality but no elementary functionality should belong to more than one component module.

The Systems Breakdown Structure (SBS)

Functionalities are performed by systems. The latter therefore represent translations of functionalities into hardware, software and human elements. In civil aeronautics, systems are classified according to Air Transport Association (ATA)-chapters, which were already introduced in Chapter 8. Each ATA-chapter system has to satisfy certain functionalities.

Components of systems like equipment units and interconnecting elements are assembled at different Assembly Stages (ASs) such that the physical completeness of the system is usually only achieved when the aircraft build is fully completed. Despite this, systems need to be designed and developed as a whole. What is therefore needed is a system view, which, for example, can be realised on an ATA-chapter by ATA-chapter basis. A zonal view as provided by the PBS is not adequate to represent a blueprint for the breakdown of a system as many of the ATA-systems stretch over some or many aircraft zones. Thus, looking at a specific zone is usually not sufficient to comprehend an entire system with all its sub-functionalities delivered by equipment units, pipes, tubes, harnesses and so on which penetrate various zones of the aircraft.

A generic breakdown logic for the SBS could be as follows – again from top to bottom:

- aircraft;
- ATA-chapter (such as fuel ATA 28);
- ATA-sub-chapter (such as fuel storage ATA 28-10);
- ATA-sub-sub-chapter (such as fuel tanks ATA 28-11);
- and so on.

The generic Work Breakdown Structure (gWBS)

The generic Work Breakdown Structure (gWBS) describes the breakdown into work categories of all work which needs to be done to design, develop, produce, test, certify and refurbish the aircraft. For the Development sub-phase, high-level work categories may include, among others:

- specific design work, including all schemes, drawings and modifications work necessary for development of specific aircraft zones;
- non-specific design work, comprising activities, which have to be carried out for the overall aircraft and cannot be allocated to a certain aircraft zone. For example, these activities may relate to:
  - design and certification philosophy;
  - configuration development;
  - aerodynamics and aeroelastics;
  - performances;
  - weights and balance;

- environment;
- loads;
- stability and control;
- systems;
- overall structure design;
- lightning protection;
- airport compatibility;
- aircraft integration; as well as
- maintainability, reliability, supportability (MRS);
- structural tests (such as static and fatigue tests);
- tests of individual ATA-systems;
- jigs & tools;
- production of development aircraft;
- flight test;
- ground support equipment;
- technical publications;
- refurbishment, if any;
- delivery of first aircraft;
- production, transport and delivery of spare parts for first aircraft; as well as
- management activities.

Note that there may be different work categories for the Series Production phase. For example, it may be necessary to distinguish between specific design work during Development as opposed to specific design work during Series Production phase.

High-level work categories may be further broken down into lower-level ones. Lowest-level work categories require dedicated methods and tools to be performed. In other words: activities described by work categories on the lowest level of the gWBS can be differentiated from each other by the different methods and tools they require for their execution. A breakdown logic for the generic WBS could therefore look like this:

- total work to be performed;
- high-level work category;
- ...
- low-level work category;
- lowest-level work category.

The specific Work Breakdown Structure (sWBS)

Most projects, however, do not apply the generic WBS, but rather a WBS which for example may also take into account aspects of the PBS, FBS, or SBS. This WBS, specifically developed for a project, is called the specific Work Breakdown Structure (sWBS), but is generally referred to as simply the Work Breakdown Structure.

The WBS features work packages (WPs) on each of its levels. Work packages are Project Management tools, necessary to set-up, monitor and control a project. However, most work packages are just 'nodes' of the WBS, introduced for the purpose of consolidation of lower-level work packages. Only work packages on the lowest level really contain the relevant information. The latter is described by Work Package Descriptions (WPDs) consisting of:

- a description of the work to be performed such as in the form of a list of individual activities;
- the list of sufficiently well-described inputs and output deliverables from and to other work packages, respectively;
- an estimation of the necessary workload (for example in manhours, costs and so on) broken down to a phasing (such as a monthly basis) compatible with the required reporting frequencies;
- the start and end date of the work; as well as
- signatures for approval of content and authorisation of work go-ahead.

Note that a work package can contain activities representing more than one work categories.

There also are WPDs on higher levels of the WBS. As work packages on higher levels represent consolidations of lowest-level work packages, descriptions for higher-level work packages represent the consolidation of descriptions of lowest-level work packages. To make the distinction, a WPD on a higher level of the WBS is often called Statement of Work (SoW). As during the earlier development phases a more crude description of the work to be performed is usually sufficient, it makes sense to introduce Statements of Work at that point. In any case, detailed work packages and their descriptions are usually not available by then.

The Generic Process Architecture (GPA)

All activities of a company do not only belong to work categories but also to processes. For example, 'Design', 'Develop', 'Produce', 'Test', 'Certify' and 'Refurbish' are generic processes which contain individual sequences of different activities of work, which in turn may each represent different work categories. A company's generic processes represent a great deal of its established and proven know-how. Aircraft development can therefore be regarded as a set of processes required to develop the aircraft. Quality Assurance checks the adherence of an organisation to its established and proven processes in order to secure product quality. Generic processes therefore provide one important link between the activities of the Quality department and other Functions such as Engineering or Manufacturing. However, due to the long cycle times encountered in aircraft development, there is not only the aspect of processes' application and adherence, but also of their continuous improvement.

There are therefore valid reasons to also create a Breakdown Structure for the company's processes: the GPA. A GPA breakdown logic could look like this:

- highest-level processes;
- medium-level processes;
- lower-level processes;
- detailed generic processes.

For example, Airbus calls some of its highest-level processes 'Sell', 'Develop', 'Fulfill', 'Support', 'Enable'.[3] Detailed generic processes could comprise Engineering activities, such as 'Release of Drawing', as well as Manufacturing operations, such as 'Assemble'.

As the list of established company processes should cover all activities, there is a definite matrix-like relation between the GPA and the WBS: each task or activity belongs to a work category and a work package as well as to a generic process, see Figure 11.3. Generic processes should be well documented, approved and accessible for use. Management must ensure that they are understood by everyone who is expected to use them.

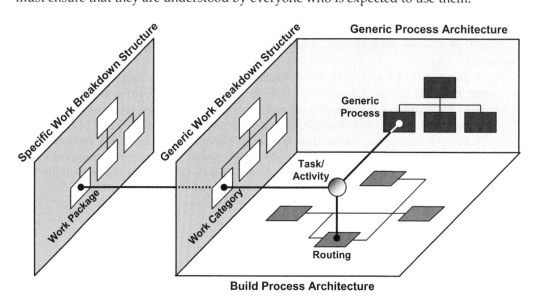

Figure 11.3   Relation between Work Breakdown Structures, Generic Process Architecture and Build Process Architecture

The Build Process Architecture (BPA)

The BPA outlines the sequence of the assembly process from the manufacture of elementary parts to the gradual build-up into larger and larger units. Thus, the BPA belongs to the category of Sequence Architectures. Its structure takes into account aspects like:

- the workshare between collaborating companies;
- the constraints set by the existence of distinct and different manufacturing sites, within the aircraft company as well as with suppliers; as well as
- the logic of on-site assembly sequences.

The BPA is broken down into Manufacturing Stages (MSs) and ASs, respectively, consisting in turn of one or more routings. A routing is a generic work plan for a

---

3   Smyth, R. and Roloff,G. (2006), *Best Practices for Systems Development and Integration* (Helsinki: FINSE Fall Seminar), 26 October 2006, p. 10.

Manufacturing Stage Operations (MSO) or an Assembly Stage Operation (ASO) to be executed at a MS or an AS, respectively. It describes sequenced Manufacturing activities, whereby each of the latter is associated with data such as workload, standard times, type of task and so on.[4]

A generic BPA breakdown logic could be:

- final aircraft assembly;
- assembly of major components to be delivered to Final Assembly Line (FAL) (such as the wing);
- assembly of major sections to be delivered to site of major component assembly;
- and so on.

The breakdown continues until those operations are reached which describe the manufacture of elementary parts.

BPA components, which are delivered from one manufacturing site to another, in particular if different companies or different legal entities of the same company are involved, are called Constituent Assemblies. They are very important because this is where conformity between 'Design' and 'Build' must be documented to gain the approval of the airworthiness authorities.

Note that in the case of Manufacturing processes, there is also a close relation between the BPA, and the WBS and the GPA, respectively. This is because routings consist of activities. This relation is also depicted in Figure 11.3.

The Infrastructure Breakdown Structure (IBS)

The development of a new commercial aircraft necessitates a dedicated infrastructure, the financial volume of which is considerable. It consists of elements like:

- land;
- streets;
- run-ways;
- buildings;
- transport means like lorries, ships, barges, and even airplanes (see for example Figure 11.4);
- machinery of different kinds;
- test equipment;
- jigs;
- tools;
- and so on.

---

4   A routing is converted into a specific routing, called work order (WO), when manufacturing and assembling a specific, single aircraft. The activities in the work order may be used to record the times spent. The link between routings and work orders on the one hand and work packages on the other will be described in Chapter 13.

Figure 11.4   Vehicles used for transport of Airbus aircraft components
Source: Airbus, with permission.

Although not aircraft parts, all of these are products, many of which need to be newly developed. They require the application of professional Project Management techniques as much as the aircraft parts in order to secure the overall success of the aircraft development project. It therefore helps to not only break down the aircraft into elementary parts or functionalities, but also to create an architecture which puts the individual infrastructure elements in a relation to each other. This architecture is called Infrastructure Breakdown Structure (IBS).

It can, for example, be structured along a non-aircraft FBS. But, of course, it is also heavily linked to the way the aircraft is built, as described for example by the WBS, the GPA and the BPA. It may also be part of an enlarged PBS. Thus, there are many ways to associate or incorporate the IBS.

The Organisation Breakdown Structure (OBS)

Once the decision has been taken about how the generic overall structure of the organisation for the aircraft development project is to be laid out (for example, as a matrix organisation based on multi-functional Design-Build Teams (DBTs) as described in Chapter 5 and 6), the specific OBS can be established. It allocates team units to the various organisational levels and defines their respective lines of reporting. However, for a variety of reasons this breakdown is different from the other breakdowns discussed so far. Reasons include any of the following:

- OBS-units are often combined organisationally for reasons of responsibility or practicability. This may lead to DBTs managing component modules, which do not share common interfaces;
- for reasons of 'political correctness', organisational units sometimes need to appear higher up in the OBS although the components, for which the units are responsible, can for example only be found on lower levels of the PBS;
- the dimension of time is completely missing in an OBS, which, for example, makes it difficult to match the OBS with the BPA;
- one OBS team could manage many work packages. Thus, the WBS and the OBS differ from each other, but are structurally interlinked; and
- there could be more than one cost category (see Chapter 10) per OBS unit.

As a result, the individual structures of the OBS and the other architectures cannot be matched so easily. This is why the OBS is often a separate architecture. But there are links to the other Breakdown Structures. For example, the link between WBS and OBS is established in what is traditionally called the Task Responsibility Matrix. Each task or activity contained in a work package is to be performed by one unit of the OBS, see Figure 11.5. Each task or activity is also related to a component or part described by the PBS (or FBS or SBS, whichever applies).

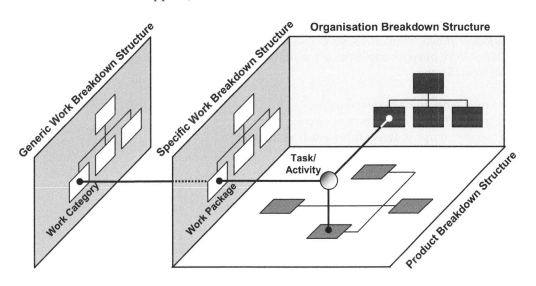

Figure 11.5   Relationship between Work Breakdown Structures, Organisation Breakdown Structure and Product Breakdown Structure

However, every attempt should be made to match the OBS wherever possible with corresponding parts of the PBS and BPA, for example. This eases greatly the communication about component responsibilities with regard to time, cost and quality objectives as well as the association of team members with 'their' components.

An OBS breakdown logic could look like this:

- Head of Company;
- Head of Programmes;
- Head of specific programme;
- Head of high-level multi-functional team;
- ...
- Head of lowest multi-functional DBT;
- Engineering key representative therein;
- Individual member of the Engineering community therein.

The Cost Breakdown Structure (CBS)

All activities generate costs. Any WBS is therefore also strongly linked with the CBS. The latter represents a breakdown of cost into cost categories which are suitable to satisfy the needs of project control. The breakdown also often has to satisfy the needs of the Cost Estimation department, which largely draws its cost estimates for future aircraft development projects from the experience with past projects.

Breaking down the costs of an aircraft development project may be required to be according to completely different criteria compared to the Breakdown Structures mentioned above. Thus, the CBS is usually not just a copy of, for example, the PBS or the WBS. Very often, though, there are similarities. For example, a CBS breakdown logic could look like this:

- total work to be performed;
- production of development aircraft;
- major component assembly (such as the wing);
- company;
- section (such as the Fixed Trailing Edge)
- team;
- processes used;
- work category;
- cost category.

Thus, it consists of a mixture of architectural levels found in the specific WBS, the BPA, the OBS, the GPA, and maybe others. As a result, some of the non-financial attributes mentioned in Chapter 10, which were introduced to be able to aggregate project-related costs according to a multitude of queries, are replaced by the modules of the CBS. In any case, the lowest level of a CBS established to help controlling projects only consists of the two cost categories described in Chapter 10, that is, (a) the category, which is based on reference units, such as hours, booked against a certain activity or milestone, and multiplied by the applicable charging rates as well as (b) the category describing costs directly charged against cost units.

STEPS LEADING TO AN INTEGRATED ARCHITECTURE

Introduction

Historically, each Function participating in a project used its own set of architectures. For example, Project Management usually used the WBS and OBS as a starting point, Management in general the OBS alone, Engineering the PBS or FBS or SBS, Manufacturing the BPA and IBS, Finance the CBS and Quality the GPA.

However, with the above descriptions of the various architectures it should have become clear that there is not a single way of grouping and structuring project and product information. Different data and information is still required to cater for the needs of Engineering, Manufacturing, Procurement, Finance, Project Management Office (PMO) and so on.

> 'The many different functional specialists who contributed to a project each broke the project down in different ways, to suit their particular needs. Many differing frameworks were developed, and still exist, for planning, estimating cost and other resources, budgeting, cost accounting, financial analysis, assigning responsibilities, purchasing, scheduling, issuing of contracts and subcontracts, material handling and storage, and many others. Often these frameworks are themselves different for the different functions of Finance, Marketing, Engineering, Procurement, Manufacturing, …, and Operations, to name a few. Bitter experience on many large projects showed that it was impossible to properly correlate and integrate the planning and control information on a large project in this hodgepodge of differing definitions of the same project.'[5]

Despite the varying needs of the different functions, if we want to promote an integrative management approach to aircraft development we still need to look at possibilities to integrate the traditional architectures into one architecture. For any given project, what is needed is a common project Breakdown Structure that all Functions can understand and agree to, and to which all the traditional frameworks can be linked to. In this and the following chapters it will be explained step by step how this can be achieved.

However, what can already be said at this point is that the variety of architectures clearly demonstrates that architectural decisions are not simply technical decisions that should be left to the Engineering department. They impact on the very essence of business and should only be made through integrative management drawing on the full expertise of Engineering, Manufacturing, Procurement Finance, Project Management, and so on . In other words: Project architectures should be discussed in multi-functional teams.

> 'Unfortunately, even in … sophisticated companies architecture is too often seen as an Engineering issue. It becomes a technical decision made to maximise product performance or minimize product cost. In reality, this fails to exploit the true potential of good architecture to accelerate development. Managers must see architecture as one of their primary management tools.'[6]

---

5 Archibald, R.D. (1992), *Managing High-Technology Programs and Projects*, 2nd Edition, (New York: John Wiley & Sons), pp. 194–5.
6 Smith, P.G. and Reinertsen, D.G. (1998), *Developing Products in Half the Time*, 2nd Edition, (New York: John Wiley & Sons), p. 115.

Note that not only an integrated architecture needs to be set up but that changes to it need to be kept under control too. This important aspect of project architecting will be explained in Chapter 12. For the time being, there is enough to be discussed about how to integrate architectures even without the aspect of change control.

The Concept of Views

Right from the start of an aircraft development in today's technological environment, the intention is that all relevant Functions such as Engineering, Manufacturing and Procurement work together in order to create as soon as possible an integrated project architecture. An architecture developed purely by Engineering without consulting, say, Manufacturing, would not be helpful in supporting this intention. With the help of today's Information Technology (IT) capabilities, the modern approach towards an integrated architectural design is to regard the above mentioned classical architectures (as well as their more sophisticated modern variants) as different *views* of one and the same integrated architecture.

The different views are mainly used:

- to provide different logical navigation routes to access project and product information, such as requirements, specifications, definitions, drawings, models, Conditions of Supply (CoS), test and justification information, operational and maintenance information, schedule plans, cost information, organisational information and so on;
- to handle the necessary data appropriately and according to the viewing needs of users; as well as
- to ensure data traceability.

Let us use the example provided in Figure 11.6 to illustrate this. View 1 may show a part of the BPA which could, for example, describe the situation in the aircraft's FAL.[7] In this example, the wings are attached to the centre fuselage before the forward and aft fuselage are added to complete the fuselage/wing assembly of the aircraft.

However, the design engineers would not use this breakdown. They would first start with the entire aircraft to sort out overall issues like aerodynamics, weight and so on before breaking the aircraft down into its major assemblies like fuselage, wing and so on. They would, for example, use something like View 2 applying the classical product breakdown in a top to bottom approach, which lacks the dimension of time. Designers need a different view compared to the Manufacturing people!

The Costing people who created the Business Case as well as the project's cost controllers might need yet another view. With View 3 in Figure 11.6 it is assumed that costs need to be controlled individually for the wings, the centre fuselage as well as the aggregate of the forward/aft fuselage. View 3 is again different from the other views.

View 2 drawings and View 3 costs, respectively, are associated with certain modules, which *could* represent real component assemblies of the aircraft. This is an important point: In order to generate Views 2 and 3 we needed to have something in mind to which

---

7   Although the BPA is a Sequence Architecture it is represented in Figure 11.6 in the format of a Breakdown Architecture for reasons of better comparison.

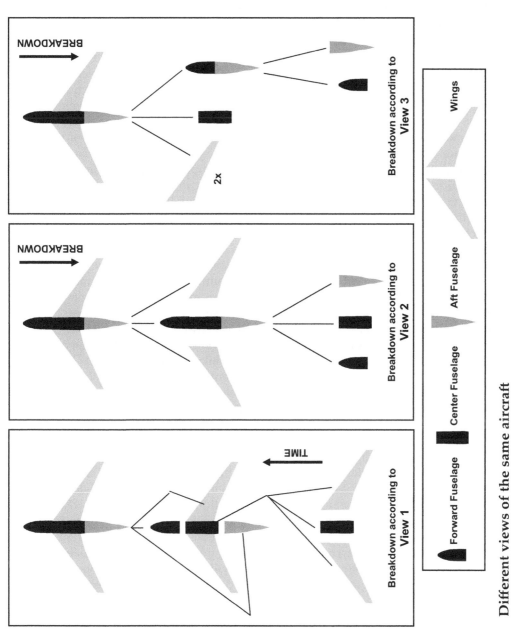

Figure 11.6   Different views of the same aircraft

we could attach data like drawings and costs, although in reality this something may perhaps never exist as a physical piece of hardware. The complete fuselage alone as in View 2, for example, will never exist as real, stand-alone hardware, as can be seen in View 1 of Figure 11.6. The lesson learned from this is that views need to be based on something which easily captures the imagination of the human mind, in particular when related to the product, but that this something most probably has no physical representation as a stand-alone object.

However, all views in Figure 11.6 make use of the same basic modules forward, centre and aft fuselage as well as left and right wing. Whatever the view, the trick is that it is constructed using certain of the same modules (and their lower-level structures) which in fact do have a physical stand-alone representation. Figure 11.7 describes how the different views of Figure 11.6 could be generated starting from common modules. Note that common modules must be selected such that there are not too many of them and that they are robust against (design) changes. Both requirements ensure that an already complex architecture does not become even more complex. This is a difficult challenge which requires a significant amount of experience. In fact, the identification of harmonised constituting modules represents important company know-how.

Note that each new view requires additional higher-level modules in the common architecture. For example, View 1 requires the module 'Wing plus Centre Fuselage', while View 3 requires 'Forward & Aft Fuselage', see Figure 11.7. But within the same architecture, the top module (here: 'Complete Aircraft') remains the same, whatever the view. Note that for the concept of views – requiring additional higher-level modules – it is necessary to accept that there can be more than one incoming *and* more than one outgoing link per module. This is, for example, required if the same higher-level module is called by different views. One now also finds modules with exactly one incoming *and* one outgoing link per module. In other words: with the concept of views, Breakdown Architectures and Sequence Architectures become structurally very similar indeed.

Type I attributes attached to the modules will identify which module belongs to which view. With the help of these attributes, different architectures can be selected, and users can select and view specific information and data in a structured way and according to their viewing needs.

With sufficient detailed granularity, that is, with sufficient lower-level modules existing, other new and higher-level modules can be generated, to which any of the existing lower-level modules can be interlinked. This allows for the generation of yet again new views for the purpose of different aggregation or summation of data attached to the modules. Thus, if one wants to view information in an entirely new way, one has to create new, view-specific modules, to which the existing common modules are linked to. These new modules are always on levels between the existing common modules and the one module at the top. Alternatively, existing common modules could be further sub-divided. In this case, the so far common modules will no longer be common. Instead, the newly created lower-level modules become the common ones. Or a mixture is chosen involving both new view-specific modules as well as breaking down already existing common modules. Whatever needs to be done to integrate the classical architectures into one, it should always be possible to view the resulting single architecture from different perspectives.

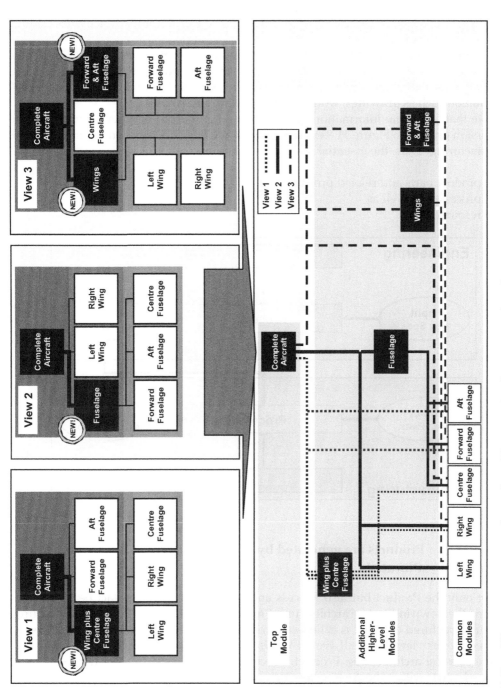

Figure 11.7   How to generate views (see Figure 11.6 for comparison)

The Building Blocks of an Integrated Architecture

If we want to integrate architectures into one single architecture while enabling different views we need to reconsider what the building blocks are of all company activities leading to the generation of products:

Outputs (such as assembled products) are generated from inputs (such as parts) by processes, to which resources (people, money, machines and so on) are allocated using adequate information (general process descriptions, specific 'input → output conversion' information such as drawings, work orders, software and so on).

Note that 'adequate information' usually is itself an output of other input → output conversion processes. Figure 11.8 explains this for the interlink between Engineering and Manufacturing. Thus, the essential building blocks are:

- product parts and related product data;
- processes; as well as
- resources.

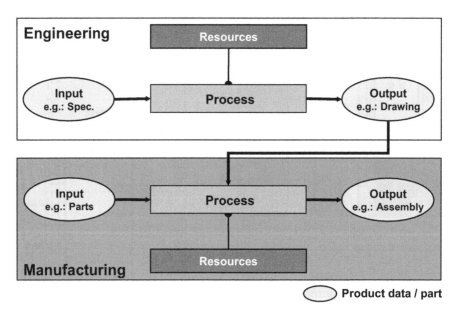

Figure 11.8   Products are generated by Processes to which Resources are allocated

Not only the Product but also Process and Resource data can in fact be structured, yielding their own individual architectures.[8] In principle, if Product, Process and Resource data are structured on what is called an Architecture Base, one can extract the classical architectures as views on the 'Level of Views'. This is depicted in Figure 11.9. Note, that for the classical architectures, Product, Process and Resources data are not interlinked with each other within the Architecture Base.

---

8   For the time being, it has not yet been described what these architectures consist of and how they look like. This will be added below.

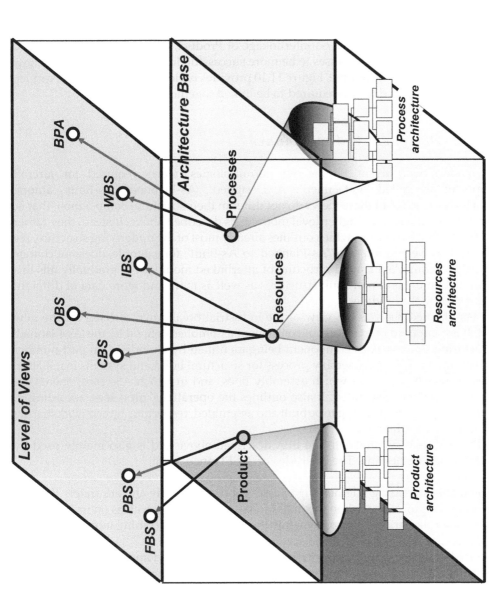

Figure 11.9   Classical architectures comprehended as views of Product, Process and Resources architectures

Now, instead of trying to find *on the Level of Views* a blend of:

- Product-related Breakdown Structures like the PBS or FBS with
- Resources-related Breakdown Structures like the OBS, CBS as well as with
- Sequence Architectures where the dimension of time introduces a dynamic element, such as the BPA,

an integrated architecture based on interlinkage of Product, Process and Resources data *within the Architecture Base* promises to be more successful. In a moment we will see how this can be done. In the meantime, Figure 11.10 provides typical examples where Product, Process and Resources data are required to be linked together.

From 'As-Designed' to 'As-Planned'

Using modern Concurrent Engineering terminologies, views required for aircraft development are called 'As-Defined', 'As-Designed', 'As-Planned', 'As-Built', among others. However, some of these views do not differ in their structural composition, that is, call for different lower- and higher-level modules as described above. Instead, they rather evolve from a first basic view, which outlines already most of the underlying structure, see Figure 11.11. The evolution from 'As-Planned' to 'As-Built', for example, does not change the already existing 'As-Planned' structure of interlinked modules but gradually fills this structure with more and more sub-structures as well as more and more data of different types attached to the modules.

The 'As-Planned' view is the view of the industrial breakdown selected for the new aircraft. It is composed of a cascade of component assemblies defined by the 'As-Planned' manufacturing process, each component being identified by an individual part number (P/N). This manufacturing assembly process for structural build and systems installation outlines, in which plants, on which assembly lines, and in which ASs components are going to be built and assembled. It also outlines the operations' processes, according to which the components are going to be built and assembled, respecting agreed worksharing between companies, if any.

'As-Planned' is very similar to the BPA described above and is also mainly used by Manufacturing Engineering. The basic idea of 'As-Planned' is to:

- help Manufacturing Engineering to assess and plan deeply and accurately the industrial process based on the definition of the aircraft as well as on its build and manufacturing philosophy while this definition is becoming more and more mature; as well as to
- put Manufacturing engineers in the position to be listened and to discuss the 'As-Planned' structure with Engineering before start of the Development sub-phase.

The 'As-Planned' proposed by Manufacturing Engineering is a key capability to support Concurrent Engineering and to reduce cycle time.

On the Engineering side, it is the 'As-Defined' view, which, in fact, represents the starting point of the evolution of Engineering views. It comprises a first set of Engineering information such as space allocation, frontier models, Master Geometry and design principles used to investigate different aircraft concepts.

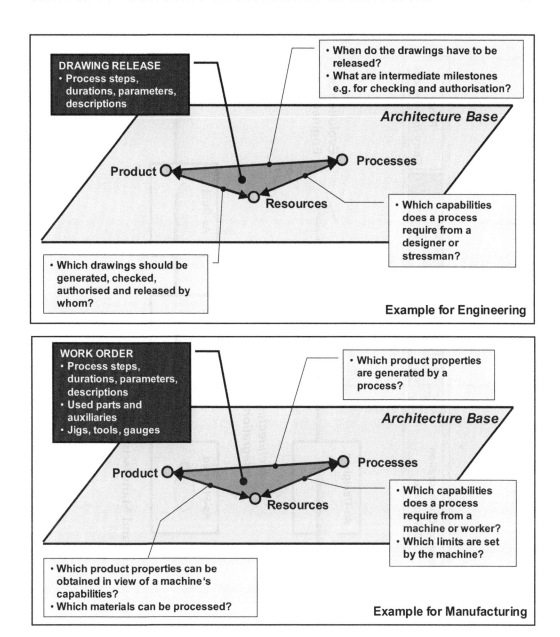

Figure 11.10   Examples of how Product, Process and Resources data is required to be linked together in Engineering (above) and Manufacturing (below)

General idea and Manufacturing example adopted from: Klauke, S. (2002), *Methoden und Datenmodell der Offenen Virtuellen Fabrik zur Optimierung simultaner Produktionsplanungsprozesse* (Düsseldorf: VDI Verlag, Fortschritt-Berichte VDI) Reihe 20, Nr. 360, p. 50, with permission.

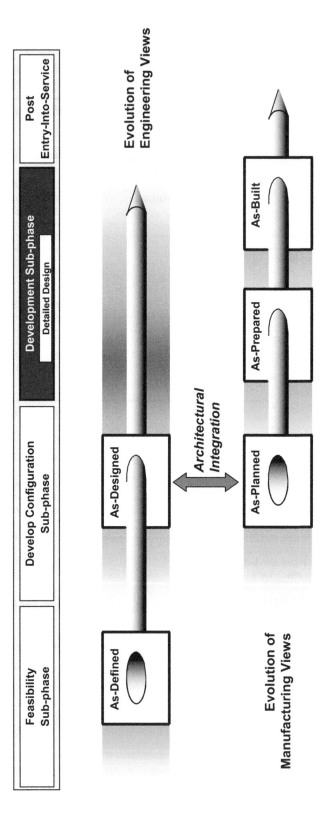

Figure 11.11 Evolution of Engineering and Manufacturing views

However, as the establishment of the 'As-Planned' view takes time, architectural integration of the Engineering and Manufacturing streams of views can only happen at a later than 'As-Defined'-stage. In addition, compared to 'As-Planned', 'As-Defined' has significantly evolved in terms of components and parts on its lower architectural levels, and it is only at these levels where integration with Manufacturing aspects really becomes both necessary and possible. This is why integration usually happens between the 'As-Designed' view on the one hand and the 'As-Planned' view on the other, see Figure 11.11.

The 'As-Designed' view is the view of Engineering for structural as well as systems installation design. It is used to manage the Digital Mock-Up (DMU) introduced in Chapter 2 and allows navigation through the DMU to individual components or aggregates of components. 'As-Designed' and 'As-Planned' are the two most important views for achieving an integrated architecture for the entire aircraft life-cycle. The following explanations therefore concentrate on these two views only.

It is important to note, that both 'As-Designed' as well as 'As-Planned' are not used with the same intensity or frequency at all times. Figure 11.12 shows the principle of how the intensity usually changes during the Development sub-phase.[9] There is a limited 'window of opportunity' for engineers from Engineering and Manufacturing (and maybe other Functions) to work concurrently to establish the founding blocks of an integrated architecture. It occurs just before and at the beginning of the Development Sub-phase.

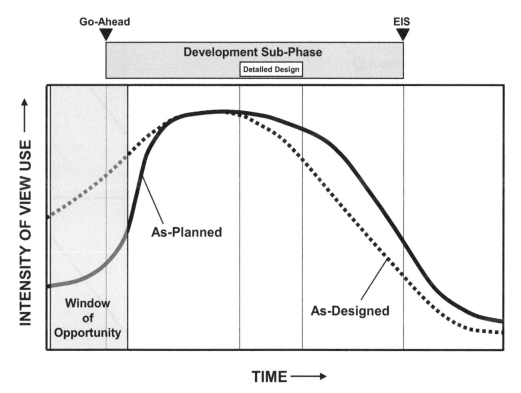

Figure 11.12   The use of 'As-Designed' and 'As-Planned' views during the
                Development sub-phase (principle)

---

9   Similar curves could be drawn for all other views in use.

Integrating 'As-Designed' and 'As-Planned'

Neglecting that Engineering usually has created earlier (that is, more upstream) views than 'As-Designed', the structure of the Product architecture within the Architecture Base as shown in Figure 11.9 would in fact initially be created by Engineering according to its 'As-Designed' considerations. This is shown in Figure 11.13. As a result, this 'As-Designed' view would be identical to the Product architecture in the Architecture Base. However, let us denote this view as 'As-Designed^P' to emphasise that it contains for the time being only Product information, that is, no Resources or Process data can be extracted from this view.

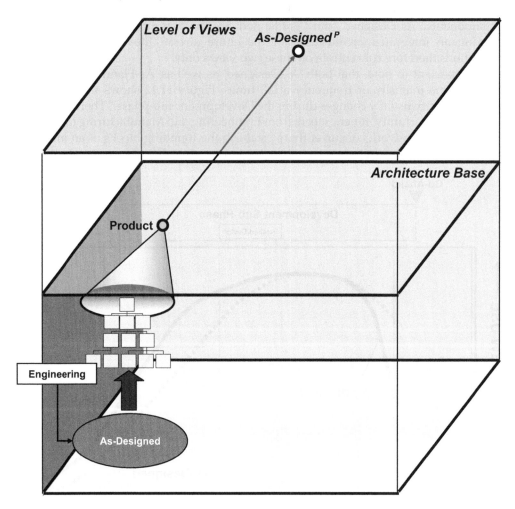

Figure 11.13   Initial Product architecture based on As-Designed considerations

This 'As-Designed^P' view can now also be used by, say, Manufacturing Engineering to scan, search and consult the Product structure in order to check whether it fits to its Processes and Resources. It probably does not, as the Engineering view is more based on a product or functionality breakdown while the Manufacturing view is based on how to assemble the aircraft components. But they need to work together to be able to continuously exchange data along the maturity cycle of the aircraft. So, something has to happen.

To exploit the benefits of Concurrent Engineering, Design engineers and Manufacturing engineers need to synchronise with each other, see Figure 11.14. Besides discussing and sorting out other important topics, they also must discuss within the multi-functional DBTs how the Product architecture should look like to satisfy both their needs. Engineering is in charge of defining all the design aspects of the aircraft. Manufacturing engineers have to define the way the aircraft can be built, checking, among others, the technical and industrial feasibility as well as the sequence of individual manufacturing step. Manufacturing engineers also have to define how it can be built efficiently with appropriate resources (facilities, skilled staff, jigs & tools, transportation means and so on).

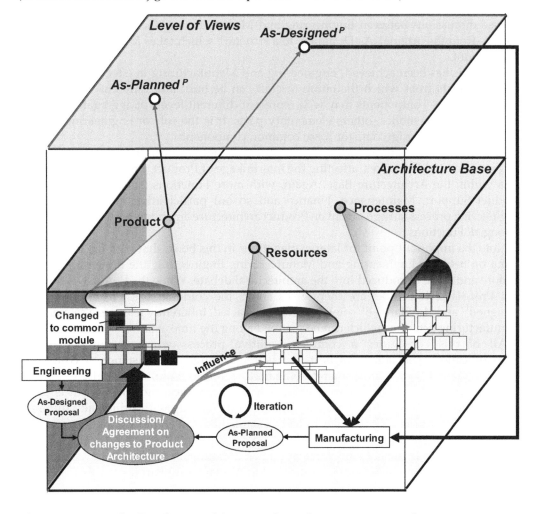

Figure 11.14   **The Product architecture based on agreements between Engineering and Manufacturing**

After having investigated Engineering's initial 'As-Designed$^{P'}$ view, which is identical to the initial Product architecture, Manufacturing Engineering will propose changes to the initial 'As-Designed'-based Product architecture in the Architecture Base. The motivation for its proposal stems from the need to be able to extract views which reflect the sequence of individual aircraft manufacture steps. Engineering must respond to this proposal und might come up with a new Product architecture compared to the initial one, which may now feature changed or added modules. However, the new Product architecture would still allow Engineering to extract its 'As-Designed$^{P'}$ view. Manufacturing Engineering will analyse the new architecture and its viewing possibilities and may still propose new changes after all.

This iteration process needs to continue until both the 'As-Designed$^{P'}$ as well as a new 'As-Planned$^{P'}$ view can be generated from a common base. As a major result of all the intense discussions between Engineering and Manufacturing, the Product architecture evolves from the original 'As-Designed' to a structure which takes Manufacturing needs into account.

Once this has been achieved, Engineering and Manufacturing in effect agree on a set of components from which the entire aircraft can be built and different views can be extracted. These components may well represent different levels of aggregation, that is, some may be assemblies, others elementary parts. It is the role of Engineering then to generate detailed design data for these common components.

In addition, the iteration process might result in changes to the initial static architectures for Processes and Resources, affecting the interlinkage of Product, Process and Resources data within the Architecture Base. Again, with more Functions (such as Procurement, Product Support, Maintenance, Finance and so on) participating in the Concurrent Engineering process, an agreement on Product architecture amendments has to be reached among *all* Functions.

Note the important point for later explanations in this book, that with the discussions going on between Engineering and Manufacturing Engineering, the dimension of time is more and more introduced into the architectural debate. While the Product, Resources and Process architectures are themselves static, the common base, from which 'As-Designed$^{P'}$ and 'As-Planned$^{P'}$ views are to be extracted, inherently features a sequence of manufacturing activities which can be planned along the time axis.

All of this constitutes a complex iterative process which requires excellent communication. The latter can best be assured through the existence of co-located multi-functional teams which therefore again turn out to be fundamentally important for integration. But what is also needed is a set of architecting rules which is to be provided to the teams. Otherwise each team could come up with principally different architectural solutions. These rules must be established and communicated before start of the development project.

To provide an idea of what Engineering and Manufacturing would have to discuss to find a common base an example is attached in Appendix A3 Part I.

COMMON SUB-STRUCTURES: PRODUCT & ASSEMBLY TREES

Introducing the Product & Assembly Trees

The product definition of the aircraft is now linked to the way it will be manufactured and assembled. In other words: the 'As-Designed$^{P'}$' and the 'As-Planned$^{P'}$' do have many but not all individual modules of the Product architecture in common.

> *The assembly processes described by the 'As-Planned' is usually based on past experiences and therefore consists of a relatively known series of phases, steps, tasks and activities that the organisation uses in creating the various components of the project. If done by a knowledgeable engineer, the 'As-Designed$^{P'}$' will therefore anyway not be done without considering the assembly process, wherever possible. It would not be wise to establish the interfacing boundaries between the elements and components of the 'As-Designed$^{P'}$'-view completely independent from the interfaces created by the assembly process.*

If carefully constructed, the agreements between Engineering and Manufacturing will lead to many sub-structures in the Product architecture where modules are common to both the 'As-Designed' *and* the 'As-Planned'. Where elements are common, the sequencing of the manufacturing operations can be allocated to the product data (such as drawings) and vice versa. Where they are uncommon, they are used to satisfy the different users' viewing needs.

Each of the common sub-structures is therefore more a *joint* Product and Assembly Breakdown Structure or – perhaps more imaginative – a Product & Assembly Tree (PAT).[10][11] The top module of a PAT could be located anywhere within the Product architecture, that is, wherever it would make sense to ensure that below this module the modules for 'As-Designed$^{P'}$' and 'As-Planned$^{P'}$' are common. This is schematically depicted in Figure 11.15.

> In Figure 11.15 the examples of Figure 11.7 are used. The top modules, below which one could image individual PATs common to all three views of Figure 11.7, are Forward, Centre and Aft Fuselage as well as Left and Right Wing.

It is essential for the establishment of an integrated architecture to create as many PATs as necessary to have sufficient sub-structures common to both 'As-Designed$^{P'}$' and 'As-Planned$^{P'}$', respectively.[12]

With the introduction of PATs, views requested by different Functions are no longer isolated from each other (as they were traditionally) but build themselves up from the common sub-structures using Type I attributes. In addition, as it will still be described below, the PATs also represent the platforms where the interlinkage of Product, Process and Resources data is taking place.

---

10 Although the 'tree' is upside down.

11 Note, that in some companies the PAT is called ADAP as it uses modules common to 'As-Designed' and 'As-Planned'.

12 The same applies to other views, for example views needed by Procurement, Product Support, Maintenance, and so on. The PATs should be common to all of them.

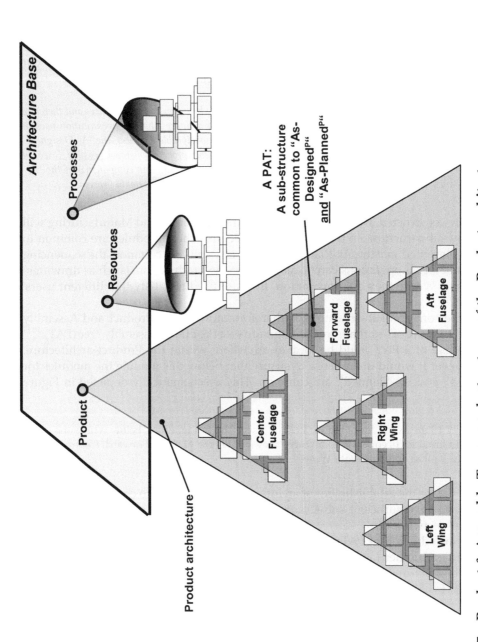

Figure 11.15 Product & Assembly Trees as sub-structures of the Product architecture
Note: They are common to both 'As-Designed[P]' as well as 'As-Planned[P]' view. Compare this figure
with Figure 11.7.

Once all necessary Process and Resources data has been linked to the common Product architecture, Manufacturing can generate an 'As-Planned^{PPR}' view – also called 'As-Prepared' – see Figure 11.16. 'As-Planned^{PPR}' is a Manufacturing view combining Product, Process and Resources information maturing along the development of the aircraft. It describes manufacture and assembly activities in all details for each stage. While it does not encompass scheduling and logistics aspects, but purely Manufacturing constraints only (such as sequence of tasks or activities to be performed, Resource data and so on), it enables the calculation of routings for specific aircraft to be scheduled in the future. As a result, individual MSOs and ASOs of the 'As-Planned^{P}' view, respectively, have now evolved to specific routings into 'As-Planned^{PPR}'.

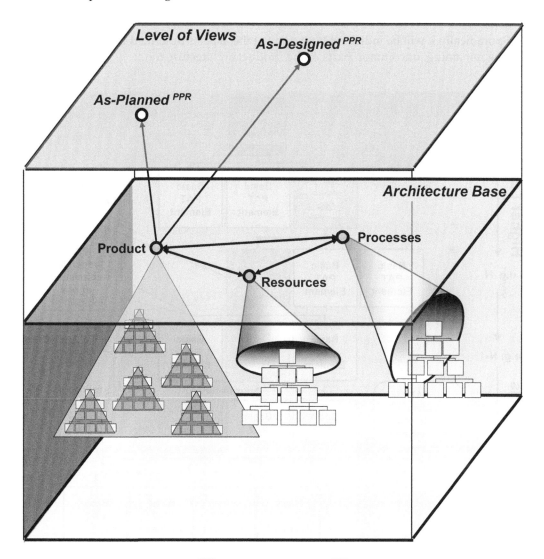

Figure 11.16   'As-Designed^{PPR}' and 'As-Planned^{PPR}' made possible through linking together Product, Process and Resources data within the Architecture Base

With the capability to allow for different views *and* for interlinkage of Product, Process and Resources data, PATs really become the backbone of integrative management structures for complex product development, with *all* Functions using the same common PATs and their attached data.

If the PATs are the 'secret' to integrated architecting, then it is worth learning more about them. This is why the remaining part of this chapter as well as Chapters 12 to 14 are to a great portion dedicated to them. It will be described in more detail:

- what PATs consist of;
- how the Product, Process and Resources data are interlinked; as well as
- how they can be used for an integrated management approach.

Sporadically it will be indicated how some of the aspects described for the PATs apply for the remaining, uncommon parts of the Product architecture too.

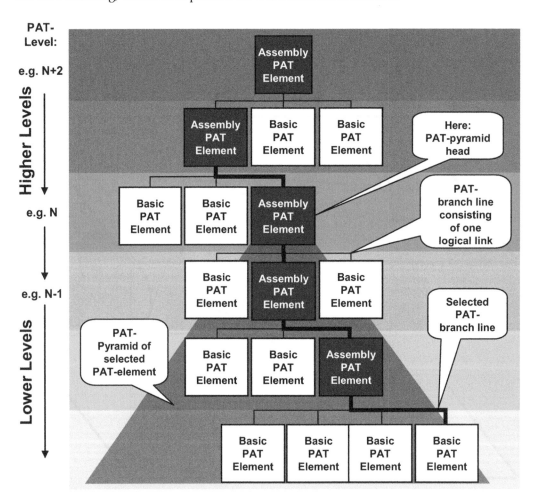

Figure 11.17   A Product & Assembly Tree in Breakdown Architecture
                format

Description of a Product & Assembly Tree

Figure 11.17 depicts a PAT in the format of a classical Breakdown Architecture: the PAT is a top-down structure, with a single PAT-element represented by a box on the top and a cascade of PAT-elements beneath. However, in order to better understand the principle of PATs, a few more definitions need to be introduced:

- PAT-elements, which represent basic (that is, single, non-assembled) components (such as machined parts, fasteners, other standard parts), are called *basic PAT-elements*. In principle, the entire final product can be broken down into basic PAT-elements.[13] However, a PAT established at one company will usually not include the breakdown into all basic PAT-elements of assembled components procured from other companies. The concept of a virtual single PAT stretching out to all suppliers ('Extended Enterprise' concept) should nevertheless be in the minds of all stakeholders, even if today it seems impossible to fully implement this concept in practice.[14]
- PAT-elements, which represent assemblies (or assembled components), are called *assembly PAT-elements*.
- Each element resides on what is called a *PAT-level*. All elements necessary to assemble a PAT-element on level N must be located on the lower level N-1.[15]
- A PAT-element on level N-1, forming part of the assembly of a PAT-element on level N, is logically linked together with the latter. This link is represented by a connecting line.
- The link between any PAT-element on the lowest PAT-level and any other PAT-element on a higher level (or vice versa) – even if there are many levels in between – is called a *PAT-branch line*. A PAT-branch line therefore consists of one or more individual logical links. Within the logic of an Extended Enterprise concept, PAT-elements on the lowest PAT-level of a given PAT-branch line are always basic elements.
- The pyramid shaped by the envelope of those branch lines, which start downwards from a selected PAT-element, and include those PAT-elements which are located on all the branch lines culminating in the selected PAT-element, is called *selected PAT-pyramid*. The selected PAT-element is the *head* of this PAT-pyramid.
- PAT-elements representing components which are subject to a testimony that they are produced as specified, are called Constituent Assemblies (CAs).[16] The latter are of high importance when considering component deliveries from one manufacturing site to another, in particular if two different companies or different legal entities of the same company are involved. They are very important PAT-

---

13 It is worth remembering that the entire architecture required during aircraft development may not only consist of the components required for the flying aircraft such as aerostructures, systems' elements and equipment, but also of jigs & tools, buildings, test and training facilities, ground equipment and so on.

14 In future, web-based IT solutions may offer the potential to extent the concept of an Extended Enterprise architecture to at least some of the major suppliers of a company.

15 As a result, many of the assembly PAT-elements on level N comprise PAT-elements on level N-1 which are assembly PAT-elements on their own.

16 The vast majority of Constituent Assemblies consist of assembled components, although basic components could also be Constituent Assemblies if delivered from one site to another.

elements because this is where conformity between 'Build' and 'Design' must be documented to gain the approval of the airworthiness authorities.

Note that every PAT-element on level N has a potential *physical* interface to any of the other elements – not only with those on level N, level N+1 and level N-1.

About Product & Assembly Tree elements

In complex projects such as commercial aircraft development, large amounts of data are generated and need to be managed. This data can come in many forms such as requirements and specifications, drawings, graphics, databases, schedules and so on. If it is not organised and managed, the multi-functional teams will soon be lost in chaos. But with the power afforded by today's computer systems navigating through an agreed architecture, it is much easier to manage large amounts of data on behalf of the entire project.

Thus, while the principal power and diversity of modern IT systems was described in Chapter 2, we now have to look into the possibility of using the PAT as a navigation system for all kinds of project-relevant data. As every PAT is a sub-structure of the Product architecture enabling – among others – 'As-Designed' as well as 'As-Planned' views, the data attached to its elements can be relevant to design *and* production of parts and assemblies, respectively.[17]

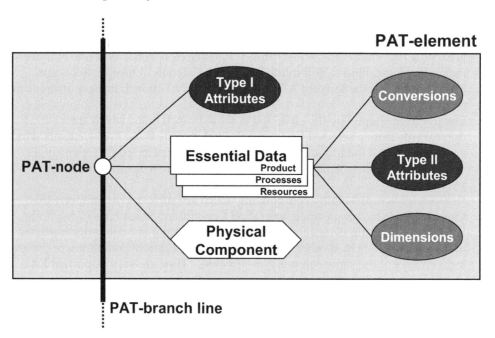

Figure 11.18   Attachments to a Product & Assembly Tree node

---

17 Does this mean that data cannot be attached to the uncommon modules of the Product architecture? No, it does not. It only means that data attached to the view-specific modules cannot be accessed by all functional views.

In the PAT, each element is nothing more than an empty box – a node, to which data can be attached. Thus, the PAT does not store the data itself but provides the access path to it. The PAT only provides a navigation path from any direction to the PAT-elements. As can be seen in Figure 11.18, each PAT-element consists of:

- a point-like and content-free node representing the architectural aspect of the PAT-element; to which can be linked:
  1. the physical component, which the PAT-element represents. The PAT can, thus, be used to address individual physical components;
  2. Essential Data linked to the component represented by the PAT-element (such as 2D drawings, 3D models of the DMU, Condition of Supply data, requirements, reports, quantifiable data such as weight, cost and so on);
  3. Essential Data describing the processes of how to design, manufacture, test and so on the component (for example schedules, Engineering process descriptions and so on);
  4. Essential Data describing the resources needed;
- data describing the links to other nodes in the architecture (Type I attributes), thus defining the precise position of the PAT-element within the architecture.[18]

In principle, higher-level PAT-elements, into which lower-level PAT-elements culminate, will show the compilation, aggregation, summation and so on – whichever applies – of the lower-level PAT-elements' Essential Data.[19]

Next to the Essential Data, there might be additional data about Essential Data, called meta-data. One should be able to distinguish here between the following meta-data:

- conversions;
- attributes; as well as
- dimensions.

Conversions allow for the transformation of quantifiable Essential Data from one format or unit into another, for example, weight in kg converted into lbs, cost in $ converted into €, and so on. Changing the formulae to calculate conversions does not change the Essential Data or attributes.

Adding attributes to an Essential Data neither changes the data itself nor the conversions. They rather add information to and around the Essential Data, usually for the purpose of filtering, searching and aggregation. Attributes attached as meta-data to Essential Data are called Type II attributes. They must be distinguished from Type I attributes. As was said before, Type I attributes are needed to generate views such as 'As-Designed' and are therefore not attached to Essential Data but to the PAT-nodes directly, see Figure 11.18.

---

18 In this book, PAT-elements are mostly represented by boxes whenever it is important to look at the entirety of the PAT-element (that is, the node, the component and the data attached to the component together). Where the differentiation between the node on the one hand and the component and its attached data on the other is necessary for the understanding of the text, these items are depicted separately.

19 However, one has to distinguish between those sets of data where there is a one-to-one relationship between each PAT-element's node and its attached Essential Data and those where there is a many-to-one or a one-to-many relationship. This is because they have different consequences when
- compiling non-quantifiable data (such as text, drawings, elementary 3D models);
- aggregating non-quantifiable data into larger units (for example elementary 3D models into larger DMU aggregates); and when
- summing-up quantifiable data (weight, cost, and so on).

However, the difference between Type I and Type II attributes is that Type II attributes comprise Product, Process or Resources meta-data while Type I attributes do not. In addition, filtering by Type II attributes leaves the selected view of the architecture unaffected while changes to Type I attributes changes the view.

Examples of Essential Data with different dimensions include cost information for various years, drawings in various release status and so on. The dimensions in these examples are the year and the release status, respectively. All attributes and conversions being equal, moving from one dimension to another could bring about changes to the Essential Data. For example, the project's cost for year x could be different from the one of year y.

In Appendix A3 Part II it is depicted in 'PAT terminology', what Engineering and Manufacturing each proposed for the Architecture Integration Example mentioned above. One will notice that the two Functions require different numbers of PAT-levels. The Manufacturing approach turns out to be more complex compared to the Engineering approach, but it does reflect its build process in all detail.

Having explained now the structure of the PATs as well as what PAT-elements consist of it is important to understand that both the structure of any architecture, as well as the attachments to individual nodes can (and very often do) change as part of the normal development evolution. This change needs to be kept under control otherwise the whole concept of an integrated architecture used by multi-functional teams becomes useless. And every time the Product architecture needs to be amended, a concurrent way of working between Engineering and Manufacturing is again required. Change control is therefore an essential part of Integrative Management. That is why we will now turn our attention to it.

MAIN CONCLUSIONS FROM CHAPTER 11

- Classical Breakdown Architectures and Sequence Architectures allow for an *organised* collection and sharing of business and technical information related to a project and product.
- What is needed is a common project architecture that all Functions can understand and agree to, and to which all the traditional frameworks can be linked to.
- Architectural decisions are not simply technical decisions that should be left to one Function only. As they impact on the very essence of business, they should be taken by multi-functional teams.
- The modern approach towards an integrated architectural design is to regard the different classical architectures used in projects as different *views* of one and the same integrated Project architecture. Views represent snapshots of an as-is situation a viewer wants to see.
- If Product, Process and Resource data related to a project are structured on what is called an Architecture Base, one can extract the classical architectures as views on the Level of Views. Then, instead of trying to find *on the Level of Views* a blend of the Product, Resources and Sequence Architectures, an integrated architecture can be achieved by interlinking Product, Process and Resources data *within the Architecture Base*.
- A multi-functional iteration process needs to be set up to identify a set of common modules, from which all required views can be generated.
- During this iteration, the dimension of time is more and more introduced into the architectural debate as the common base from which views are to be extracted brings with it timely sequenced activities. The Engineering-dominated product definition of the aircraft is then linked to the activities required to design, manufacture, assemble, test and deliver it.
- Sub-structures containing common modules are called Product & Assembly Trees. Each Product & Assembly Tree consists of a point-like and content-free node, to which a variety of data and attributes can be attached.

12

Controlling Configuration Change

INTRODUCTION

Why the Control of Change is so Important

Aircraft are built to contractually binding specifications. These include:

- the standard definition of the aircraft;
- any options which the customer may have selected from the options catalogue; as well as
- specific customer requirements to be embedded in the delivered configuration.

As we saw earlier, commercial aircraft consist of hundreds of thousands, if not millions of parts. In addition, there are numerous related tooling, fixtures, gauges, templates, test equipment and so on. Thus, with a different mix of options selected, but certainly with additional specific customer requirements embedded (for example in the cabin), it may well be that no aircraft is identical to another, even if it were designed to the same standard definition.[1]

In addition, there are numerous other reasons why an aircraft configuration changes. These changes occur frequently during the aircraft's Development sub-phase as well as – to a lesser extent – during Series Production. Root causes include – among others:

- change of specifications;
- change of regulatory requirements;
- corrections of drawings or other Engineering documents;
- corrections of usability, reliability or safety problems;
- removal of bug or product defect;
- improvement of performance or functionality;
- new producibility aspects;

---

1   Boeing, for example, has introduced what it calls Tailored Business Streams (TBS) to arrive at simpler, reusable and more cost-effective processes and solutions. TBS distinguishes between parts and processes that go into every airplane to manage product variation (TBS 1), parts and processes that are reusable (including options that are available for a customer to order; TBS 2), as well as parts and processes that are unique, custom designed or need special tooling, whose design is not meant to be reused (TBS 3). The idea is that Engineering does not need to get involved again for TBS 1 and 2 parts and processes, once they have been set up initially and have proven to be stable.

- necessity to reduce cost;
- enhancement of installation, service, or maintenance; as well as
- change of materials.

Thus, the standard definition as well as the options catalogue may evolve over time.[2] By consequence, an aircraft's configuration must be defined as one of a series of sequentially created variations of the aircraft's standard, whereby each of the aircraft's performance, functional and physical attributes may vary.

For an analysis of the implications of change it is important to understand the two different classes of change. Changes which affect an item's form, fit or function can be denoted as Class I changes. They affect specifications, weight, interchangeability,[3] reliability, safety, schedule, cost and so on of an item. Changes to correct documentation or changes to hardware not otherwise defined as Class I are denoted as Class II changes. Class II changes could for example include changes to the sequence or to the calculated times of Manufacturing activities producing and delivering components, as this does not change their form, fit or function. All Class I changes affect in the end drawings and components as well as the architectural structure of the Product & Assembly Trees (PATs). Very often physical interfaces are affected also.

With a product as large as a commercial airliner, there are many thousands of changes during its design and build. Yet, over the entire life-cycle of the aircraft, the aircraft manufacturer must assure that the as-designed configuration at any point in time will satisfy all requirements and that the actually delivered hardware or software as-built configurations correspond to the approved as-designed configurations. As a result, the effort required to manage the configuration of an aircraft is tremendous.

Overview on Configuration Management

The discipline which manages this effort in a structured way is called Configuration Management. It is the Management discipline that:

- applies technical and administrative direction and surveillance to identify and document the functional and physical characteristics of the aircraft;
- controls changes to those characteristics;
- records and reports change processing and implementation status; and
- verifies compliance with specified requirements.

---

2   Note that a commercial aircraft may well be in production for 30–40 years as it is financially such a huge challenge to launch a new aircraft programme. Every aircraft-producing company is therefore sooner or later forced to adapt the aircraft's standard configuration in an attempt to keep it competitive during that long period. This is very different to, say, the automotive industry where required adaptations can be implemented a few years later when launching a new car programme with a new standard.

3   'Interchangeability is defined as when two parts possess such functional and physical characteristics as to be equivalent in performance, reliability, maintainability, etc. These parts should be able to be exchanged one for another without selection for fit or performance and without alteration of the item itself or of adjoining items.' From: Crow, K. (2002), *Configuration Management and Engineering Change Control* (Palos Verdes, CA: DRM Associates), http://www.npd-solutions.com/configmgt.html, accessed 10 October 2009.

Its purpose is therefore to establish and maintain consistency of an aircraft's performance, functional and physical attributes with its requirements, design and operational information throughout its life.

It is international practice to distinguish four different Configuration Management disciplines, see Figure 12.1. Configuration Identification (1) is the Configuration Management activity that assigns and applies unique identifiers to an aircraft, its components, and associated documents, and that maintains document revision relationships to the aircraft configurations.

Configuration Control (2) involves controlling the release of and changes to baselined products throughout their life-cycle. It is the systematic process to manage the proposal, preparation, justification, evaluation, coordination and disposition of proposed changes, approved by an appropriate level of authority. In addition, it ensures the implementation of all approved and released changes to the baseline into:

- the applicable configuration of the aircraft;
- the associated product information; as well as into the
- supporting and interfacing products and their associated product information.

Configuration Auditing (3) verifies that the aircraft is built according to the requirements, standards or contractual agreements, that the aircraft's design is accurately documented, and that all change requests have been resolved according to Configuration Management processes and procedures.

Configuration Status Accounting (4) involves the recording and reporting of the change control process. It includes a continuously updated listing of the status of the approved configuration identification, the status of proposed changes to the configuration, as well as the implementation status of approved changes. This configuration information must be available for use by decision-makers over the entire life-cycle of the aircraft.

In an environment where a large number of multi-functional teams operate concurrently it is particularly important to understand how Configuration Control works. This aspect of Configuration Management will therefore be the focus of the explanations below.

Control of Modifications

Not all changes affect the configuration of an aircraft directly as some are the consequences of others. Some are more upfront, others are more downstream. For example, change to an aircraft design might be a downstream change resulting from upfront changes like the change to a specification. Physical changes to parts and assemblies can be downstream changes resulting from changes to an existing aircraft design.[4] Arguably, design changes – also called modifications – lie in the middle of the whole chain of change, which stretches from the root causes of change to implementation of changes on the aircraft. Because design changes to the aircraft configuration are so important in terms of their technical and financial repercussions, and because they occur so frequently during the Development sub-phase, we will in this chapter concentrate on the management of the changes to the aircraft's *design*.

---

4   Other changes to physical parts or assemblies can, of course, result from production errors, such as misdrillings. However, these changes do not result from design changes and therefore are treated in a different way, usually by analysing, correcting and documenting them as concessions.

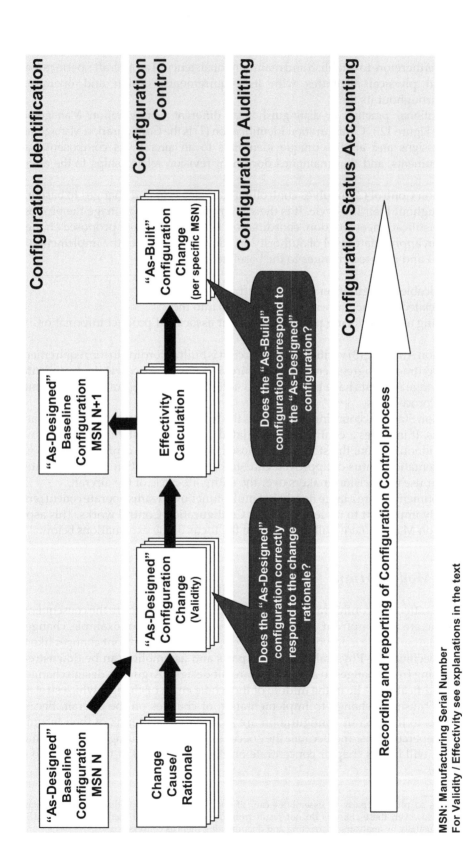

MSN: Manufacturing Serial Number
For Validity / Effectivity see explanations in the text

Figure 12.1   Simplified process chart for Configuration Management disciplines

Concentrating on the control of design changes is also justified by the fact that its management becomes very complex indeed in a concurrent working environment. This is because in a more concurrent process access to drawings must be given to many departments while Engineering is still working on it. In order to avoid complete chaos, different access hierarchies need to be introduced so that different people are not changing the same drawing at the same time. This control problem is a natural application for computerised data management tools, which is why these tools are so commonly used in concurrent design processes.

Typically, during the Development sub-phase, many thousands of engineering changes must be evaluated and processed. Controlling the resulting large number of changes to the aircraft configurations is a major challenge for the integrated management of any development project based on multi-functional teams. During the Development sub-phase, it essentially can be met by processing a proposed modification through two major phases:

- the phase leading to the approval of a modification;
- the implementation of the modification ensuring that the right parts go onto the right aircraft.

These two steps will now be described in more detail.

ACHIEVING MODIFICATION APPROVAL IN A MULTI-FUNCTIONAL TEAM ENVIRONMENT

Identifying Affected Stakeholders

The first essential step in achieving modification approval is the communication of the intended change to all stakeholders. It is to be identified whether a proposed change is actually justified on its own and whether it makes sense in the context of other implemented or proposed changes. This communication needs to reach *all* stakeholders and not only stakeholders who are perceived to be affected. Perceptions may well be wrong, even if they are expressed by senior or experienced individuals.

In the multi-functional team environment this communication therefore needs to reach *all* teams. With the complex matrix organisation in place, this is a big communication task and a leadtime challenge at the same time. It requires a preliminary description of the change, which on the one hand is detailed enough for the stakeholders to decide whether they are indeed affected or not. On the other, the description should not be too detailed in order to avoid unnecessary delays. With the tough pressures existing on development leadtimes, one has to stress that only a short amount of time can be made available for checking affectedness. Strict discipline and use of Information Technology (IT) tools are required to get the necessary feedback from all teams within a short period of time.

Where teams are affected, they must be given the opportunity to challenge the change. As teams are accountable for component delivery to time, cost and quality, any change potentially puts the achievement of these objectives in jeopardy. Successful mitigation of an intended but in the end unjustifiable change is obviously the most time and cost efficient option in controlling change. Unfortunately, changes are only too often taken for granted instead of being challenged. Team leaders and key functional representatives have here the possibility to control the project more efficiently but frequently fail to make use of this opportunity.

If more than one team will be affected by a change, then there may be conflicts arising from the fact that some teams are in favour of implementing the change and others are not. This classical conflict must be resolved and this can only be done at highest project level. Usually, a Change Control Board (CCB) is established for the project. It consists of the highest-level Project Manager, who may act as the chairman, as well as key representatives of the multi-functional teams, including representatives from Engineering, Manufacturing, Procurement and possibly other areas. This CCB takes all important decisions during the change control process. For example, it collects the feedback from all stakeholders to identify who is affected in order to decide whether the repercussions of a change need to be analysed in detail or not.

Analysing and Evaluating Change Repercussions

In fact, before the CCB can give the final 'go-ahead' for a change, the detailed repercussions need to be known. However, detailed repercussions can only be derived on the basis of a technical description of the change, which after all must be more detailed than the preliminary description used earlier.

'When a change is being evaluated, the following must be considered:
- How many of the old items are in inventory? Must they be scrapped or can they be used on other products or reworked? What is the cost to rework or scrap? Is the new item in inventory? ...
- How many of the old items are in work-in-process? Can they be reworked to the new configuration considering their current stage of completion, completed and used up before the change is effective, or must they be scrapped? Has production of the new items begun? What is the leadtime and cost for production of the new item? What is the additional leadtime for building tooling, fixtures and test equipment? ...
- Is the old item on order? Can it be cancelled or reworked? At what cost? What is the leadtime for procuring the new item? Are new suppliers required? ...
- [concerning] the impact on ... customers ... : What notification is required? How long will the process take? What documentation, manuals and catalogs need to be updated? What are the implications on spare parts requirements? ...
- Are the changes significant enough to require retesting? What testing needs to be performed? Does the product need to be recertified? What regulatory approvals are required?'[5]

The Engineering community within that multi-functional team, which proposed the change, kicks-off the in-depth repercussion analyses by describing the change in all its technical details. In particular it needs to describe what a component or assembly looked like before the change and how it should look like after implementation of the change. Let us call this technical description the Primary Technical Change Description (PTCD). It usually also includes a proposition for the intended Point of Embodiment (PoE) and the

---

5   Crow, K. (2002), *Configuration Management and Engineering Change Control* (Palos Verdes, CA: DRM Associates), http://www.npd-solutions.com/configmgt.html, accessed 10 October 2009.

modification's validity. The PoE identifies the *first* aircraft while the validity identifies *all* individual aircraft, where the change is supposed to be implemented.[6]

The PTCD is then communicated to all teams and Functions who had earlier declared their affectedness. In each of them, again the Engineering community needs to establish the detailed change description of their components and assemblies which change as a result of the primary technical change. Eventually, a Complete Technical Change Description (CTCD) describing the change for all affected areas of the aircraft is compiled. This compilation should again be done by the team who proposed the change. It is now possible to estimate the costs for the design changes, in particular the costs associated with the creation and release of new drawings and documents.

Concurrently – and also on the basis of the PTCD/CTCD – the Procurement communities within the multi-functional teams can estimate the repercussions of the change in terms of leadtime and cost if procured items are affected. This may involve contacting suppliers to get their quote on the repercussions. However, suppliers may not be contacted on a change-by-change basis as this approach may just too easily 'open the doors' for financial claims by the suppliers.[7] Therefore, Procurement often prefers to provide its own estimate of the repercussions on a change-by-change basis, and only enters into negotiations with a supplier once a batch of changes affecting the same supplier has emerged. The right strategy for managing suppliers with regard to design changes and their resulting claims is difficult to chose. In any case it depends on the actual change situation, the supplier as well as the contractual obligations, which both supplier and contractor have already entered.

With the PTCD/CTCD of the change *and* the potential input from Procurement at hand, Manufacturing can then analyse its cost and leadtime repercussions too. Again, this should happen within all affected multi-functional teams. Most importantly, Manufacturing will identify the earliest PoE, where the change could be implemented at (almost) no additional cost to Manufacturing and with no outstanding work at the next assembly site (in particular if it is located at another company or at the Final Assembly Line (FAL) of the aircraft, respectively).[8] If this Manufacturing PoE is later than the one suggested by Engineering (as mentioned above), and if the Engineering PoE must still be met, Manufacturing will have to estimate:

- its additional effort and cost required to meet the Engineering PoE while still trying to avoid outstanding work;
- or – where an internal recovery is no longer possible – the additional effort and cost for performing outstanding work at the site of the next Assembly Stage (AS);
- or – if time has already too much advanced for any recovery possibility prior to aircraft delivery – the cost for a post-delivery retrofit solution.

---

6   Note that the validity describes individual aircraft independent of whether these aircraft are produced in a strictly sequential series (one after the next) or not.

7   This very much depends on the contracts with the suppliers. Ideally, from a contractor's point of view all design changes should be absorbed by a supplier without any additional claims. However, in reality this will not be possible as usually only a certain amount of additional cost resulting from design changes is agreed to be absorbed by the supplier. Amounts above that agreed level will result in justified claims, although admittedly it needs some analysis to really find out what is justified and what not. The 'devil is in the detail'. However, negotiating a contract with a supplier is in any case a bet on the magnitude of design changes!

8   Performing outstanding work at assembly sites other than the one where the work was originally planned to be performed can cost 3–6 times more. This is because of the limited time available and the non-optimal arrangements for such work during the already tightly scheduled work at the other assembly site.

In the latter two cases, agreements must be reached with the teams or functional units who have to absorb the outstanding work or retrofit, respectively, in their area of responsibility.

As the change had been originally proposed by one team, it should be this team who gathers all the estimated repercussions and outstanding work agreements from all affected stakeholders, aggregates them and compiles them in a Change Repercussion Document (CRD). In addition, the team should develop an agreed and integrated change plan for the implementation of the modification. This plan should identify all of the required actions, the responsibilities, the timing and schedule as well as the associated resources. Ideally, it is a section of the CRD. In fact, less detailed versions of the plan should be developed much earlier as all affected teams need to have visibility on important milestones – and consequently work towards them – such as the date at which the CCB is expected to approve the modification.

On the basis of this document, the CCB is then able to take a judged decision whether to go ahead with a change or not, and when the change is to become effective. If it provides the 'green light' then the change enters the phase of modification implementation.

Challenges to the Control Process

The process described here is complex due to the large number of teams participating in the development effort. It therefore inherently bares the risk of adversely affecting leadtime. Different affected and possibly geographically distant teams have to work concurrently to achieve the modification approval. For all affected teams this means disruption of their mainstream activities. It is therefore difficult for the team who raised the change – and who consequently should steer the whole process as a modification owner – to push the modification through the process. This is, for example, because resources needed for configuration control may be under the leadership of other team leaders. They are therefore not so easily accessible to do work in the interest of the modification-owning team. The latter will encounter multiple conflicts of priority with other teams, which will be the more difficult to resolve the more teams are affected.

A theoretical solution to the management of the configuration change control process in a complex multi-functional team environment could be to appoint a dedicated expert modifications team. However, in practice this often does not work: the knowledgeable specialists needed for describing all technical, cost and leadtime repercussions of a given change are inevitably among those members of multi-functional teams already required to make them operate well.

The better solution is to ensure right from the beginning that the multi-functional teams are adequately staffed in terms of both quantities as well as skills. Already when the Business Case described in Chapter 4 is set up, it should be taken into account that a higher numbers of resources will be needed to perform the complex change control activities compared to the classical, functionally divided and therefore limited management practices. Well-prepared Business Cases do not only check a project's financial viability but also provide evidence that sufficient and sufficiently skilled resources can be made available for the project.

However, even with enough resources, there may be situations where there is not enough time during the Development sub-phase between raising a change and getting it approved by the Configuration Control Board. For example, modifications of utmost importance to ensure aircraft safety may sometimes have to be implemented at the last minute.

The assembly of aircraft components may by then already be very advanced. In such a case, the modification must go ahead even without an in-depth analysis of all cost and leadtime repercussions. But even if it is acceptable in some cases that detailed analyses of change repercussions are not performed, this must under no circumstances become an invitation for all other changes to also by-pass careful analyses. By-passing this aspect of the company's configuration control processes must be limited to exceptional cases and must be authorised by a senior person in the multi-functional organisation. In any case, as diverting from the process represents a high risk, a risk mitigation plan should be established. However, no compromises must be made when it comes to the proper documentation of the change, even for time-critical modifications (as it can still be done later).

Finding the right balance between insisting on process adherence and being flexible in exceptional cases is a major leadership challenge. This applies to all kinds of processes but not respecting the processes in the case of Configuration Management could really have serious consequences. In this case, therefore, full management attention is required. All in all, achieving modification approval in a multi-functional team environment remains a major challenge for every team and project leader.

MODIFICATION IMPLEMENTATION

Configuration Items

Once the modification has been approved by the CCB, it needs to be implemented. As was mentioned before, drawings, which affect the form, fit or function of the aircraft, must be attached to PAT-elements. Thus, every approved modification will affect the content of one or more PAT-elements or even the structure of the PATs themselves.

If, for example, the design of a basic component is modified, a new drawing needs to be established, that is, an attachment of the corresponding PAT-element changes. If assemblies change, not only do the drawings for some or all involved components change but potentially also the interface drawings between them, the assembly drawings as well as the drawings describing the interfaces between this assembly and other assemblies. In other words: the change propagates through the Product architecture. The number of affected PAT-elements, and the PAT-levels on which they can be found, depend on the nature of the change itself. A very careful analysis is required in order not to miss out any PAT-element. Clearly, this needs to be supported by the use of three-dimensional Digital Mock-Ups (DMUs) (see Chapter 2).

To manage the changes in the PAT-architectures, a standard methodology is applied. It is based on PAT-solutions (PS) attached to Configuration Items (CIs), see Figure 12.2. A CI is an aggregation of hardware (or software or both) that is designated for Configuration Management: it is treated as a single entity in the Configuration Management process. In a PAT, a CI is a node introduced on a PAT-line between two adjacent levels of the PAT-structure where the configuration status of:

- the PAT-pyramid below needs to be change controlled and is subject to Configuration Management; and where
- the PAT-structure above remains unaffected.

A selected CI represents a barrier which signals the finish line to any change tempting to propagate further upwards in the PAT-structure.

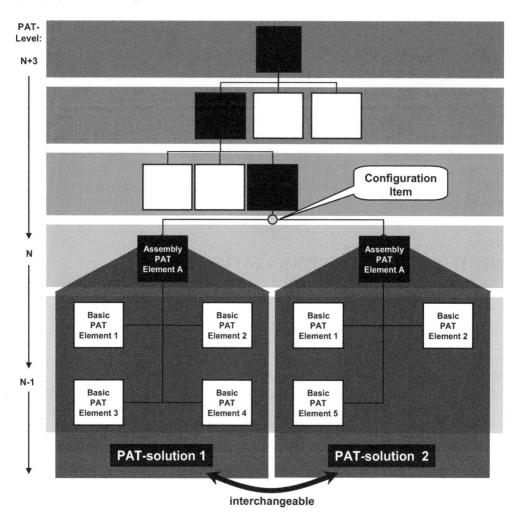

Figure 12.2   Different Product & Assembly Tree solutions attached to one Configuration Item

This does, however, not mean that CIs only appear between specific adjacent PAT-levels. It may well be that a configuration change is after all so fundamental that it turns out to be impossible for lower-level CIs to stop the change propagation. For example, a change to the aerostructure material of the fuselage could affect the specified form, fit and function of this major component. This would result in part numbers changing all the way up the product structure to the level of this major component. In this case, a CI at a higher level than the major component would have to be selected to stop the propagation. But this clearly is a matter of judgement. As a result, there may be CIs between many PAT-levels on many different PAT-branch lines embedded in the Product architecture. We will discuss in a moment how to select the right CI for a given change.

The fundamental aspect of CIs is, that different, interchangeable and mutually exclusive PS can be attached to them, see Figure 12.2. Interchangeability requires that the form-, fit- and function-interfaces with surrounding components are the same for each PS attached to the same CI. Every PS is based on agreements between Engineering and Manufacturing (and maybe other Functions), as described above. A PS represents a PAT-pyramid, which has an assembly PAT-element as the dedicated pyramid head. It is to be *configured*, that is, it can be allocated to an individual aircraft. It consists of those data and meta-data, such as drawings, Engineering documents and Manufacturing routings, as well as the components themselves, which belong to all basic or assembly PAT-elements of the *configured* PAT-pyramid.[9]

It is at the level of the head of this PAT-pyramid that one needs to be able to demonstrate conformity to the airworthiness authorities between the product's (that is, an assembly component's) form, fit or function on the one hand and design data on the other. Below this PAT-pyramid head the components constituting a PS remain unconfigured. There is design data available for them, and the Quality department checks conformity between this design data and the manufactured component. However, the conformity check on levels below individual, configured PAT-pyramid heads is not relevant to airworthiness authorities from a Configuration Management perspective.[10]

The envelope of all PSs under the same CI – thus covering all manufactured or to-be manufactured aircraft – represents a PAT-pyramid in its entirety. A PS and its associated Bill of Material for a *specific* aircraft can be extracted using what is called the effectivity calculation. It defines for which aircraft a specific PS should be switched to 'active' while leaving other PSs inactive (more explanations below).[11] Note that different PSs can make reference to the same lower-level PAT-elements, see for example the 'Basic PAT-Element 1' in Figure 12.2.

To implement a modification to the aircraft configuration, a new PS must be generated at the correct PAT-level, replacing an earlier one. The modification implementation methodology therefore consists of three steps:

1. Once it becomes clear which PAT-elements are affected by a specific change situation, one CI must be selected which is high enough up in the PAT-structure to avoid change propagation to parent levels further up.
2. For each change a new PS is established.
3. The effectivity of the modification needs to be calculated.

Let us examine these three steps in more detail.

Selecting Configuration Items

During the gradual build-up of the PATs during the course of the aircraft's development, CIs should already be introduced independent of specific change situations. Introduction of higher-level CIs should start at the early stages of an aircraft development project,

---

9   Note that not all parts which go on a commercial aircraft need to be configured. Standard parts, for example, do not. But there may also be assembly components which do not need to be configured.
10 This approach leaves flexibility with regards to those changes to manufacturing processes which do not change the form, fit or function of the design solution, thus offering the potential for later productivity improvements during Series Production.
11 Note that the link between the CI and the PAT-solution needs to contain the correct effectivity information. Once the effectivity is known, a PAT-solution is completely identified by its part number and the effectivity.

that is, during Feasibility and Develop Configuration phases. Lower-level CIs should be introduced early in the Development sub-phase. But even after termination of the Development sub-phase, new CIs may still need to be created.

The main criterion for introducing and selecting CIs is the ability to manage a PAT-pyramid's performance parameters and physical characteristics separately and independently from the higher-level PAT-structure, so that lower-level changes can principally be stopped from propagating to higher levels. Other criteria for CI selection, which should be applied, include:

- criticality in terms of high risks, safety and mission success;
- new or modified technologies;
- procurement conditions;
- logistics and maintenance aspects; as well as
- robustness of interfaces with other components.

CIs should be implemented in the PAT such that each PS represents one Manufacturing work order representing the activities for one Assembly Stage Operation (ASO).[12] The effectivity of work orders is then equal to the effectivity of the corresponding PSs. In other words, a CI ensures the link between product and processes Configuration Management. If the product's form or fit changes, Manufacturing processes have to be amended. And if Manufacturing processes require changes to the product's form or fit, design data has to change.

Also, a work order should be designed commensurate with the working conditions on the shop floor. Ideally, it comprises the work of an individual or team which can get completed during a single shift at most. In return, this sets limits to the size of a PS. All changes need to happen in a controlled, concurrent way under the 'umbrella' of a CI.[13] If engineers adhere to this principle, the concurrent incorporation of Engineering *and* Manufacturing needs in the PAT will be even more successful. In addition, fewer problems with linking drawings to work orders will be encountered during industrialisation and Series Production.

As the PATs evolve, there will be CIs on many different PAT-levels depending on the assembly processes. This often results in lower-level CIs, which can be found directly below upper-level PSs. This CI-PS-CI cascade might be necessary all the way down to a level where components do no longer need to be configuration change controlled (such as standard parts), see Figure 12.3a. However, instead of this cascade, some companies prefer a flat architecture like the one in Figure 12.3b. The latter is a result-oriented PS, where the structure no longer corresponds to the assembly and build process. The PS only represents the *result* of one or several ASOs.

---

12 As was said earlier, a work order is a sequence of Manufacturing activities featuring data such as workload, standard times, type of task and so on to be followed-up for a specific, single aircraft. A work order is data which is specific to the 'As-Built' view, and it corresponds to a routing in 'As-Prepared' as well as to an ASO in 'As-Planned'. Using IT tools, a work order can be generated automatically once the correct PAT-solution with its attached data has been selected for a given aircraft according to the effectivity calculation.

13 If form, fit or function are not affected by a process change, the latter can be implemented without being subjected to Configuration Management procedures. Thus, for Class II changes, work orders could still be changed by amending process steps, for example to cash cost saving opportunities. It is, however, recommended to also control and document these changes.

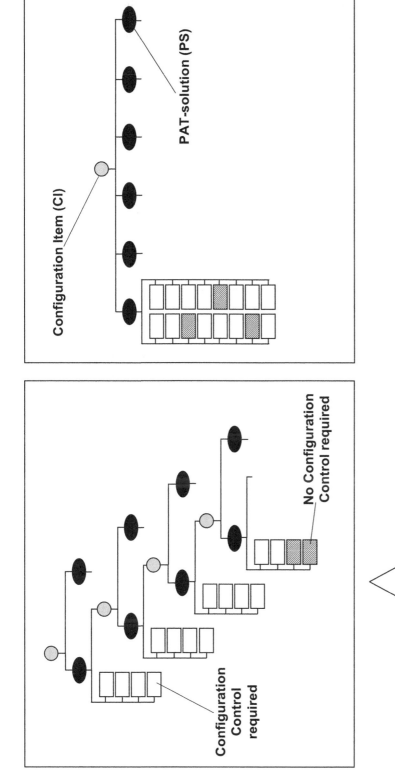

Figure 12.3   Configuration Item – PAT-solution structures:
Cascaded, operations-oriented (a) verrsus flat, result oriented (b)

If one Manufacturing work order corresponds to one PS, individual work orders for (b) must be of bigger size (that is, cover a lot more activities) compared to (a). But we saw above that the size of individual PSs should be limited in order to be able to accomplish at least one work order during one shift. Also, as (b) features significantly less CIs compared to (a), there is the risk that insufficient decomposition creates architectural logistics and maintenance difficulties and limits management possibilities.

But Manufacturing will also need the possibility to potentially change the content and sequencing of activities of a routing or work order at a later stage, for example in order to improve productivity. For this, Manufacturing would prefer not to change the corresponding PS, provided the result in terms of form, fit and function remains unchanged. This need would speak in favour of a flat structure. In addition, a result-oriented, flat structure also makes it easier to measure the quality of the components built. With a Process-oriented structure it is more work-intensive to quality-measure the result of manufacturing activities. One also has to bear in mind that the introduction of too many CIs affects product visibility, hampers management and increases cost. Finally, the choice between (a) and (b) is also depending on the company's IT capabilities to process Computer Aided Design (CAD) data: data configured using (b) can be processed faster compared to (a).

Whatever the choice, introducing the right number and location of CIs in PAT-structures is a difficult job and requires a significant amount of multi-functional experience. Once appropriate CIs have been incorporated in the PATs to the satisfaction of all involved Functions, CIs can be selected individually for specific change situations. Experienced experts should be entrusted with the task to identify the change-specific CIs out of the pool of all possible CIs.

The interested reader may at this point wish to return to the Architecture Integration Example introduced earlier to understand what Engineering and Manufacturing have agreed upon and how this links to configuration control requirements. This can be found in Appendix A3 Part III.

Protecting Interfaces

As different PSs can be attached to a given CI and because the physical components represented by the PSs will have physical interfaces with neighbouring components, interface protection as a design principle to support the management of aircraft configurations becomes very important. Wherever interfaces in the aircraft can be made stable during design and development, they will be candidates for becoming CIs. If, then, components change as described by alternative PSs, it is less likely that the changes propagate across the interfaces. In other words: non-propagation of change across a CI in the PAT often has the meaning of non-propagation of change across a physical interface on the real aircraft.

> 'Experience indicates that many of the difficulties in building large systems show themselves in the interfaces between the subsystems and between the system and external systems with which it must interoperate. For this essential reason, we include interface control as a key element of Systems Engineering.'[14]

---

14 Eisner, H. (2002), *Essentials of Project and Systems Engineering Management*, 2nd Edition, (New York: John Wiley & Sons), p. 209.

Robust CIs do not come for free. Proper interface protection is key to generate them. It can for example be achieved by:

- providing margins to space allocations;
- ensuring redundant functionalities;
- using standard parts; as well as by
- providing reserve factors when dimensioning the strength of aerostructures resisting loads.

Standard parts have the advantage that their quirks have been eliminated, which saves time. Also, they are understood by the individuals developing interfacing modules, thus easing communication. Reserve factors stop the penetration of changes in the calculation of stress and strain across the interface. Margins in space allocation avoid propagation of geometrical changes.

Unfortunately, protecting interfaces adds weight to the system, which is highly undesired in aircraft development. During the early stages of design and development, engineers are often forced to reduce design margins in an attempt to reduce weight. While weight reductions can, of course, be achieved in this way:

- later stress and loads changes can no longer be absorbed by any reserve factors; and
- resulting configuration changes across critical interfaces cannot be absorbed any more by, for example, space allocation margins.

Late changes to an as-built configuration will therefore happen more frequently and will generate high cost and delays. But as lowest possible weight is also an important objective to achieve, this is a serious conflict, which needs to be resolved.

It is suggested here that for the first prototype aircraft weight-saving activities should not encompass the cannibalisation of interfaces' robustness. If an empty aircraft with no passengers and cargo on board does not take off because critical interface margins have not been reduced, then the aircraft has a fundamental weight and design problem in the first place. Usually, the prototype aircraft is not sold on to customers. If, for example, aircraft No. 4 is the first to be sold to a customer, then this gives designers three more aircraft to save weight. As the aircraft design will become more and more mature during the development of these three aircraft, margins in the interfaces' design can also be reduced successively. This has to be done in a very controlled way using the principles of configuration control outlined in this chapter. If properly controlled, this approach would protect the programme schedule, while also achieving the weight target where it really matters, that is, for the first customer aircraft. However, this strategy is easier to be implemented for metal aircraft compared to aircraft with a larger composite material content. For the latter, a change in design requires more changes to tools compared to metal aircraft, thus significantly increasing costs.

Protecting critical interfaces has another important aspect. If teams have to completely redesign a component because of change of interfaces, they will be very demotivated. The motivation will remain high if they can proceed quickly with the component development of the module: 'People will often give 110 percent when they are properly motivated. However, nothing destroys the motivation to give extra effort as much as seeing work that is almost completed made totally useless because the requirements have been changed. This is like trying to run a 100m dash, only to be stopped 90m down the track and told that you are running in the wrong direction. It should be no surprise that after several such episodes designers simply slow down the pace. They will have learned that there is less wasted effort by waiting for everybody else to finish before completing their module. Unfortunately, if no one wants to finish first no races will be won.'[15] The expected Time-to-Market (see Chapter 7) will not be met.

Generating a New Product & Assembly Tree Solution

As a starting point for configuration control, a complete design baseline, including both the Engineering and Manufacturing Bills of Material, must be established. This baseline serves as a reference against which any change is defined, designed, managed, monitored and reported. It becomes impossible to properly change control the aircraft's design if the design baseline is not complete and not thoroughly documented. In aircraft design and for a given aircraft, any evolutionary change to a baseline n creates a new baseline n+1. The reference for the next change is then baseline n+1, not baseline n, see Figure 12.1.

In order to alter the PS of baseline n and to generate baseline n+1, all affected PAT-elements below a selected CI – including their attached data – are revisited to check whether they have to be amended or not. For Class II changes it usually is sufficient to only amend the data attached to the PAT-elements, for example, by replacing an old document by a new one.

However, if any of the PAT-elements below the CI is affected by a Class I change, it will need a new part number. As by definition the change affects form, fit or function, a unique identifier is required for all kinds of purposes, for example planning, procurement, stocking, manufacture and support, because it is distinct and different in how it can be used.

> 'The usual practice when a drawing is changed is to reissue it with a new revision number. If, however, a change results in a manufactured component or assembly being made different from other items with which it was previously interchangeable, it is not sufficient merely to change the drawing revision number. The drawing itself (and therefore the part number) must also be changed. This is a golden rule to which no exception should ever be allowed.'[16]

Thus, it would not be sufficient to only amend the data attached to the PAT-elements of the original PS. The change in part numbering requires that a new PS – maybe comprising new PAT-elements – is created. Figure 12.2 provides an example for this. Controlling design changes therefore not only leads to amendments of the data attached to already existing PAT-elements but in fact also and primarily leads to changes of the PAT-structure itself.

The combined set of PAT-elements below a CI which:

15 Smith, P.G. and Reinertsen, D.G. (1998), *Developing Products in Half the Time,* 2nd Edition, (New York: John Wiley & Sons), p. 108.
16 Flouris, T. and Lock, D. (2008), *Aviation Project Management* (Aldershot: Ashgate), p. 236.

- remain unchanged compared to the original PS;
- require amendments to the attached data only; and those which
- need to be newly generated because the change under consideration is in fact a Class I change (with new attached data such as drawings, manufacturing data and so on);

constitutes the new PS.

Any proposed change to a previous PS must be reflected in the CRD mentioned earlier. Prior to any decisions taken, it must become clear to all stakeholders – including the CCB – what the implications of a change are in terms of amendments to the PAT-structure or the data attached to it.

To better understand the implications of creating a new PS, the interested reader may at this point wish to return to the Architecture Integration Example described in Appendix A3 Part IV.

Implementing Options

While so far the control of design changes was in the focus of this chapter, it is nevertheless worth mentioning that the method of using CIs and PSs for configuration control is also applied to the management of options ('Customisation'). Commercial aircraft manufacturers create options (that is, PSs), which are offered to the customers. They are listed in an Options Catalogue from which every customer can chose its specific options selection.

When developing a new commercial aircraft, it is modern practice to provision for all standard options which are to be offered to customers using the Options Catalogue. This may, for example, require that some harnesses must be designed and installed on board the standard aircraft to cover the requirements of all standard options. This clearly adds weight but allows a response to customer requests much more quickly.

> 'Boeing ... [had identified] about 200 standard options that covered the majority of individual customer requirements. Provisions for these 200 standard options were incorporated into the basic design of the 777 to reduce change, error, and rework costs for individual customers.'[17]

Thus, it is not only important to master all steps of the configuration control process for managing design changes but also to ensure efficient and effective customisation as well as rapid response to customer requests. The corresponding PSs, which are verified and tested and which can be called-up again and again, must be available.

However, despite each individual option being well tested individually, it is of course by no means trivial to ensure that there are no clashes between parts or unfavourable interferences between systems whatever the mix of options and whatever the stage of the standard aircraft's evolution. The difficulty increases with each additional option added to the catalogue. However, the clash risk can be reduced if each option is designed as one interchangeable PS. Only the strict application of Systems Engineering principles in

---

17 Spitz, W., Golaszewski, R., Berardino, F. and Johnson, J. (2001), *Development Cycle Time Simulation for Civil Aircraft*, NASA/CR-2001-210658, pp. 3–13.

conjunction with an integrated architecture provides a chance to master this complexity in a cost efficient manner.

Validity and Effectivity

Commercial aircraft customers do not usually order a single aircraft but rather batches of aircraft having few or many aircraft in each. Aircraft are therefore uniquely identified by the batch number and a sequence number within the batch. If a planned change to a configuration has to be implemented on a certain number of aircraft, they can be identified by a series of batch numbers and sequence numbers, respectively. This set of numbers is called the validity of the modification. It describes the set of all aircraft where a change needs to be implemented.

An aircraft allocation table identifies which batch and sequence number corresponds with which Manufacturing Serial Number (MSN). This conversion is necessary as the numbering needs of a customer are different from the ones of Manufacturing. However, using the aircraft allocation table a modification's validity can be expressed as a set of MSNs too.

For each modification, there is a validity identifying the aircraft where the change needs to be implemented. Also, for each modification there are one or more CIs in the PAT-structure, to which new PSs are attached to represent the modification in terms of drawings, Bill of Material and so on. However, different modifications have different validities and different sets of affected CIs. Thus, for a given CI, there may well be an overlap of different validities resulting from different modifications. To identify which PS is to be embodied at which aircraft, one has to apply what is called the effectivity calculation.

> 'The effectivity of the configuration change typically will be specified through one of two basic techniques: date effectivity or serial effectivity. ... [With date effectivity], a change implementation date is used as the basis for planning when the new item will be phased into the Bill of Material and the old item phased out of the Bill of Material. Dates can be associated with the start of production lots to control the configuration of the lot ... Serial effectivity works in a similar way, but the change is tied typically to the end-item serial number. Serial effectivity is sometimes the preferred effectivity technique because the planned configuration of each end-item serial number is pre-defined and not subject to shifting schedules.'[18]

Let us look at the following example to illustrate this, see Figure 12.4: Modification Mod1 approved at time T1 may affect a CI with a validity 1-9999 (that is, it affects MSN 1 to MSN 9999). The corresponding PAT-solution PS1 is effective from MSN 1 to MSN 9999. Modification Mod 2 at T2 may affect the same CI with a validity 371-9999 (that is, it affects MSN 371 to MSN 9999). The corresponding PAT-solution PS2 is effective from MSN 371 to MSN 9999. However, this changes the effectivity of PAT-solution PS1, which is now effective from MSN 1 to MSN 370 only. The effectivity has to be recalculated. Modification 3 at T3 may affect the same CI with a validity 391-9999 (that is, it affects MSN 391 to MSN 9999). The corresponding PAT-solution PS3 is effective from MSN 391 to MSN 9999. The effectivity of PAT-solution PS2 changes and is now only effective from MSN 371 to MSN 390. The effectivity for PS1 remains unchanged.

---

18 Crow, K. (2002), *Configuration Management and Engineering Change Control* (Palos Verdes, CA: DRM Associates), http://www.npd-solutions.com/configmgt.html, accessed 10 October 2009.

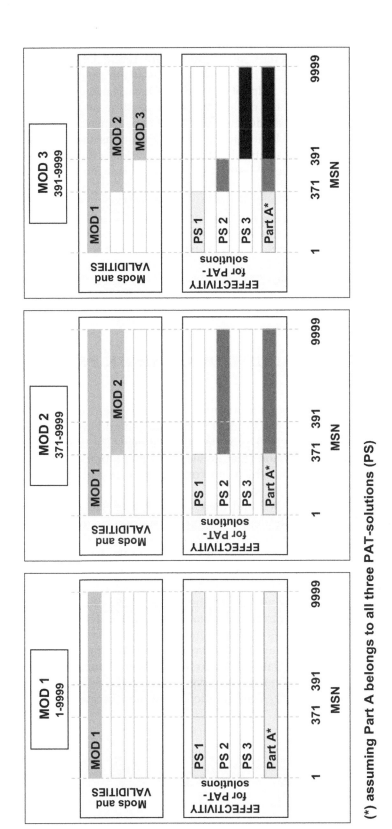

(*) assuming Part A belongs to all three PAT-solutions (PS)

Figure 12.4   Calculation of effectivity based on a given validity

The effectivities for the different modifications are recalculated with every new modification affecting the same CI. They are attached as an attribute to the link between the CIs and the PSs. With the help of a Product Data Management software tool (PDM), individual PSs are activated or de-activated depending on the effectivity of the modification. This then triggers all downstream data management, such as the change of Bill of Material, planning for Manufacturing Stage Operations and so on. The concept of effectivity is fundamental for controlling the configuration within the constraints of a given PAT-structure: it links the PAT-structure with the management of configuration change.

> *The power of the effectivity calculation can be described by the following example: imagine that for many years shop floor workers have riveted the same brackets to the fuselage shell of many aircraft. According to a new modification, those brackets get replaced by new ones, which are also attached at different positions. Thanks to the effectivity information applied to the DMU data attached to the PAT-structure, workers can 'dive' into any of the DMU's zones of interest where the change is supposed to be implemented. There they can see in 3D the new brackets attached at the new locations which helps them understand the changes.*

WHAT HAS BEEN DISCUSSED SO FAR ON PRODUCT ARCHITECTURE

Let us stop for a moment at this point to summarise what has been described so far in this and the previous chapter on how the Product architecture should look in order to enable integrated management. Figure 12.5 provides an overview. The Product architecture consists of common and uncommon (that is, view-specific) modules, respectively. The former are represented by PATs, which are common to different Functions (such as Engineering, Manufacturing and so on).

PATs are configuration controlled and therefore consist of flat or multi-level CI-PS-CI cascades or a mixture of both, depending on the specific circumstances. They also may contain unconfigured components, usually on lowest level of the product structure. The number of PATs, their precise location and size within the Product Architecture, as well as their flatness or depth, respectively, depend on agreements between Functions. Individual PSs become effective for individual aircraft using validity information and effectivity calculations. The effectivity information is carried by the link between the CI and the PS. Ideally, one PS results in one Manufacturing work order.

Additional, uncommon modules outside of the PATs are required to enable Function-specific views. These modules may carry data too. However, their data must not be subject to configuration control. Data related to the uncommon modules and generated by the aggregation, summation and so on of data attached to PSs (such as higher-level DMU-aggregates) is automatically configuration controlled because of the PSs being configured.

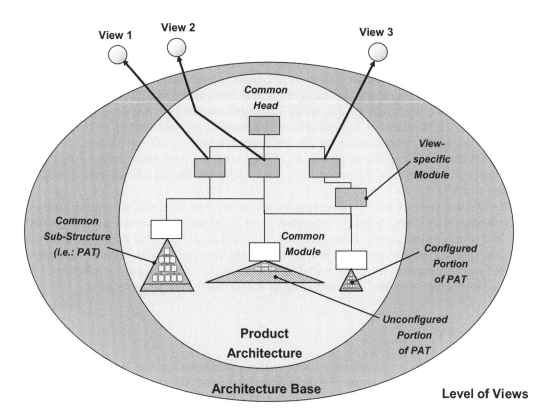

Figure 12.5   Specifics of the Product architecture

THE TEAM(S) FOR PRODUCT ARCHITECTING

Depending on the number of PATs, views and PSs, the Product architecture becomes very complex indeed. Some people have to control this complexity. As was mentioned earlier it is not only necessary to change control data attached to the PAT-elements, but in fact the PAT itself. With the introduction of the CI design changes are no longer independent of changes to the structure of the PAT.

Who should be in charge to control the Product architecture and its PATs? We learned that the Configuration Control Board (CCB) is in charge to decide when and how to change the PATs. But we also need an authority which has the power to propose changes to the PATs and to ensure their implementation once approved by the CCB.

Let us call this authority the Team for Product Architecting (TPA). For reasons outlined above it should be a multi-functional team. In fact, it consists of the same people who are needed for achieving the architectural agreements required for the creation of common Product architecture sub-structures. This was explained in the previous chapter. However, the TPA is not a Design-Build Team (DBT). It rather analyses possible PAT changes as well as their repercussions from the perspectives of the different Functions.

Should each multi-functional team have a TPA? Given the large number of multi-functional teams required for commercial aircraft development, this would not be realistic as the experienced experts needed to staff such a team are a scarce resource in any company.

In addition, in order for the TPAs to be able to keep control of the PATs and their changes, a certain degree of centralisation is required.

However, to limit the number of TPAs to only one would also not be wise. It would be a serious bottleneck if all architectural decisions would have to be taken by a single TPA. In conclusion, therefore, it is suggested that there should be TPAs for each major component assembly, such as the wing, fuselage, empennage, landing gear and so on. Each TPA is fundamentally important in achieving project objectives and should therefore be led by a person with significant aircraft development experience. Members of the TPA should be capable to represent architectural requirements of their Functions. In case of conflict between TPA-members, the TPA leader should be authorised to have the last word.

The TPA must ensure that architectural decisions are taken first, that is, before drawings, documents and so on are going to be amended. Thus, the structural environment is provided first and is only subsequently filled with the amended data. Note that this represents the opposite approach to the approach practiced by many companies, where data is amended first before deciding how to accommodate it in any kind of architectural environment. For the purpose of better control, the former approach is clearly the preferred one.

SUMMARY

Leaving changes uncontrolled may cause serious problems for the aircraft producing company, resulting, for example, in development delays and customer dissatisfaction. If it comes to the worst, failure to know the exact configuration of an aircraft may result in aircraft accidents. Or an accident has happened as a result of other reasons but the accident investigation by the relevant authorities is hampered by the unavailability of precise configuration information. In both cases, an aircraft developer has failed to comply with airworthiness requirements and might be sued.

Whatever the repercussion of a change it will generate additional cost, whereby late changes are always relatively more expensive than early changes, as was already mentioned in Chapter 4. It therefore comes as no surprise that failure to properly control changes is the most frequent cause of cost overruns. In addition, insufficient change control may well cause a degradation of the reputation of the company as an excellent aircraft developer, resulting in considerable revenue losses. Finally, if the configuration of individual aircraft is not well managed this also has major drawbacks for delivering an efficient in-service support.

To ensure the management of configuration control within a multi-functional environment:

- procedures for controlling configuration changes must be in operation;
- a frozen baseline configuration must be established;
- CIs need to be introduced to the PATs and need to be well selected to manage specific changes from an architectural point of view. Robust interfaces support this;
- subsequent changes to the configuration (that is, modifications) must be controlled, approved and documented by a dedicated CCB, which also determines the validity of each change;
- the effectivity for new PSs needs to be calculated;

- new PSs need to be established and attached to the CIs;
- team leaders must ensure that team members adhere to the change control processes in a disciplined way but also allow for controlled process by-passing in exceptional cases.

There can be no doubt about this: proper and effective change control is cardinal for the well-being of a project and a 'life insurance' for the company as a whole. All in all, it is vital to manage and to control changes affecting the configuration of individual aircraft in order to ensure conformity of the product with its contractual specification and customisation.

However, tight configuration control can only be justified once a first set of DMU-models or drawings has been released, such that there exists an initial aircraft configuration. Until then, only a much lighter change control process should be applied to limit the administrative burden associated with tight configuration control.[19] Everyone attached to the processes developing a commercial aircraft should be aware at what point the switch from the lighter change control process to a more stringent configuration control can and should happen.

MAIN CONCLUSIONS FROM CHAPTER 12

- For a commercial aircraft to be developed successfully it is essential to control two classes of changes: Class I changes affecting any aircraft part's form, fit or function as well as Class II changes comprising changes to documentation or changes to aircraft items not covered by Class I changes. All Class I changes affect in the end drawings and components, perhaps even the architectural structure of the Product & Assembly Trees. This is why they need special attention, in particular in complex product development.
- The application of professional Configuration Management techniques ensures that attention. It consists of Configuration Identification, Configuration Control, Configuration Auditing as well as Configuration Status Accounting. However, in an environment with large numbers of multi-functional teams working concurrently, it is Configuration Control, which is the most challenging.
- In this context, there are two types of teams, which have a significant impact on the professionalism and efficiency of Configuration Control: the Configuration Control Board, which decides upon a change to be implemented or not, as well as Teams for Product Architecting, which have the power to propose changes to the Product & Assembly Trees and to ensure their implementation once approved by the Configuration Control Board.
- The Configuration Control Board takes its decision on the basis of a series of change repercussion analyses which have to be prepared by

---

19 The notion of 'Configuration Control' assumes the fact that at least a first defined configuration must exist before control can start.

the affected multi-functional teams. However, it is a matter of balance to ensure that these analyses are not too detailed, which would consume too much precious time, but also not too limited, as this could delay the decision process. One also needs to ensure that not already the *identification* of teams impacted by a potential change takes too long. Thus, there are some significant process challenges to be met to ensure that Configuration Control can keep pace with the rate of change during the Development sub-phase.

- Teams for Product Architecting work with the concept of Configuration Items. A Configuration Item is an aggregation of hardware (or software or both) that is designated for Configuration Management: it is treated as a single entity in the Configuration Management process. A selected Configuration Item represents a barrier which signals the finish to any change tempting to propagate further upwards in the Product & Assembly Tree-structure. The fundamental aspect of Configuration Items is that different, interchangeable and mutually exclusive Product & Assembly Tree-solutions can be attached to them. The Teams for Product Architecting must ensure that architectural decisions are taken first, that is, before drawings, documents and so on are going to be established or amended. For many companies, 'Architecture first' represents a radically different approach.
- Effectivity-calculations are applied to identify, for which aircraft a specific Product & Assembly Tree-solution should be switched to 'active' while leaving other Product & Assembly Tree-solutions inactive.
- Finding the right balance between a disciplined process adherence and being flexible in exceptional cases is a major leadership challenge. This applies to all kinds of processes, but not respecting the process in the case of Configuration Management could really have serious consequences. In this case, therefore, the full attention of team leaders, Configuration Control Board and Teams for Product Architecting is required.

13

Integrated Project Planning

THE NEED FOR GOOD PLANNING

There have been many examples and it is often cited that projects with inadequate planning have failed miserably. This is why for complex product developments it is usually necessary to create a comprehensive Project Plan which represents a blueprint for the work to be performed. It describes precisely how to meet the external as well as the company-internal requirements and how the entire project organisation is supposed to work.[1] It generally is established prior to the launch of the Development sub-phase and includes essential aspects such as:

- goals, objectives, and requirements;
- project needs;
- architectures to be used, including project structure and organisation;
- staffing of organisation;
- roles and responsibilities within the project;
- rules for reporting;
- determination of ways and means of communication;
- methods and processes to be used;
- technical approach;
- schedule plan(s);
- interdependencies;
- determination of how to measure quality and performance;
- Change Management;
- budgets and their breakdowns; as well as
- analyses of risks.

As R.D. Archibald states: 'Without an adequate plan, the required resources cannot be assured and committed at the proper time, the team members cannot be fully committed to the project, monitoring and control will not be effective, and success will be a matter of luck.'[2] Creating a sound Project Plan is therefore extremely important to project success.

---

1   The Boeing Project Plan used to be called 'Design Objectives and Criteria'. 'Boeing has produced these volumes for every airplane type from the earliest days onward. Each new airplane builds on the most recent one before it and benefits from everything learned up to that point.', in: Sutter, J. (2006), *747: Creating the World's First Jumbo Jet and Other Adventures from a Life in Aviation* (New York: Smithsonian Books, HarperCollins Publishers), p. 128.

2   Archibald, R.D. (1992), *Managing High-Technology Programs and Projects*, 2nd Edition, (New York: John Wiley & Sons), p. 180.

Many experts on multi-cultural Project Management, such as H.-E. Hoffmann et al., strongly suggest preparing the initial planning of a multi-national project – such as commercial aircraft development – on an international level first, followed by a subsequent working along purely national rules.[3] According to them, it is also important to leave the overall project control in the hands of an international team. This is because the planning process works differently in different cultures:

- in Japan, for example, significantly more time is spent planning and explaining objectives than in western countries;[4]
- in the US everything will be prepared in a very detailed way, while
- in China a Project Plan is regarded as a general agreement to formulate a relationship based on confidence;[5]
- in-depth planning is also important in Germany. In contrast, it is regarded quite normal in France and Italy to work with crude planning only.[6]

It is quite important to be aware of the different cultural biases towards project planning in order to achieve a plan which is really accepted by all stakeholders. In any case there should be more time allocated to the kick-off of an international project compared to a purely national project, as it takes more effort to find the common basis for the project needs.[7]

However, as a result of the development dynamics, Project Plans typically become obsolete six to 12 months after they have been written. Does this mean that for commercial aircraft developments the Project Plan should not be established in order to save time and costs? It depends. On larger aircraft development projects it is certainly suggested to still lay out a Project Plan. It definitely serves as a good means to allow all project team members, in particular newly assigned personnel as well as upper corporate management, to comprehend the most important aspects of the project. However, the Project Plan should be kept as short and concise as possible. Having said this, it is important to note that resourced schedule plans must be vital elements of any Project Plan, whatever the size.

During the early phases of aircraft development, detailed schedule plans are not required as only a limited number of resources are allocated to the project. But there is a point in time where this drastically changes: at the beginning of the Development sub-phase. Very quickly, a tremendous ramp-up of resources needs to happen, the work of which requires management and coordination based on schedule planning. However, what is needed is not only a schedule plan, but an *integrated* schedule plan. This is why in the following we will concentrate on this part of the Project Plan.

The major advantage of integrated planning is that the activities of each functional area are being planned and carried out for the benefit of the overall needs of the project rather than the functional needs. To lever this benefit, the causes behind inadequate planning need to be well understood and addressed. They include, among others:

---

3   Hoffmann, H.-E., Schoper, Y.-G. and Fitzsimons, C.J. (2004), *Internationales Projektmanagement* (München: Beck-Wirtschaftsberater im Deutschen Taschenbuch Verlag), p. 54, free translation by the author.
4   Hoffmann, H.-E., Schoper, Y.-G. and Fitzsimons, C.J. (2004), *Internationales Projektmanagement* (München: Beck-Wirtschaftsberater im Deutschen Taschenbuch Verlag), p. 44, free translation by the author.
5   Hoffmann, H.-E., Schoper, Y.-G. and Fitzsimons, C.J. (2004), *Internationales Projektmanagement* (München: Beck-Wirtschaftsberater im Deutschen Taschenbuch Verlag), pp. 70–71, free translation by the author.
6   Hoffmann, H.-E., Schoper, Y.-G. and Fitzsimons, C.J. (2004), *Internationales Projektmanagement* (München: Beck-Wirtschaftsberater im Deutschen Taschenbuch Verlag), p. 47, free translation by the author.
7   Hoffmann, H.-E., Schoper, Y.-G. and Fitzsimons, C.J. (2004), *Internationales Projektmanagement* (München: Beck-Wirtschaftsberater im Deutschen Taschenbuch Verlag), p. 10, free translation by the author.

1. the aversion of many people to do planning in the first place ('Do you want me to do planning, or do some productive work?');
2. the reluctance of many people to expose their knowledge of the job (or lack thereof) to others in plans; as well as
3. the sheer difficulties in planning highly complex projects.

The first two causes reinforce the necessity to expose adequate leadership behaviours which recognize the vital contribution of planning data for project success. We covered these aspects in Chapter 6.

Schedule planning for commercial aircraft development projects is very complex. This is why the third cause needs significant attention too. What needs to be avoided is that the processes, methods and tools used for planning purposes force planners and managers to make inherently complex plans even more complex. This would lead to a situation where monster plans would be created, which no one will use. On the contrary, what is needed is an approach which allows the condensation of complex planning data into something which satisfies the needs of project teams as well as of general managers at all levels in the organisation. However, as a minimum one could list the following fundamental requirements for good schedule planning:

- plans need to correctly reflect what is going on in terms of work progress, resources, cost and so on. It is of no use constructing plans that do not reflect the work of the teams;
- there should be *one* integrated aircraft development plan across all components, teams and functions, across 'Make' and 'Buy', compatible with the Product architecture and its Product & Assembly Trees (PATs);
- for communication purposes it should be possible to aggregate detailed tasks or activities (and their associated resources, costs and so on) to different levels of aggregation (such as project level, team level, senior management level). Plans should be set up in a way which makes aggregation and status reporting easy and understandable.[8] This will also satisfy the constraint of the rather limited resources, which in commercial aircraft development can be made available for planning and monitoring purposes – for the reasons outlined above.

Around the start of the Development sub-phase, a (specific) Work Breakdown Structure (WBS), as introduced in Chapter 11, is established to enable Project Management during this intense period, see Figure 13.1. At this point, higher levels of Product-related architectures, such as a Product Breakdown Structure (PBS), Functionality Breakdown Structure (FBS) or Systems Breakdown Structure (SBS), whichever applies, are already in place. However, at this point one is still far from having common modules leading to PATs. It is part of the development activities during this sub-phase to build up – step by step – the integrated architecture described in Chapter 11.

---

8   The communication aspect of schedule planning is usually not exploited sufficiently well, neither from those who should make use of it, nor from the companies, which develop software tools for planning purposes. Those companies usually give preference to the scheduling functionalities compared to communication functionalities. However, in very complex projects like aircraft development the communication aspect becomes very important as well.

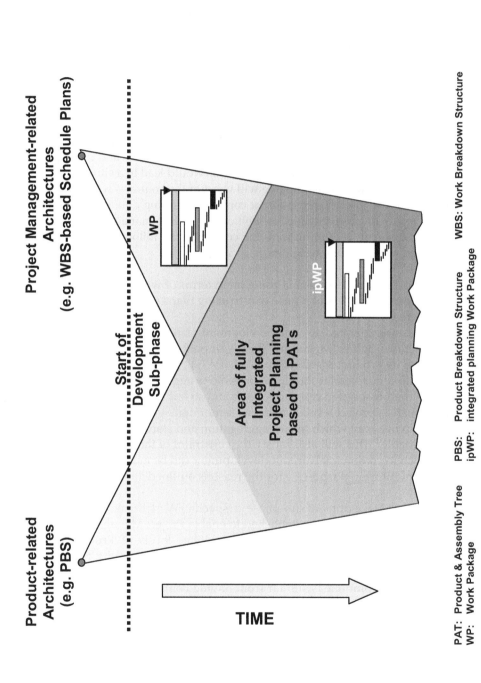

Figure 13.1   Interference of Product Breakdown Structure- and Work Breakdown Structure-architectures to enable integrated planning based on Product & Assembly Trees

However, before this can be achieved, there should already be a WBS-based schedule plan established which is integrated across all work packages. To achieve this for complex projects like commercial aircraft development is already a major undertaking. To allocate resources to the activities contained therein to build up a proper Earned Value Management system is an even bigger challenge. However, even at that level there would still be no real link to any of the Product-related architectures. Only once all relevant architectures have evolved sufficiently well to allow for PATs to be generated can the architectures of the product- and project-worlds be interlinked.

Clearly, the first PATs do not yet contain all the details which are required at a later stage. But they are sufficiently detailed to reflect the following:

- the major downstream stages of the assembly process, thus ignoring the smaller component assemblies at this point in time;
- the major component assemblies to be transported from one manufacturing site to another, in particular to the Final Assembly Line (FAL);
- the major parts of the Master Geometry;
- known elements where 'Make/Buy' decisions have been already made or are obvious;
- the organisational structure of higher-level multi-functional teams, to which time, cost and quality (TCQ) responsibilities have been entrusted for the design, development and production; as well as
- the aircraft zones, for which design concepts are to be established.

At this stage, that is, at the beginning of the Development sub-phase, the size of any PAT-pyramid is small. Later, the PAT-pyramids will grow and expand, and more and more elements will be added.

Before looking into some pragmatic guidelines for schedule planning in Chapter 14, a suggestion for an integrated approach to schedule planning will be presented below. It is based on the integrated Project architecture introduced in Chapter 11, that is, once PATs have been established. To the knowledge of the author this integrated planning approach has not been implemented anywhere yet. It rather presents an outlook of what might be possible a few years from now. However, the quick reader may prefer to move on to Chapter 14.

WAYS TO INTEGRATED SCHEDULE PLANNING

The Product & Assembly Tree as a Sequence Architecture

The PAT has so far been introduced as a top-down pyramidal structure like any other Breakdown architecture. However, turning a PAT clockwise by 90° will yield a format more similar to a Sequence Architecture, see Figure 13.2.

In this format, one can perhaps more easily imagine that each PAT-element could also be sub-divided into three *temporal* phases, see Figure 13.3:[9]

---

9   While being a node, PAT-elements are represented here as boxes for the purpose of better comprehension.

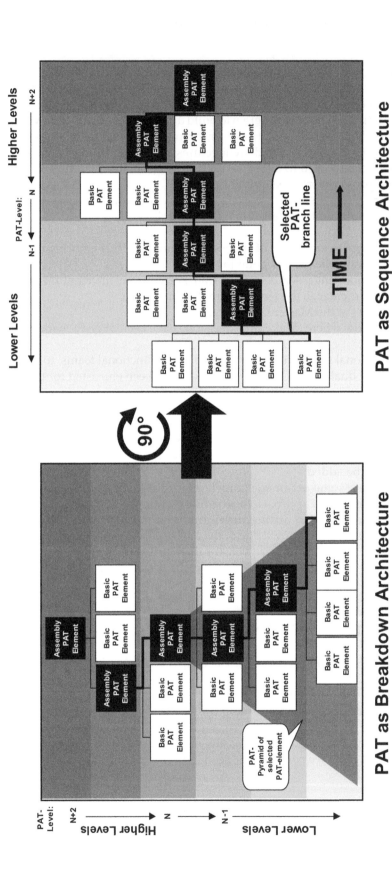

Figure 13.2   Turning the Product & Assembly Tree by 90°

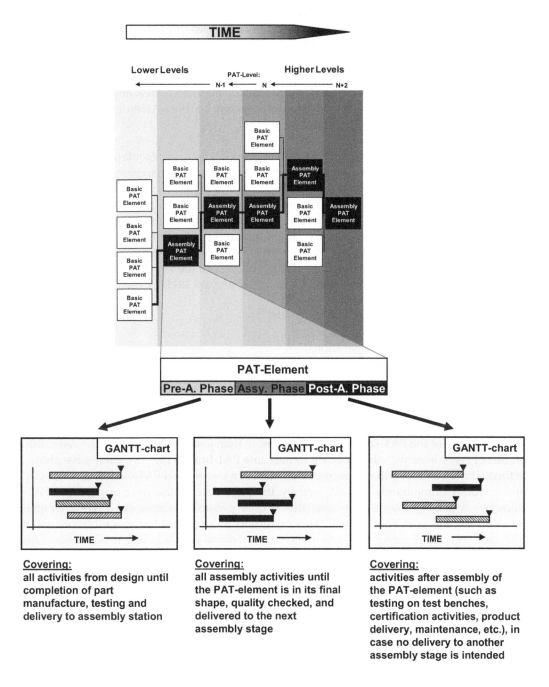

Figure 13.3   The three phases of each Product & Assembly Tree element
(Pre-Assembly, Assembly, Post-Assembly)

- one 'Pre-Assembly' phase covering all activities from design until completion of part manufacture, testing and delivery to assembly station;
- one 'Assembly' phase covering all assembly activities until an assembly PAT-element is in its final shape, quality checked and delivered to the next Assembly Stage (AS);
- one 'Post-Assembly' phase covering activities after assembly of the PAT-element (such as testing on test benches, certification activities, aircraft delivery, maintenance and so on). Note that there can only be 'Post-Assembly' phases where there is no delivery to another AS.

Seen from this perspective, each PAT-element is not only a representative for a physical component (as introduced in Chapter 11), but also for the *processes* necessary for its creation and operational use, that is, for its life-cycles.

Note that for a basic PAT-element there is no Assembly phase as basic PAT-elements do not represent assemblies. For an assembly PAT-element there is no parts' manufacture during the Pre-Assembly phase. Parts' manufacture is only considered for basic PAT-elements.

One can now also imagine that data of the kind, where the *dimension of time*[10] plays an essential role, can be attached to the PAT-elements' phases, such as:

- planning of Engineering activities;
- manufacturing planning;
- planning of assembly operations;
- detailed parts planning;
- parts ordering planning;
- maintenance planning;
- and so on.

The turning of the PAT-structure by 90° is more than just a 'mental trick': the Assembly phases of PAT-elements, which belong to the same PAT-branch line, are successive phases in time. That is: in a schedule plan, the finish date of a lower-level PAT-element's Assembly phase would immediately be followed by the start date of the next upper-level PAT-element's Assembly phase. Note that this is only possible because the PATs are equal for 'As-Designed' and 'As-Planned'. This possibility to sequence the Assembly phases of PAT-elements is not only a fundamental characteristic of the PATs but also an important backbone in achieving integration of architectures.

GANTT-Charts

The three phases introduced above are the 'starting points' for more detailed schedule plans. GANTT-charts play a key role in this.[11] A GANTT-chart is the classical tool to represent planned activities and milestones. It is also used to:

---

10 That is, the 'first dimension' of time. A 'second dimension' of time will be introduced later.
11 GANTT-charts are named after the industrial engineer Henry Gantt (1861–1919) who was a management consultant revolutionising the managing of large, complex projects such as construction, when he introduced his GANTT-chart in around 1910.

- capture any dependencies between activities;
- perform scheduling;[12]
- perform Critical Path Analysis (CPA); as well as to
- share views on the status of the project (aspect of communication).

As it is shown in Figure 13.3, it is suggested here to consider GANTT-charts more as an attachment to architecture elements rather than a starting point for setting up a project. This is a very different concept compared to the classical planning approaches because it requires that architectures or modules thereof are established prior to the start of any planning activities! It cannot be emphasised enough that this really represents a radically different approach to schedule planning.

Of course, the GANTT-charts can also be accessed in the PAT's Breakdown architecture format – once they are attached to the PAT-elements, see Figure 13.4. However, in this format it is more difficult to understand how time dependent information is linked to the Product architecture. This is where the 'mental trick' with the 90° turn helps.

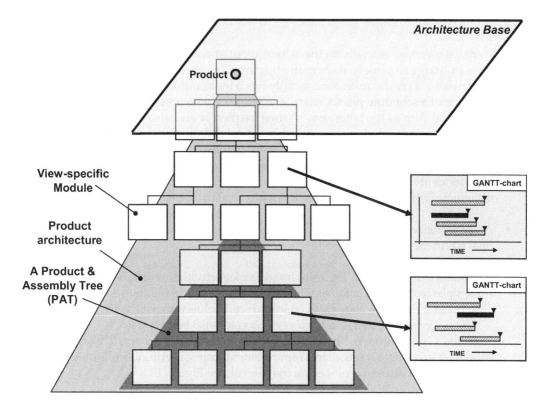

Figure 13.4   **GANTT-charts attached to modules/elements of the Product architecture**

---

12 'Scheduling' is a planning activity, which calculates the correct location of activity bars along the time axis in a GANTT-chart according to the interdependency logics (such as Start-to-Finish, Start-to-Start, and so on) as well as the dates or durations associated with those activities.

To better understand the concept of integrated GANTT-charts within the context of architectures, some useful definitions for GANTT-chart elements should be made:

- A *task or activity* is the smallest planned unit with a non-zero duration (in hours, days or months).[13] In the GANTT-charts, tasks/activities are represented by horizontal bars;

- A *milestone* is a zero-duration event, the achievement of which can be clearly recorded and validated.

    'An event is an occurrence at a point in time that signifies the start or completion of one or more tasks or activities. To be useful for project planning, scheduling, and control, an event must be understandable to all concerned, clearly and unambiguously described in precise terms, and its occurrence must be immediately recognizable. The occurrence of an event ... must clearly be pinpointed to one specific day, month, and year; in other words, a calendar date.'[14]

    Milestones are digital (or binary) in the sense that they are achieved or not achieved – for example they cannot be 50 per cent complete.

- A *critical milestone* is an event, the achievement of which is regarded critical by a team in charge to achieve its overall objectives. In other words: it is anticipated that missing a critical milestone usually has a measurable effect on the overall leadtime of a schedule plan. Critical milestones are not necessarily milestones on the Critical Path as the latter may change frequently and also because a Critical Path of a lower-level plan, to which this critical milestone belongs, may not be identical to the Critical Path of a higher-level plan. However, by definition all milestones on the Critical Path are critical milestones. Examples for critical milestones are: delivery of requirements and specifications, of documents and drawings, critical interfaces, delivery of physical components, and so on.

- A *payment milestone* represents an event in a schedule plan where a stage, part or final payment is due to be made.

- A *thread* is a unique set of logically linked tasks/activities, which belong to:
    - the *same booking code*, against which workload (such as manhours), and costs can be booked (booking codes will be introduced below); as well as to
    - the *same generic aircraft development process* (see also below).[15]
    In the GANTT-charts, threads are also represented by bars. Threads always culminate in a single (critical and reportable) output- or terminal-milestone, which, of course, must also exist on task/activity level, see Figure 13.5.[16]

---

13 Some sources prefer the terminology 'task', others 'activity' Throughout the remaining part of this book, the term 'task/activity' will be used.

14 Archibald, R.D. (1992), *Managing High-Technology Programs and Projects*, 2nd Edition, (New York: John Wiley & Sons), p. 211.

15 For example, activities related to the verification of requirements belong to the generic Verification & Validation process.

16 Planning tools usually allow for viewing individual levels of a given plan. For example, in one view all activities can be seen, in another only threads. Good tools maintain the logical links across the border around those tasks/activities, which are to be aggregated into threads, see Figure 13.5. Also, by default, the critical milestones on tasks/activities remain visible on thread level.

Threads play an important role when it comes to applying the Earned Value method to complex, commercial projects. This will be described in Chapter 15.

- An *aggregated thread (A-thread)* is a unique set of logically linked threads. A-threads are also represented as bars in the GANTT-charts. A-threads representing groups of threads follow exactly the same aggregation principles as threads representing groups of tasks/activities, see again Figure 13.5. However, A-threads may cover more than one booking code and/or more than one generic process. There is exactly one A-thread per phase per PAT-element. The length of an A-thread represents the overall duration of a PAT-phase.[17]

For a PAT's Assembly phase's A-thread planning, any two of the following three subjects need to be estimated[18] and provided by Manufacturing (or the Manufacturing Engineering community within the multi-functional teams, respectively):

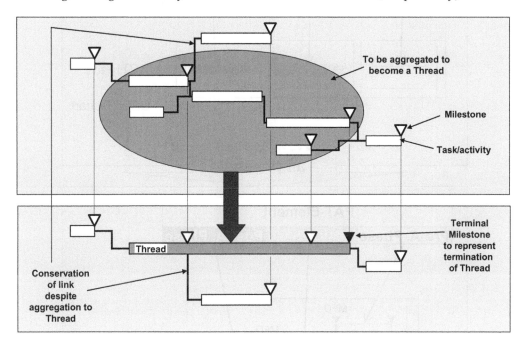

Figure 13.5   Aggregation of tasks/activities (top) to a single thread (bottom)

  - the (perhaps at this stage estimated) duration of the envelope of all assembly tasks/activities necessary to finish the Assembly phase of the PAT element;[19]
  - a Must Start Date (MSD);
  - a Must Finish Date (MFD).

---

17 As basic PAT-elements do not feature any Assembly phase, the duration of this phase can be regarded as being of zero value. As a consequence, for basic PAT-elements there are no A-threads for this phase.

18 Based on past experience, extrapolations and so on.

19 Note that this duration is equal to the time between on-dock delivery to the assembly point on the one hand and delivery to the next assembly point on the other. Thus, it includes not only the actual durations for assembly activities but, for example, also all activities related to logistics and transport between assembly points. Finally, any margins deemed necessary may also be included. At this point the Assembly phase of *each* PAT-element is only regarded as a 'black box' with a certain duration, without looking into more details.

For a given PAT-element, the three A-threads of its three phases are connected by Finish-to-Start links. Thus, it is possible to view the entire schedule planning for each individual PAT-element in one GANTT-chart, using MSD and MFD as connectivity milestones. The principle of this is shown in Figure 13.6. Post-Assembly A-threads are only available for Assembly PAT-elements, which are to be delivered to another AS.

In addition, A-threads of the Assembly-phases of PAT-elements on the same PAT-branch line are connected to each other by Finish-to-Start links, see Figure 13.7. As mentioned above, this is possible because PAT-elements on the same PAT-branch

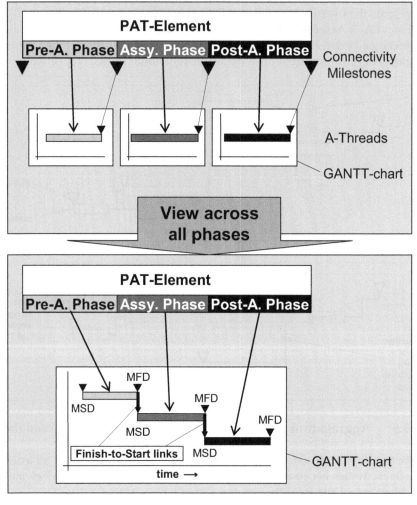

MSD: Must Start Date   MFD: Must Finish Date

Figure 13.6   Single GANTT-view across all three phases of a Product & Assembly Tree element

line are sequenced according to the assembly process, following the results of the discussions between Engineering and Manufacturing described in Chapter 11.[20]

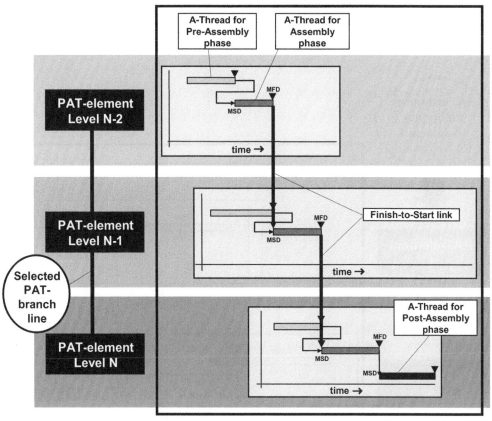

Figure 13.7 **Linked A-Threads of the Assembly phases of three elements on a selected Product & Assembly Tree branch line**

- *Summary threads (S-threads)* are representatives for the time span between the start date of the earliest and the finish date of the last of a number of selected A-threads. If S-threads are used in GANTT-charts attached to PAT-elements they are representatives for the time span between the start date of the earliest and the finish date of the last of all A-threads, which belong to the same type of phase (such as Assembly phase) of *all* PAT-elements, which culminate into the one selected PAT-element, for which the S-thread is to be shown, see Figure 13.8. In other words: S-threads represent the time span of individual phase types of selected PAT-elements. 'Selected PAT-elements' could, for example, be all those elements which belong to an entire PAT-pyramid.

---

20 Note that Figure 13.7 represents an idealisation of production reality. For example, the assembly processes may require some time for assembly preparation, starting well before the parts arrive. However, this diversion from reality does in no way compromise the statements made in this chapter, as it will become clear in a moment.

This concludes the introduction into some necessary definitions. In the following, we will concentrate on the PATs as sub-structures of the Product architecture in order to understand how integrated schedule planning can be performed based on integrated architectures.

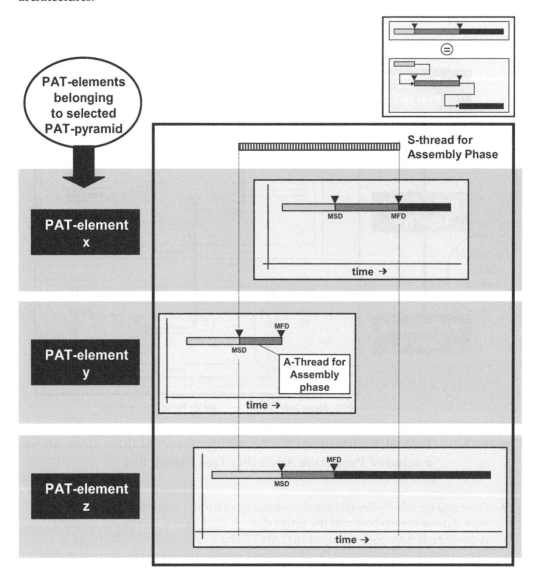

Figure 13.8   **S-Thread for Assembly phase of three selected Product &
Assembly Tree elements**

Aggregated-Threads: Gateway to Detailed Planning

The A-threads of the various PAT-phases can be regarded as gateways to lower-level GANTT-charts, which allow to plan in more detail, see Figure 13.9.[21] This represents again a very different approach, namely a top-down-approach, compared to the widely used planning practice where tasks/activities are planned first and then, subsequently, grouped together to be represented by summary bars.

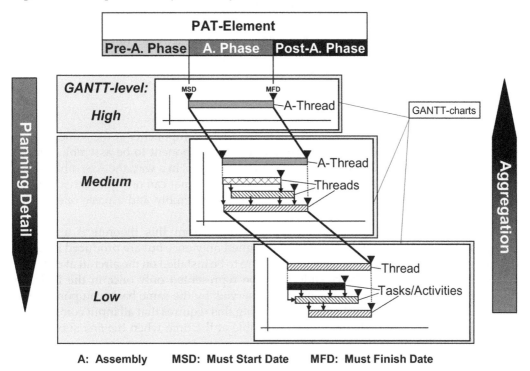

Figure 13.9   The cascade of GANTT-charts
(here shown for the Assembly phase only)

The detailed plans themselves consist of threads. Such threads are logically linked together and individually finished by a critical milestone. Detailed plans must have an overall envelope of start and finish dates of threads, which corresponds with the start and finish dates, respectively, of the selected A-threads.

From each such thread another, even lower-level GANTT-chart may be populated with tasks/activities and milestones. As a result, for a given phase of a selected PAT-element, there may be as many as three levels of GANTT-charts. These levels are called GANTT-levels as they are fundamentally different from the PAT-levels described in Chapter 11. GANTT-levels represent grades of schedule planning details, while PAT-levels represent details in the product and assembly structure. Because GANTT-levels are available for each PAT-element, and because there is also a whole cascade of PAT-elements, there is

---

21 The A-threads only represent groupings of threads and, thus, tasks/activities within the same PAT-element. If a more detailed breakdown of the PAT-element in lower-level elements is required, this must be done using the PAT-structure itself.

overall sufficient granularity available to allow for rather detailed planning. According to the author's personal experience there is therefore no need for more than three GANTT-levels per PAT-element.

For most PAT-elements – and for at least one of the three phases of them (that is, Pre-Assembly, Assembly and Post-Assembly) – eventually a detailed plan needs to be established. Each of those plans can be aggregated to the one A-thread on highest GANTT-level, see Figure 13.9. However, when starting to create the first PAT-elements, only sufficient information to generate the A-threads for the Assembly phases are usually available, because only the durations for these phases have already been estimated by Manufacturing. The starting point for top-down planning activities of the majority of all PAT-elements should therefore be their Assembly phases.

Planning the Assembly Phase

It is obviously impossible to assemble more than two components at the same time at precisely the same aircraft location. Any additional component to be assembled in the same detailed zone would require a new assembly step. So, in a way, the Assembly phase of every element of the PAT in Sequence Architecture format can only just consume two incoming components – either two basic ones or an assembly and a basic one or two assembly ones, see Figure 13.10, part a).

However, in practice there are some deviations from this theoretical approach. Components, for example, which are to be designed only once but are produced in larger quantities (for example in batches), and which are to be installed on the aircraft at different locations, can nevertheless be considered to be represented only once in the PATs. In this case, different assembly components are 'served' by the same basic component, see Figure 13.10 part b). In terms of schedule planning this requires that all input components belonging to the same batch are already available at the time when the *first* subsequent assembly component needs one of them.

In practice, too, more than two incoming components are considered to be schedule planned for one and the same Assembly phase. The latter would than include assembly tasks/activities on a lower GANTT-level to describe the precise sequence of assembly steps. It is a matter of choice whether an Assembly phase contains more or fewer sequential assembly steps and what the resulting complexity of the phase will be. This is what Manufacturing brings into the negotiations with Engineering when discussing 'As-Designed' and 'As-Planned', among others. Figure 13.10 part d) provides an example for a different structure containing the same basic elements than Figure 13.10 part c), but with more than two basic components consumed by two assembly components.

The practice to accept more than two components for the same Assembly phase leads to an important assumption for schedule planning purposes (and later availability of real parts): no components are added to the assembly process *during* any individual Assembly phase. This means that *all* required incoming components are considered available *at the start* of the Assembly phase – even if in reality they are to be 'consumed' one-by-one in a sequential order.

Once the PAT 'skeletons' have been generated, the process to add planning data to it commences by populating the Assembly phases of the (by then already existing) PAT-elements with the following data for the first production aircraft:

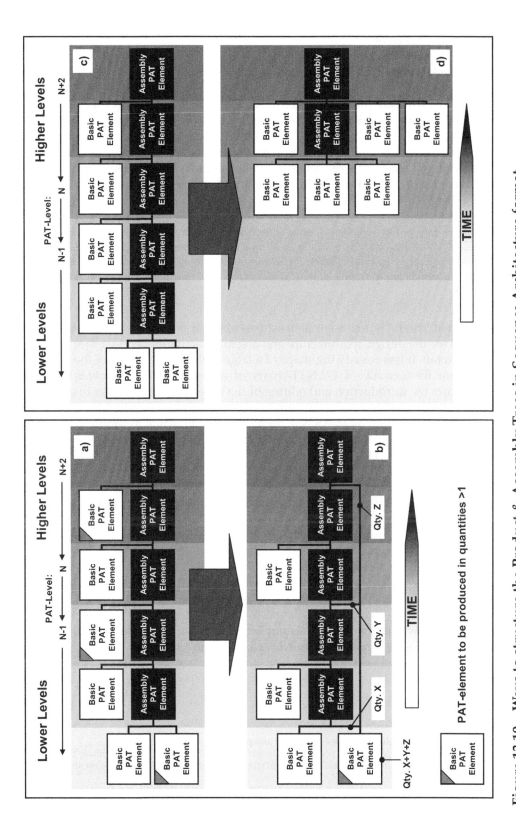

Figure 13.10 Ways to structure the Product & Assembly Trees in Sequence Architecture format

- a MSD *and* a MFD; or
- a MSD *and* the overall duration; or
- a MFD *and* the overall duration.

As all PAT-elements are linked together in the PAT-structure, and because Assembly phases of PAT-elements on the same PAT-branch line are linked in a sequential order, the detailed shape of the PAT in Sequence Architecture format can be calculated with this information – defining the PAT-elements relative position to each other.[22] This calculation (called 'scheduling') rearranges the relative positions of all the PAT-elements along the time axis according to the provided MSD, MFD and duration information as well as according to the PATs' logical links.

> *If, for example, only the top-level element, which is furthest to the right in Sequence Architecture format, would have a MFD, the calculation would arrange all the other PAT-elements along the time axis following a right-to-left sequencing approach. If some PAT-elements would be attributed with MFDs, others with MSDs, the calculation would arrange all the PAT-elements around these dates while respecting the PAT logical links.*

In this format, the PATs represent a structured starting point for a high-level target schedule plan for the design, manufacture and assembly activities necessary for the first production aircraft. It has already the shape of a large-scale schedule plan of the project.

'Diving' into the cascade of GANTT-charts of all relevant PAT-elements via the A-threads allows the introduction and editing of the necessary schedule details. For the Assembly phase, detailed plans should contain all tasks/activities required between start of the actual assembly process at AS N-1 up to start of the actual assembly process at the subsequent AS N. This includes, among others, the periods and dates for the tasks/activities of (see Figure 13.11):[23]

- the actual assembly process steps;
- any required inspections (such as First Article Inspection (FAI)) or testing before delivery to the next AS;
- any transport or logistics efforts prior to the next assembly site;
- any duration(s), during which specific and often very expensive jigs & tools are needed for the component assembly;[24]
- any AS de-activation or cleaning; as well as
- any additional assembly preparation, if any, prior to the Assembly phases' MSD.

As it will be shown in a moment, it is important to group the tasks/activities of the actual assembly processes under the same thread, see Figure 13.11.

---

22 Any schedule conflicts would have to be highlighted and need to be resolved.
23 Note that the manufacture of basic components is to be planned during the Pre-Assembly phase rather than the Assembly phase. In addition, for basic PAT-elements the Assembly phase is of zero duration, that is, MSD equals MFD.
24 Times and durations for the use of jigs & tools are important constraints to the industrial ramp-up, for which the planning will be described below.

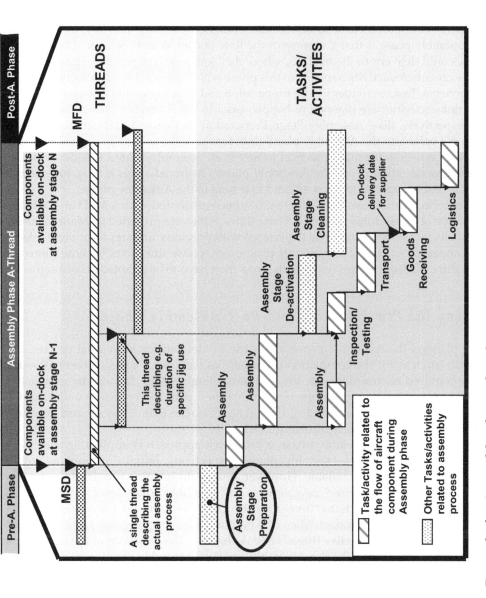

Figure 13.11   Example for Assembly phase planning

Note that while it was said earlier that the length of A-threads represents the overall duration of a PAT-phase, in the example of Figure 13.11 this does not seem to be the case.

For example, the task/activity 'Assembly Stage Preparation' happens prior to the MSD of the A-thread. Thus, the actual Assembly phase seems in fact to last for longer than the A-thread.

It is therefore important to be somewhat more precise about the earlier statement. The idea of the Assembly phase is that it represents the flow of components generated by assembly processes until delivery to the next AS, where they join new components. However, there might be other tasks/activities related to this phase which are not directly affecting the flow of components. Tasks/activities like jig preparation and cleaning would be examples of this. If such tasks/activities are planned to happen prior to the Assembly phase's MSD or post MFD, respectively, (like 'Assembly Stage Preparation' in Figure 13.11), strictly speaking, they do not belong to the Assembly phase (but, for example, to the Pre-Assembly phase). But it is nevertheless far more practical to access all assembly-related tasks/activities via the same gateway (that is, via the Assembly phase's A-thread). This is why, for example, 'Assembly Stage Preparation' is considered to belong to the Assembly phase.

As described earlier, each PAT-element has been attributed with a MSD and MFD for the envelope of its assembly activities. These dates, which are provided by Manufacturing, may initially be estimates based on experience with previous aircraft programmes. As the Development sub-phase of the aircraft progresses, phase durations become more and more mature. Tasks/activities (or threads) may then have to be adapted accordingly.

Planning the Pre-Assembly and Post-Assembly Phases

Having planned the Assembly phase, dates can be defined for individual process start and finish (such as for the completion of a shell), as well as for parts delivery. Based on this, the required release dates for the individual Engineering data can be established. Engineering (and maybe other Functions) must then plan the processes and resources required to ensure these release dates. Thus, it now also has to link Process and Resources data to the Product data, see again Figure 11.16.

When planning the Assembly phase, a top-down approach is applied. In the case of the Pre-Assembly and Post-Assembly phase, respectively, initial durations as well as MSDs and MFDs are not available. However, one could imagine introducing 'dummy' A-threads, which can be used as gateways to 'dive' into lower-level GANTT-charts, as well as 'dummy'-threads to 'dive' into the GANTT-charts on the lowest GANTT-level. Here, individual tasks/activities can be planned. Assigning the latter to threads converts the 'dummies' into live threads and A-threads. The durations of the threads and A-threads (and therefore of the Pre-Assembly and Post-Assembly phase, respectively), are calculated automatically by a planning software tool once the tasks/activities and critical milestones, respectively, have been entered into the detailed GANTT-charts.[25] Thus, also the

---

25 As for the Assembly phase, tasks/activities of the Pre-Assembly and Post-Assembly phase, respectively, may spill over to an adjacent phase. Formally, they would then belong to the respective adjacent phase. However, it might still be more practical to access them via the A-threads of the Pre-Assembly or Post-Assembly phase, respectively, in much the same way as described for the Assembly phase.

Pre-Assembly and Post-Assembly phases are now populated with A-threads. Note that for these two phases essentially the classical bottom-up planning approach is applied.

When 'diving' for the first time into the highest-level GANTT-chart for the Pre-Assembly phase of a selected PAT-element, only one critical milestone appears: the milestone which represents the delivery to an assembly station:

- of either a manufactured basic components (with its associated Engineering data);
- or of the relevant Engineering data (such as assembly or interface drawings) required to perform an assembly.

This critical milestone is a MFD and is equivalent to the overall MSD of the subsequent Assembly phase. It represents the connectivity milestone between the Pre-Assembly phase and the Assembly phase, see Figure 13.12.

The detailed schedule plan for the Pre-Assembly phase should – among others – contain the periods and dates for:

- all major design tasks/activities, such as for generation of
  - different schemes (evolutions) for component drawings;
  - interface drawings;
  - installation drawings;
  - assembly drawings;[26] as well as the
  - Data for Manufacture (DfM)
- Numerically Controlled (NC) programming;
- the actual basic component manufacture; as well as for
- component testing and inspection;

whatever applies.

The planning, if any, of all tasks/activities for the period after an assembly component has been generated – and is not delivered to another AS – needs to be established within the Post-Assembly phase. This can be done in a similar way as for the Pre-Assembly phase. Compared to the other two phases, fewer PAT-elements will need detailed planning for the Post-Assembly phase, though. This is because validation and testing activities are usually performed with assembly components of a relatively high level of aggregation only and because Post-Assembly planning for basic components usually is not required.

As before, when 'diving' for the first time into the GANTT-chart on the highest GANTT-level of a PAT-element, only one critical milestone appears for the Post-Assembly phase: the milestone which represents the date where an assembly component is fully completed and is delivered to the site where the initial tasks/activities of the Post-Assembly phase take place. This critical milestone is a MSD and is equivalent to the overall MFD of the previous Assembly phase of the same PAT-element. It represents the connectivity milestone between the Assembly phase and the Post-Assembly phase, see again Figure 13.12.

The detailed schedule plan for the Post-Assembly phase should – among others – contain the periods and dates for all tasks/activities required:

---

26 If the DMU is used, then there may be no need for 2D assembly drawings as a 3D visualisation is available. 3D assemblies are digitally generated from the digital component models using the relative links provided by the Product Data Management (PDM), see Chapter 2.

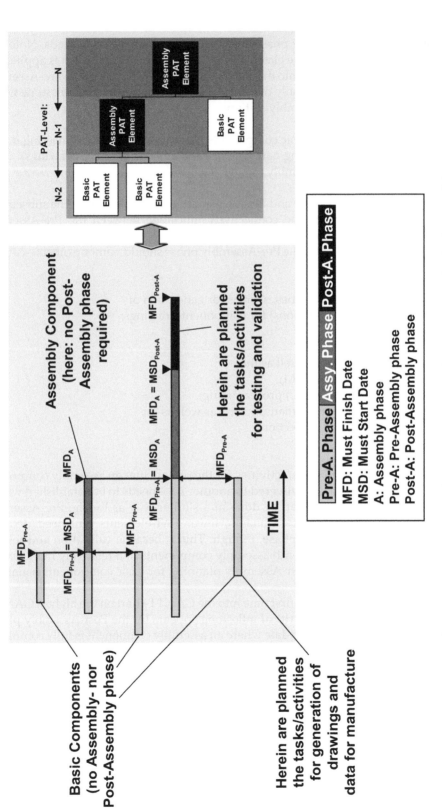

Figure 13.12　Planning of the Pre-Assembly and Post-Assembly phases, respectively (example)

- to prepare for validation tests;
- to perform the test;
- to analyse and document the results.

While the Assembly phases of PAT-elements on the same PAT branch line are successive elements in time with no overlaps between A-threads, the corresponding A-threads of the other two phases may well overlap, see Figure 13.13 for the example of Pre-Assembly phases. Also, the threads and tasks/activities in the GANTT-charts for each of those phases may, of course, show overlaps and other logical links (such as Finish-to-Finish, Start-to-Start, Finish-to-Start) than the standard Finish-to-Start applied for the Assembly phases.

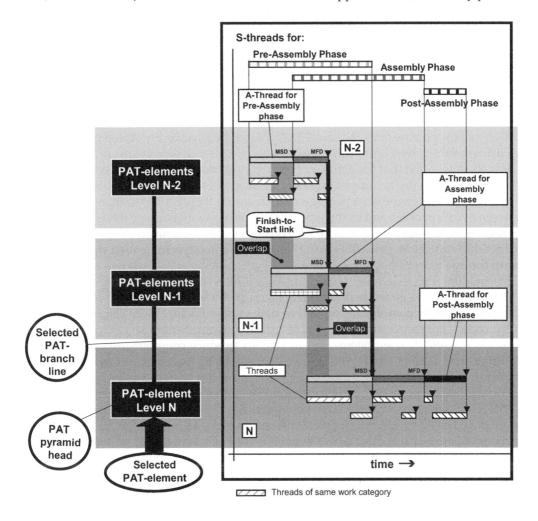

Figure 13.13   Overlapping Pre-Assembly phases on a single GANTT-chart for selected elements on same Product & Assembly Tree branch line

THE RELEVANCE AND IMPORTANCE OF WORK PACKAGES

What Integrated Planning Work Packages are all About

In Chapter 11, work packages (WPs) were introduced as units of the sWBS necessary to set-up, monitor and control a project. It is now time to analyse what the terminology 'work package' generically means in a fully integrated planning environment covering PATs with their Pre-Assembly, Assembly and Post-Assembly activities.

As was said before, every work package belongs to a single work category[27] and is usually described by an individual Work Package Description (WPD). The latter consist of a description of the work to be performed in terms of tasks/activities, among others. It would, therefore, now be suitable to establish a rule of how to schedule plan threads with their tasks/activities of both different PAT-elements as well as of uncommon, view-specific modules in relation to integrated planning work packages.

It is suggested here that no work packages are established where their tasks/activities cover more than or stretch beyond one PAT-phase or different PAT-element types (that is, basic or assembly), respectively. In other words: work packages should consist of tasks/ activities, which may belong to different PAT-elements but which belong to the same PAT-element type as well as the same PAT-phase type (that is, Pre-Assembly, Assembly or Post-Assembly), see Figure 13.14 part c.

It is also suggested that the entire scope of every work package is covered – and can be schedule planned – by an integer number of threads, see again Figure 13.14 part c. Each thread belongs to only one work package. In other words: there is a one-to-many relationship between a work package and its threads. All threads of a work package therefore belong to the same work category *and* the same PAT-phase type *and* the same type of PAT-element – as does the work package itself. Figure 13.14 part c depicts recommended ways of how to group threads to work packages.

During the Pre-Assembly phases, tasks/activities are, for example, created and schedule planned to generate design data (such as drawings, 3D models and so on) for basic and assembly components as well as data to manufacture basic components (such as NC codes). For the Assembly phases, tasks/activities are created and schedule planned to produce assembly components. But to actually produce components and assemblies, Manufacturing needs to issue routings which describe the sequence of tasks/activities to be performed.

For Manufacturing applications, it is therefore suggested here to understand the content of threads as being equivalent to the content of routings (or of work orders, which are routings dedicated to specific aircraft). This would allow to group Manufacturing tasks/ activities to threads in GANTT-charts attached to PAT-elements. In terms of planning, work orders are, thus, represented by threads. In other words: it should not take more than one work order to produce a basic component out of a design data input or to assemble an assembly component out of delivered components at the corresponding AS.[28] For the Pre-Assembly phase this means that threads describing the work orders for the manufacture of different basic components *cannot* be grouped into one work package. However, for all other, non-manufacture threads this is quite possible, see Figure 13.14 part c.

---

27 Of course, the total work scope to be performed within a work category may be covered by more than one work package.
28 Of course, more than one work order per AS are possible.

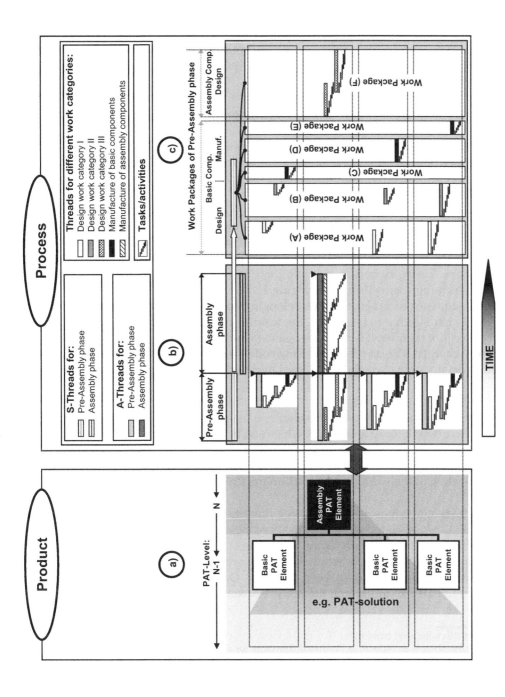

Figure 13.14   How threads link into work packages and vice versa (Post-Assembly phase not shown)

To adhere to the Manufacturing principle of 'one work order = one work package' is particularly important in cases where the assembly process belongs to a configured PAT-element, that is, to a PAT-solution (PS). It was suggested earlier that every PS should be produced by one work order only. A thread representing such a work order would then automatically also represent the corresponding PS. This is why it was stated in Figure 13.11 that all actual assembly process steps should be grouped under one thread. This is required to allow Manufacturing to issue one single work order for all individual assembly steps of a PS. Therefore:

- for a given Assembly phase, one work package corresponds to one work order, consisting of one thread;
- for the other phases, one work package consists of threads of one or more PAT-elements, provided they belong to the same work category. This is except for threads covering basic component manufacture. Here again one work package consists of one thread.

For the PATs, it is suggested that work packages should only be generated for PSs and not for lower-level PAT-elements. This is an important contribution to make the management of complex projects such as commercial aircraft development more pragmatic, as was required above. The work packages would then 'reside' on the S-thread level of the PAT-pyramid, which represents the PS, see Figure 13.14c.

When a sequence of tasks/activities belonging to the same work package in any of the GANTT-charts is due to be started, the team in charge requests a booking code, also known as account number. It is this code against which workload (such as manhours) and costs can be charged. Release of the booking code is authorised if a WPD exists and if the estimated cost does not exceed the allocated budget.[29] A booking code should identify:

- the aircraft type (or standard);[30]
- the aircraft version (where applicable);
- the aircraft rank (only for threads which can be allocated to individual aircraft);
- the corresponding work package, thus also identifying the corresponding WPD and work category; as well as
- the location of the work package within the Product architecture.[31]

Thus, a booking code is a work package identifier. By consequence, all threads of a work package are associated with the same booking code allocated to this work package. However, the company's cost accounting processes should not only just allow to record workload against a booking code but should also require to simultaneously record the thread for which this workload has been used up.[32] Only when using such a mechanism it can be ensured that also individual threads are linked to the correct booking codes.

---

29 For budget allocation see Chapter 15.
30 Assuming that there is an individual Product architecture for each aircraft standard.
31 For each aircraft standard, we can imagine a work package 'residing' on S-thread level of the head of a PAT-pyramid. One can identify the location of the work package by going downward the Product architecture into the PATs, there following PAT-levels from top to bottom to the right PAT-element (that is, the right PAT-pyramid head), continuing further down via one of the three PAT-phases, then 'jumping' into the GANTT-charts, further continuing the way down to finally arrive at the level of the S-threads.
32 Note that it is not suggested to record the tasks/activities for which these hours have been used up. This is another contribution to make the management of the complex development progmatic more pragmatic.

Finally, if for each work package an elementary cost unit as shown in Figure 10.1 is created, then booking codes represent elementary cost units too. In fact, booking codes usually are the codes identifying the elementary cost units. The latter provide the link between cost accounting on the one hand, and Project Management represented by work packages as well as work orders and their threads on the other.

Let us take a break and revisit all the suggestions which have so far been made in this chapter in order to define what work packages in a fully integrated planning environment would be all about. Figure 13.15 provides a summary.

WHAT IS A WORK PACKAGE ?		**General Rules:**	Thread $\Leftarrow \Sigma$ Tasks/Activities		A-Thread $\Leftarrow \Sigma$ Threads	
		Pre - Assembly Phase		**Assembly Phase**	**Post - Assembly Phase**
in terms of ↓		Design	Manufacture		
Basic Component(s) / Element(s)	Work Breakdown Structure	ONE Work Category	ONE Work Category	/////	/////
	Schedule Planning	Σ **Thread(s)** (of different basic components)	ONE Thread	/////	/////
	Product Output	e.g. 3D Model(s)	ONE Basic Comp.	/////	/////
	Configuration Management	/////	/////	/////	/////
	Finance	ONE Cost Unit / Booking Code	ONE Cost Unit / Booking Code	/////	/////
	Component Manufacture	/////	ONE Work Order	/////	/////
Assembly Component(s) / Element(s)	Work Breakdown Structure	ONE Work Category	/////	ONE Work Category	ONE Work Category
	Schedule Planning	Σ **Thread(s)** A-thread $\Leftarrow \Sigma$ Thread(s)	/////	ONE Thread A-thread $\Leftarrow$ ONE Thread	Σ **Thread(s)** A-thread $\Leftarrow \Sigma$ Thread(s)
	Product Output	e.g. Σ 3D Model(s)	/////	ONE Assy. Comp.	Validation Results
	Configuration Management	/////	/////	ONE PAT-solution	
	Finance	ONE Cost Unit / Booking Code	/////	ONE Cost Unit / Booking Code	ONE Cost Unit / Booking Code
	Component Manufacture	/////	/////	ONE Work Order	/////

$Y \Leftarrow \Sigma X$: Y consists of one or more of X	

Figure 13.15   Architecting rules for Work Packages.
            By type of phase and component

Threads: The Attachment Points for Additional Project Data

In Figure 11.18 it was shown how Essential Data such as Product, Process and Resources information as well as the physical components are linked to the PAT-elements. In the meantime, we have learned how the project schedule planning aspects are introduced to the PATs. It is therefore now possible to describe precisely the way in which data is attached to PAT-elements.

Figure 13.16 shows the complete arrangement of data attachments to a PAT-element. In principle, it is suggested that threads should be the platforms to which Essential Data can be attached. Attachments are generally regarded as the output of threads (Product data) or as required to perform the tasks/activities of a thread (Process and Resources data). As a rule, Product data is always attached to the thread which has generated it, with Process and Resources data required by it.

> For example, to generate a 3D model for the Digital Mock-Up (DMU), design tasks/activities are required. They are to be planned and grouped as a work package, represented by, say, one thread. The output of the thread is a 3D model (that is, an essential Product data). The thread represents one of the company's work categories (such as 'Design in 3D') as well as the resources needed to perform the design.

Note that this represents an important change from many existing architectures where Product data is directly attached to a structural node without a link to a thread which has produced it. But to regard Product data as the output of threads provides the integrative link between Product data and Project Management.

Note also that it is not suggested to attach Process and Resources data to the thousands of tasks/activities which are schedule-planned during an aircraft development project. It is far more pragmatic to only use tasks/activities to structure and schedule plan the work to be performed under a thread, while to attach most of the relevant additional data to the threads. In fact, the tasks/activities already represent most of the Process data attached to threads themselves.

Physical components, that is, produced basic or assembly components, are to be understood as the output of work orders which are represented by threads in schedule plans. They are schedule planned to be available at the next production stage at the date of the thread's finishing milestone. This finishing milestone is always equivalent to the MFD of the corresponding A-thread.

In a way, therefore, physical components can also be understood as being A-thread level outputs of the Pre-Assembly or Assembly phases, respectively. Of course, there is no direct hard link between the physical components and the integrated architecture layout described here. But the former are linked to the latter via the part numbers associated with each component. The part number provides the link between the physical world and the architecture employed with all its data attached, see Figure 13.16.

Threads are in the centre of the sequence of architecting steps which need to be taken for every new PS – during initial development as much as during later management of modifications – to achieve project architecture integration. The four steps are as follows, see Figure 13.17:

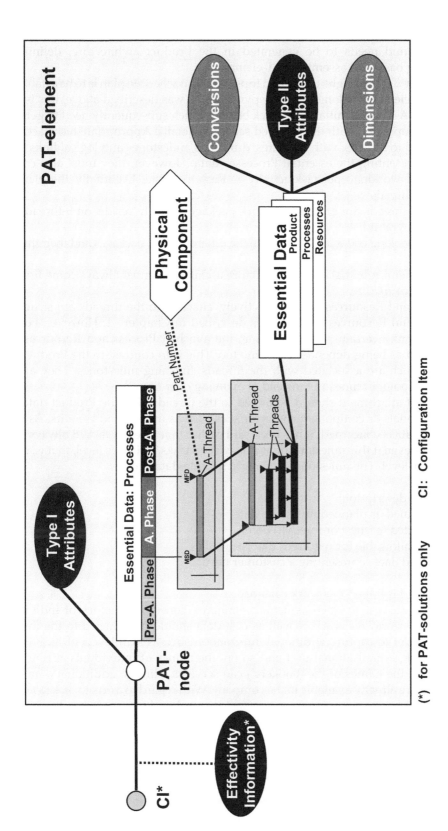

Figure 13.16 Data attachments to Product & Assembly Tree elements and their threads

I.   a PAT-pyramid needs to be generated in the Product architecture, defining its
     constituting basic and assembly PAT-elements;
II. a sufficiently detailed right-to-left and top-to-bottom schedule plan is to be established
     for all PAT elements belonging to the pyramid. As was described above, this is done
     by creating A-threads initially as 'black boxes', which subsequently need to get more
     and more populated with the required schedule details. Appropriate tasks/activities
     are grouped to threads, with the dates, durations, milestones and logical links of the
     former representing the essential Processes data. However, they must also contain
     the planned workload per task/activity as Resource data. Eventually, the plan also
     contains S-threads;
III. the various threads are grouped to work packages, which 'reside' on S-thread level
     of the PAT-pyramid;
IV. individual booking codes are allocated to the different work packages and their threads.

Thus, the general rule that Product structure architecting must always come first also
applies here, of course.

The Process and Resources data, respectively, attached to the threads are structured
in the Process and Resources architectures described in Chapter 11. However, because
threads span across certain periods of time, the attached Process and Resources data
can be regarded as being dependent on time too. This is in contrast to the Product data
attachments which are associated with the threads' finishing milestones. They appear,
thus, at a given point in time only instead of evolving over time.[33]

What kind of attachment should be linked to the threads? For the Product data, this
could be 3D models or component characteristics such as masses and weights. As far as
the Processes data is concerned, it has been said earlier that threads should always finish
by a milestone, even if this milestone does not exist on tasks/activities level, that is, on the
lowest GANTT-level. The milestone data might include data such as:

- milestone description;
- the scheduled or planned milestone date;
- the predicted (earliest or forecast) date;
- the latest allowable (or required) date; as well as
- the desired date representing a customer need.

Also, each thread is linked to a work category.

Finally, Resources data, such as actual manhours consumed, names of individuals
nominated to perform the work as well as cost data should be attached to the threads.
Note that in order to capture additional information such as skill levels of individuals,
location of work, organisational unit and so on, the names of individuals attributed to
threads (that is, the names of the 'bookers') can serve as a link to additional databases,
which usually are already available in the company. With regard to a link to organisational
units this works because employees can in fact themselves be regarded as the smallest
organisational units of an Organisation Breakdown Structure (OBS). In other words:
bookers to a thread provide the link to the OBS.

---

33 Depending on how in detail these data-sets are structured, individual threads have one-to-one, one-to-many
   or many-to-one relationships with the attached data. One-to-one: one X corresponds to exactly one Y; One-
   to-many: exactly one X corresponds to one or more Y; Many-to-One: one or many X correspond to exactly
   one Y.

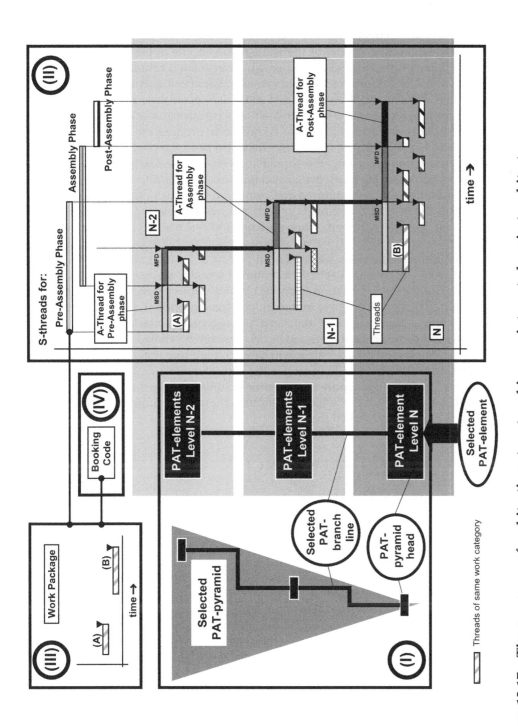

Figure 13.17 The sequence of architecting steps to achieve an integrated project architecture

The latter may then provide the links to further databases on the characteristics of organisational units, such as hourly rates, economic conditions (for example for units abroad), and so on. Whenever, therefore, an individual books his or her workload in hours (or machine hours are charged) against a certain booking code, the integrated system of databases immediately is 'aware' of the skills involved, the applied hourly rate, the organisational home of the individual, and so on. Hour-related actual costs can then be determined and analysed easily, responding to a large variety of queries.

Because of the architectural links described so far, it is possible to make comparisons between work progress on the one hand and actually consumed workload (cost) on the other. People working for a work package record their progress on tasks/activities or thread level (see Chapter 15 for more details). They also book the used workload against the booking code while recording the thread to which those tasks/activities belong. Because threads aggregate the work progress data of their tasks/activities, progress and workload consumption can be monitored and compared with each other on thread level. And because threads belong to A-threads as well as to work packages and S-threads, the same can also be done on A-thread, work package and S-thread level respectively, see Figure 13.18.

Note that with the suggestions made here in favour of an integrated architecture, a comparison of work progress and actually consumed workload is not possible on tasks/activities level. This a drastic deviation from the classical Project Management approach, where it is required that work progress and workload consumption should be comparable on the lowest level. Note also that data can be aggregated via both the thread→ A-thread→S-thread route or via the thread→work package→S-thread route.

We now have almost all architectural tools in hand to create integrated plans linked to an integrated architecture. What is still missing is how interdependencies between schedule plans are dealt with. In addition, only the schedule planning for the first aircraft has been described so far. Thus, the way in which the industrial ramp-up should be planned and prepared still needs to be described too.

MANAGING INTERDEPENDENCIES

What Interdependencies Management is all About

So far we saw how the PAT-structure can be filled with planning data. However, initially there will be many schedule mismatches. They will become apparent when looking at the many interdependencies between milestones of, for example, Pre-Assembly and Post-Assembly phases of *different* PAT-elements or between PAT-elements and uncommon modules of the Product architecture. For the Assembly phase, the PAT-structure itself provides a natural interdependency logic. For the other phases, something else is needed.

The main purpose of Interdependencies Management between the supplier of an output and the customer of an input is to achieve agreement in terms of content of a deliverable and its delivery date. Typical interdependencies between teams during aircraft design and development include the delivery of items like:

- requirements/specifications;
- model data (Finite Element Model (FEM), DMU, Space Allocation Model (SAM) and so on);

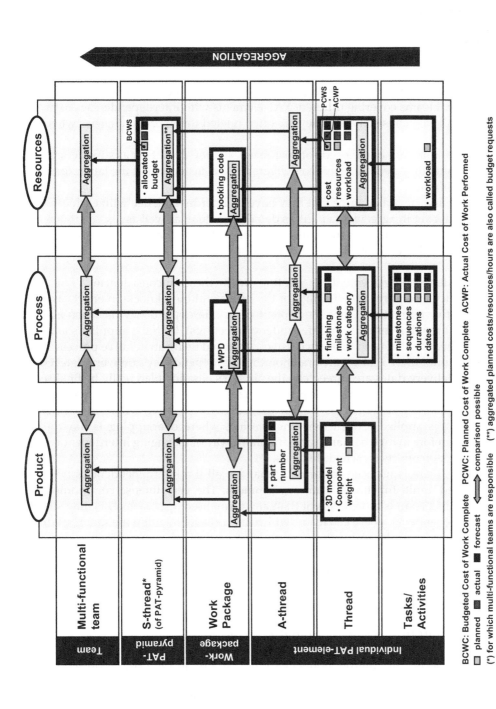

BCWC: Budgeted Cost of Work Complete   PCWC: Planned Cost of Work Complete   ACWP: Actual Cost of Work Performed

□ planned ▨ actual ■ forecast ⟷ comparison possible

(*) for which multi-functional teams are responsible   (**) aggregated planned costs/resources/hours are also called budget requests

Figure 13.18   Allocation and aggregation of important project data for a Product & Assembly Tree solution

- loads data;
- masses and weights;
- Data for Manufacture (DfM);
- assembly, installation, frontier drawings; and even
- physical parts;
- and so on.

Initially, where different schedule plans are produced independently by different multi-functional teams covering different PAT-elements with interdependencies between them, there will be inconsistencies. They are often related to interdependencies where:

- teams are unaware of the needs of customers, so they do not act as suppliers;
- teams think they need to deliver a deliverable to a customer but the latter does not agree with this;
- supplier and customer agree that they have an interdependency with each other but are not yet in agreement about its detailed content; as well as
- the need date for the deliverable as expressed by the customer is earlier than the earliest possible delivery date stated by the supplier.

Interdependency Management is dealing with these inconsistencies. For any interdependency it requires full visibility of the issues and their knock-on effects as well as their proactive mitigation. Once the content of an interdependency has been agreed (for example through a series of interdependency meetings), both the supplier's and the customer's schedule plan need to be successively adjusted using an iterative process. This must be done until there is also agreement on the interdependency's delivery date. Only once the key delivery dates are agreed can the aircraft's baseline plan really be finalised.

Thus, there is a definite link between the management of interdependencies and the schedule planning task. This is true in particular for all areas where the principle of overlapping is applied. Concurrent Engineering, where overlapping is day-to-day business, is therefore in high demand for a pragmatic way of managing interdependencies and their knock-on effects on schedule plans.

Fortunately, there is no need to formally manage all the interdependencies *within* a fully co-located, multi-functional team environment. The very intense communication going on within a team is more efficient than any formalised approach. This may still be true for interdependencies between teams which are co-located within the same building. However, research results show that the amount of desired communication reduces drastically with distance, even if teams are located close to each other.[34] As a rule of thumb, therefore, formalised Interdependency Management should become the more important the more geographically apart teams are.

Interdependencies Between Schedule Plans

In smaller projects, all interdependencies are usually graphically represented as connecting lines in a single GANTT-chart. However, complex projects such as the ones for aircraft development require the management of a myriad of interdependencies.

---

34 See for example: Allen, T.J. (1977), *Managing the Flow of Technology* (Boston, MA: MIT Press), p. 239.

Representing all these by connecting lines in a GANTT-chart (and maintaining them on a regular basis) not only binds many planning resources – which are just not available in commercial aircraft development for the reasons outlined above – but also leads to GANTT-charts which lack the clarity required by Management.[35]

While looking for a pragmatic way of managing interdependencies, another constraint still needs to be taken into consideration from a planning perspective. It is unrealistic to believe that there is a common schedule planning tool available for all multi-functional teams – in the aircraft-developing company as well as with the many suppliers in the supply chain. Instead, teams will use different tools as available. If new tools get introduced, then at best *more* teams use the same tool compared to the past – but probably never *all* teams. If data is to be exchanged between schedule plans – which is certainly the case if we want to manage interdependencies – then some semi-automatic or manual effort of translating data from one tool to the next is inevitable. Any Interdependency Management process must therefore prescribe a regular (perhaps monthly) procedure outlining the automatic as well as the manual steps.

However, as it so important to manage interdependencies across team borders properly to achieve project success, it would be preferable to use a common database for all interdependencies for which a structured management is necessary. This database should be Internet-based so that everyone participating in the aircraft development, including vendors, is capable to access and edit it.[36] It should be proactively used by both the supplier and the customer of an interdependency deliverable. This could, for example, be done at dedicated team interface meetings, during which updates of the common interdependencies would be directly fed into the database. Such tools are available and have, for example, been used by Airbus for the development of the A380.

Note that the use of a common database should be limited to interdependencies between multi-functional teams where the latter do not have access to each other's planning tools.[37] In fact, if aircraft-developing teams are not using the same schedule planning tool, an interdependency database serves as a vital linking element between them.

What could this link look like? A pragmatic approach for linking a common interdependency database with the different planning tools could be as follows: each interdependency deliverable in the database represents an input milestone in the customer's schedule plan and an output milestone in the supplier's plan, see Figure 13.19. They are represented in these plans on the lowest GANTT-level as an input and output milestone respectively, starting and finishing a specific task/activity, respectively. If different tools are used by the customer and the supplier for their GANTT-charts, the milestones in there will have different identification numbers. These numbers should be captured in the interdependency database as attributes to the interdependency. In this way, a virtual link is established between the database and the GANTT-charts.

---

35 As a result of how the PAT-structure was set up, there are two exceptions to this: (1) the interdependencies between the Pre-Assembly phase preceding the Assembly phase and the Assembly phase preceding the Post-Assembly phase, respectively, of the same PAT-element; as well as (2) the Assembly phases of PAT-elements on same PAT-branch line. Both are always graphically represented by a connecting line in the relevant GANTT-charts.

36 Access to certain areas of the database may be limited to certain users only (for example airline customers).

37 Because otherwise the common planning tool could be used as an interdependency database in the first place.

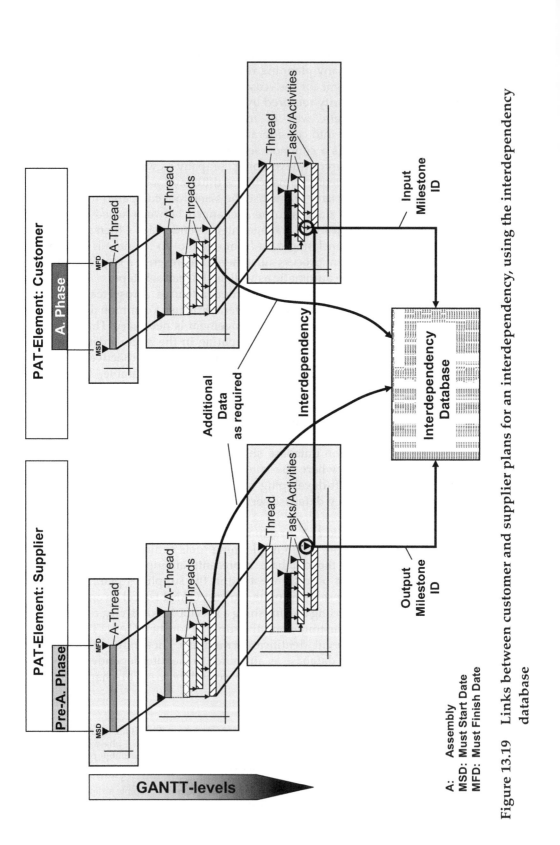

Figure 13.19 Links between customer and supplier plans for an interdependency, using the interdependency database

As tasks/activities are grouped into threads, it can now be identified which interdependency milestone belongs to which thread in both the customer and the supplier schedule plans. This offers the significant advantage of linking the interdependency milestones to other relevant data such as the associated WPD, the PAT phase, the aircraft rank, and so on. As mentioned above, every WPD needs not only to contain a description of the work to be performed by each of its threads, but also a list of required inputs and produced outputs. Thus, it can now be checked whether an interdependency – represented by a virtual link in the database or directly in GANTT-plans – is reflected in the WPD and vice versa.

As was described earlier, milestones need to be categorised according to their potential impact to the Critical Path. If interdependency milestones are on the Critical Path – or are otherwise regarded as critical to a team's project success – they become Critical Interdependency Milestones.[38] Consequently, they are depicted on Medium-level GANTT-charts. This ensures the planning clarity required by Management. Obviously, the criticality of interdependencies needs to be revisited on a regular basis.

In summary, a typical interdependency database may include any of the following data:

- the identification number (a) of the milestone where the interdependency is an output;
- the identification number (b) of the milestone where the interdependency is an input;
- the booking codes of the threads, to which the milestones belong;*
- the responsible teams* associated with (a) and (b), respectively;[39]
- the definition or description of the interdependency;
- the (hyper-) link to the actual deliverable (such as a drawing) once delivered;*
- the need, agreed,* forecast and actual delivery dates of the deliverable;
- the importance of the interdependency for maintaining the customer's Critical Path; as well as
- any additional information (such as a part number, drawing number, equipment number, and so on).

Data marked by an asterisk could be loaded into the database automatically, as it would already be attached as Essential Data to the involved threads.

In so far as interdependencies are related to the transfer of physical hardware (basic components, assemblies), the database becomes a parts tracker at the same time. It can provide valuable support when chasing parts for the build of the first aircraft.

---

38 As modern schedule planning tools calculate the Critical Path automatically, it is possible to identify the Critical Interdependency Milestones in the interdependency database too.

39 The teams in charge of threads are identified automatically by the system as threads have been linked previously to the OBS.

PREPARING FOR SERIES PRODUCTION PLANNING

Introduction

So far we have seen the establishment of all plans for all applicable phases of all PAT-elements for the first aircraft. This chapter introduces the industrial planning needed to ramp-up and then maintain the production of a series of aircraft.

Disregarding for the moment any configuration change or customisation aspects (as described in Chapter 12), as well as any learning curve effects, the planning of the industrial ramp-up can be considered as being essentially composed of copies of the schedule plan for the first aircraft which are staggered in time. For the first aircraft we learned that all A-threads, threads, tasks/activities and milestones are somehow linked to the overall MFD of that first aircraft. Knowing a new MFD for the second aircraft, all linked A-threads, threads, tasks/activities and milestones for the second aircraft can be comfortably calculated provided everything else remains constant. The same can be done for the third aircraft and so on. Thus, in principle, all necessary dates for all aircraft ranks for the basic and assembly components as well as all necessary resources for each PAT-element can be calculated using this approach.

The copies represent periods of time, which successively lie more and more in the future. They are linked together by Finish-to-Start relationships. Why is that? Many jigs, for example, will only be available once, so that the corresponding parts of the aircraft will all have to go *successively* through the same 'bottleneck' represented by these jigs. In the GANTT-charts so far described we cannot see this link because they are purposefully constructed to plan the tasks/activities for *one* aircraft. The tasks/activities are scheduled along the dimension of time, which had been introduced earlier as the 1st dimension of the time. But now we can imagine that the staggering of GANTT-chart copies happens along a 2nd dimension of time, which is perpendicular to the 1st dimension, see Figure 13.20.[40]

If we link together the copies of the schedule plans for successive aircraft along the 2nd dimension of time it is possible to view, for example, individual threads and their associated resources for the Assembly phases of selected PAT-elements *across all aircraft ranks*, see Figure 13.21.[41] This allows, for example, to identify whether the always existing constraints with regard to availability of resources or jigs & tools can be met or not. As a result, either these constraints need to be changed (that is, additional resources or jigs & tools are added) or the overall MFD for the corresponding higher rank aircraft needs to be shifted along the 2nd time dimension.

Now, the combined effects of:

- outstanding work and rework activities;
- learning curve effects;
- leadtime variations resulting from configuration changes and customisation; as well as
- shop floor line balancing;

---

40 Note that the terminology used here (1st and 2nd dimensions of time) is for better explanation only. Clearly, time has only one dimension.
41 In Figure 13.21 the two mentioned time dimensions are merged again into a single one.

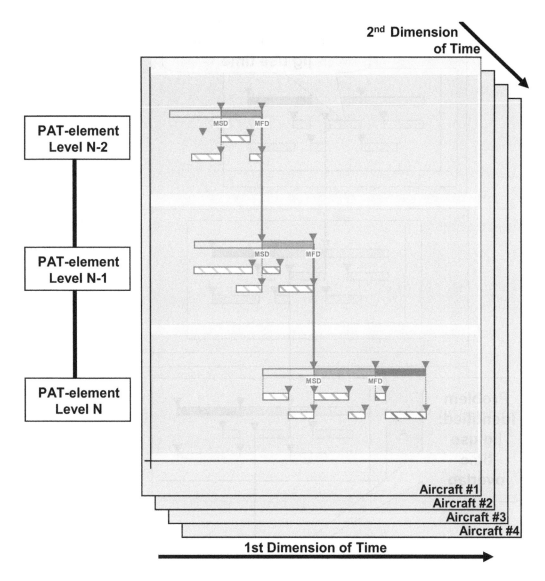

Figure 13.20   The concept of two time dimensions for schedule planning

which all result in an updated task/activity planning on the lowest GANTT-level, can be applied to all relevant PAT-elements.[42] This essentially changes the leadtimes of the Assembly phase of the PAT-elements. The resulting new durations, MSDs and MFDs for all PAT-elements and all aircraft ranks can then be calculated using a top-down and right-to-left planning approach.

For higher aircraft ranks, baseline component design is no longer necessary. Instead, design is dominated by modifications intended to change the aircraft's configuration.

---

42 When the industrial plan is set up, estimated learning curves based on historic experiences or analyses of best practices elsewhere can be applied to each individual PAT-element. However, learning curves may not necessarily follow a simple mathematical rule as described in Appendix A1, but might need to be adjusted manually on an aircraft by aircraft basis.

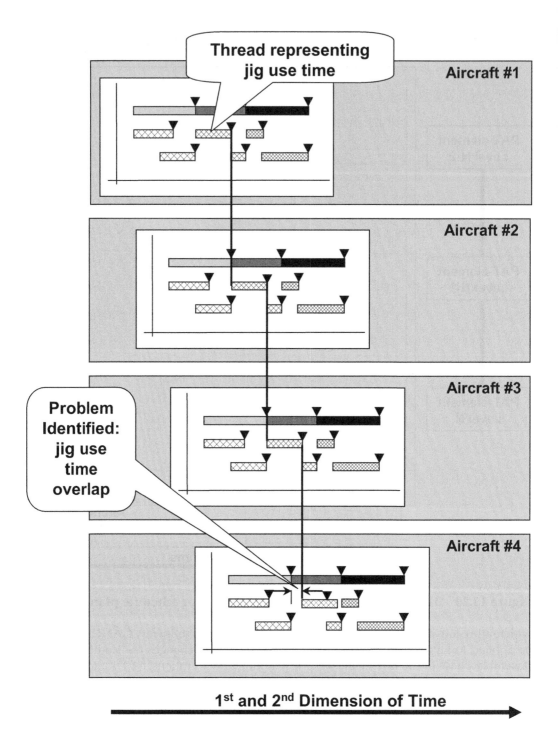

Figure 13.21   Schedule planning across aircraft ranks

The processes required to manage modifications were explained in Chapter 12. It remains to be noted here that all three PAT-phases for all PAT-elements and aircraft ranks, which are affected by modifications, need to contain the corresponding detailed schedule plans for the analyses, approval and implementation for every modification. This means that individual aircraft have in most cases individual schedule plans too, even in Series Production.

All necessary schedule plans for all Product architecture-elements for all aircraft ranks are now linked to the same, integrated architecture. This is a remarkable achievement, which fulfils the promise made in Chapter 11.

Industrial Demand

Thanks to the Engineering/Manufacturing agreement on common PAT-sub-structures of the larger Product architecture, the Bill of Material (BoM) of, for example, a PS can be easily generated. In fact, any one item on the list is associated with an individual PAT-element. Thanks to this, the complete BoM is available for each aircraft rank.[43, 44] In addition, the (industrial) schedule plan for each aircraft and each PAT-element is known, as was described above. As each assembly PAT-element consists of an Assembly phase, the MSDs are known for all Assembly phases of all PAT-elements for all aircraft ranks. Remember, the MSDs describe the dates by which parts need to be available for assembly.

The combination of links, planned dates and BoMs represents the 'demand' with regard to parts. Among others, it describes for each Assembly phase:

- which BoM items are required;
- when they are needed (that is, MSDs);
- where they will be used (that is, location of assembly station);
- which serial number (or aircraft rank, MSN) the BoM-item will be used for;
- which team is in charge to use them;
- which team/supplier is in charge to deliver them; as well as
- additional data, such as weight, geometry and so on.

The complete industrial demand represents 100 per cent of all components which are needed to build the aircraft.

As a result of the industrial demand being fully established and transparent, production lots for the same type of BoM items and for a range of aircraft MSNs can be determined. This should take into account that certain BoM items are used at quantities >1 on each aircraft already, while their architectural counterparts – the PAT-elements – may well be located at different positions within the PAT-structure. This was discussed above.

Because of all the additional attributes linked to the PAT-structures in one way or another, it is also possible to identify the batches which need to be procured from external suppliers – and where therefore Purchase Orders have to be issued – and those which are to be produced in-house.[45] Purchase Orders for all PAT-elements are usually issued and

---

43 This even includes the changes to the BOM as a result of modifications, as was described in Chapter 12.
44 There is a slight difference between an Engineering BoM and a Manufacturing BoM: the Engineering BoM consists of all components of the aircraft while the Manufacturing BoM includes all parts required to manufacture the components. However, whatever is required can be represented using the PATs.
45 In case in-house parts/assemblies are to be delivered to another company (for example to a supplier), Sales Orders have to be issued.

managed by the Procurement Function. This represents another example of how the latter is linked to the common PAT-architecture.

When production lots get delivered to Logistics (after having passed inspection at the 'Goods-Receive' station), they replenish the reduced stocks. With the demand-information at hand, Logistics can be instructed some time before start of assembly activities to prepare all BoM items required at the ASs, that is, to collect them from different lots and to commission them into kits to be transported Just in Time to the ASs. Here, they get consumed and assembled according to the schedule plans described in the GANTT-charts of the PAT-elements' Assembly phases.

AT LAST: THE LINK BETWEEN PRODUCT, PROCESSES AND RESOURCES

We now have all architectural tools in hand to create integrated plans linked to a Product architecture. This is summarised in Figure 13.22. Individual elements of the Product architecture are divided into three phases. For each of the phases a three-fold cascade of GANTT-schedule plans can be established. A-threads represent a phase on the highest GANTT-level. Threads are scheduled on the Medium level, while tasks/activities are planned on the lowest GANTT-level. The Pre- and Post-Assembly phases threads represent major gates to time dependent Product, Process and Resources data. Where tasks/activities planning would be too elaborous (such as for planning of individual drawings), databases can be attached to and accessed via the threads to facilitate schedule planning. Product data which is not dependent on time (such as data related to weights or deliverables like specifications, actual drawings once released, DMU models and so on) is attached to and can be accessed directly from the PAT-elements. An Internet-based interdependency database is used to facilitate the management of interdependencies between tasks/activities of schedule plans belonging to different elements of the Product architecture.

Wherever there is an agreement between Engineering and Manufacturing (and possibly other Functions) on how to embed an element of the classical PBS into the overall Product architecture, it becomes a common one. Common elements form sub-structures in the Product architecture and are called PATs.

Much of the aforementioned in this chapter is principally valid for both the common as well as the uncommon elements of the Product architecture. Thus, it is possible to attach schedule plans to the individual phases of uncommon elements too. They would in this case contain tasks/activities specific for a view. Pre-Assembly phases of uncommon elements are mostly used to plan non-specific or ATA-specific design activities (see Chapter 11). The Post-Assembly phase can be used to plan activities such as for flight testing.

But there are three major differences between the common and uncommon elements:

1. The aggregation, summation or listing – whatever applies – of the data attached to all PAT-elements always yields the total aircraft aggregate, sum or list, respectively. In contrast to this, the aggregation, summation or listing of the data attached to those elements which are only required to generate a specific view, yield the aggregate, sum or list, respectively, of that view.

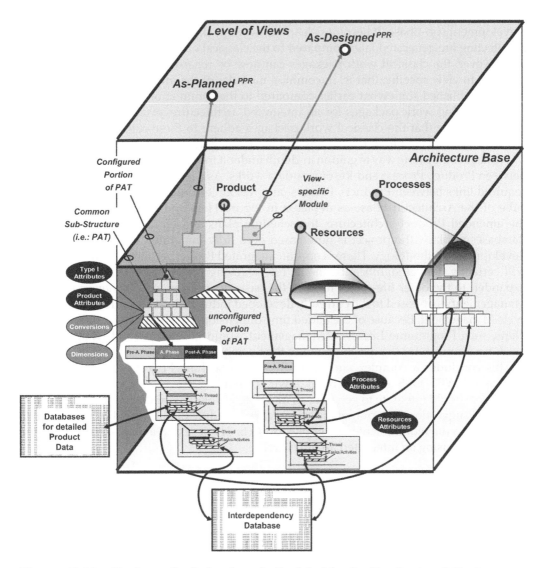

Figure 13.22   Project schedule plans imbedded in the Product architecture while enabling links to Process and Resources data

2. An integrated schedule plan of PAT-elements, which are being successively integrated into assemblies, shows the corresponding Assembly-phase A-threads connected by Start-to-Finish links. This is not the case for uncommon modules.
3. Data attached to uncommon modules is not directly configuration controlled, while data attached to common ones is, see Chapter 12.

Note, that the (specific) WBS and its constituting work packages were above introduced as a prerequisite to project manage the aircraft development with its associated complexities. It had been said that these classical work packages were to be generated around the *beginning* of the Development sub-phase. Having gone in this chapter through the full explanation of an integrated architecture, which is only to be developed *during* the

Development sub-phase, it becomes clear that work packages associated with integrated architecting are generated later compared to the classical ones, see again Figure 13.1.

However, the classical work packages can now be regarded retrospectively as to be attached to view specific, that is, uncommon modules of the integrated architecture and can be established somewhat earlier compared to the common ones. Knowing the rules for establishing work packages for an integrated architecture, every attempt should be made to ensure that the classical work packages adhere to these rules too. This greatly eases the gradual evolution from a classical to the integrated architecture.

We have come a long way to gain an in-depth understanding of how the interconnectivity between Product, Process and Resources data works. As a result, we can now say that the required links between Product-, Process- and Resources-data rather happens *within* the cube of the Architecture Base as depicted in Figure 11.16 and not on its surface. With the amended Product architecture, the schedule plans as well as the interdependency databases in place, the project is now based on a fully integrated Product and Process development methodology. There is now an integrated Project architecture in place which will certainly deliver significant benefits for achieving project success. It can be easily expanded to the other life-cycles to form the basis of an extended *Programme* Life-cycle Management as opposed to the currently developed industrial standards for Product Life-cycle Management. Because of the added time dimension, covering for example schedule plans, such Programme Life-cycle Management would go well beyond what has been used, for example, for the development of the Boeing 787.[46]

This concludes a complex and difficult chapter on how Project Management tools such as the WBS, OBS and Cost Breakdown Structure (CBS) as well as the planning of schedules should integrate to generate an integrated Project architecture. In the following chapter, a suggestion for a pragmatic approach to schedule planning will be presented. It is adapted to the requirements of complex projects developed in a multi-functional team environment using an integrated architecture.

---

46 Compare with quotation on page 43.

MAIN CONCLUSIONS FROM CHAPTER 13

- An integrated Project architecture can be achieved if the individual elements of a classical Product architecture are divided into three timely phases: Pre-Assembly, Assembly, Post-Assembly (called 'phase-types'). Each element is then no longer a representative for a physical component only, but also for the *processes* necessary for its creation and operational use.
- For each of the phases a three-fold cascade of GANTT-schedule plans can be established. A-threads represent a phase on the highest GANTT-level, threads are scheduled on the Medium level, while tasks/activities are planned on the lowest GANTT-level.
- Product & Assemby Trees are sub-structures to the integrated architecture, consisting of modules common to all defined views. An integrated schedule plan of Product & Assemby Tree-elements, which are being successively integrated into assemblies, shows the corresponding Assembly-phase A-threads connected by Start-to-Finish links.
- Threads should be the platforms, to which Product data such as Digital Mock-Up models are attached, as well as Process and Resources data required to perform the tasks/activities necessary to generate the Product data.
- Tasks/activities required to develop and build an aircraft are scheduled along the dimension of time, which had been introduced as the 1st dimension of the time. But one can imagine that a staggering of GANTT-charts describing successive aircraft builds happens along a 2nd dimension of time, which is perpendicular to the 1st dimension. The 2nd dimension of time can be used to schedule plan the project's industrialisation as well as industrial demand.
- Any work package established within an environment of an integrated architecture must adhere to some clearly defined rules, because otherwise the link between Product-data and Project Management does not work.
- It is possible to attach Product, Processes and Resources data to the individual phases of uncommon elements too. They would in this case contain tasks/activities specific for a view.
- Work packages introduced for the (specific) Work Breakdown Structure as a prerequisite to project manage the aircraft development can now be regarded retrospectively as to be attached to view specific, that is, uncommon modules of the integrated architecture. However, knowing the rules for establishing work packages for an integrated architecture, every attempt should be made to ensure that the classical work packages adhere to these rules too.
- With the amended Product architecture, the schedule plans as well as the interdependency databases in place, the project is now based on a fully integrated Product and Process development methodology. There is now an integrated Project architecture in place which will certainly deliver significant benefits for achieving project success. It can be expanded to the other life-cycles to form the basis of proper Programme Life-cycle Management.

PART V

Using Integrative
Project Architectures

14

Planning Mega-Projects in a Pragmatic Way

PROJECT TEAM PLANNING PART I

Traditionally, those in charge of schedule planning within a team had to chase other team members to get the required information in order to set up and maintain the steering plan. A series of bi-lateral meetings was therefore required. However, in a multi-functional team environment bi-lateral meetings are no longer regarded sufficient to sort out all the schedule conflicts and discrepancies. In addition, 'this process is inefficient and consumes much time of all concerned, compounding the dislike most people have for planning in the first place. This round robin [or honey-bee] approach slows down the critical start-up period of the project, and does not enhance teamwork or communication.'[1] It will also generally be based on a bottom-up view of the project. For complex projects this is not necessarily the best way to kick-off schedule planning activities as plans tend to become overloaded with details and usually include too conservative assumptions.

The combination of bottom-up and bi-lateral establishment result in schedule plans which are often poorly integrated. In addition, they may harbour unrecognised conflicts that will only be identified later. However, by then there may well be insufficient time to solve them. With only bi-lateral meetings in place, resulting in a lack of integration, the plans will not reflect how the development objectives can be achieved. Consequently, it will be difficult to achieve a sense of commitment to the plans among the people collaborating within the project perimeter. The possible result is that people are not committed to the project itself.

> [Given this], 'it is not surprising to find many situations where more than one plan exists, either for the entire project, or for many portions of it. The project manager who does not believe in and is not committed to the plans and schedules produced by a central planning department will produce his or her own plans. Many functional managers and project leaders often do the same thing.'[2]

---

1   Archibald, R.D. (1992), *Managing High-Technology Programs and Projects*, 2nd Edition, (New York: John Wiley & Sons), p. 238.
2   Archibald, R.D. (1992), *Managing High-Technology Programs and Projects*, 2nd Edition, (New York: John Wiley & Sons), pp. 237–8.

Fortunately, there is a better way to kick-off and maintain schedule planning: through planning as a joint team effort. There are many advantages when involving all team members simultaneously in the planning process, such as to:

- recognise that a commonly established plan brings the team closer together – it has a team-building effect;
- capture the group wisdom in a structured way;
- get a common understanding that in order to assure overall project success all team members need to be involved in the planning process;
- demonstrate that important contributions are delivered by every team member, and how these contributions are interrelated;
- agree on overall target master plans on a team-by-team basis;
- agree on rules about how to establish detailed steering plans;
- establish the semantics, terminologies and so on to be used among the entire community related to or in charge of planning;
- introduce robust and harmonised Project Management methods and tools;
- create a shared vision about the challenges, risks, scope and objectives of the project as well as the windows of opportunities (for example, for cycle time reduction, cost reduction and so on);
- better understand the interrelationships between cost, schedule and technical performance across the entire project;
- develop planning skills among team members;
- avoid the creation of more than one set of plans as team members realise that their contributions are respected;
- demonstrate the benefits of free and open communication among team members; as well as to
- perform 'training on the job' for less-experienced team members on all the steps it takes to develop an aircraft.

As a result, individual team member commitment to and motivation for the schedule plan will be achieved through the involvement in the planning process. In addition, individual team member commitments are made to the peer group and not just to the team leader – or the person in charge of schedule planning.

Joint team planning should take place:

- in dedicated planning workshops on all levels of the organisation, starting with the highest level first (to set up a target schedule plan); as well as
- during regular meetings of the multi-functional teams (to update and maintain the steering plan).

However, at no point should team planning sessions produce the detailed schedule plans (or any detailed workload distributions). This would certainly be a huge waste of team members' time. Team planning sessions should rather be set up to set the scene for subsequent, more detailed and efficient schedule planning (with their associated workload distributions). It is essential that planning workshops are well prepared (under supervision of the team leader) and well facilitated (ideally by a professional facilitator from outside of the team). Clever facilitation during the workshop can make life a lot easier.

A good way of living through a team planning workshop is by jointly creating a Program Evaluation and Review Technique (PERT) plan. This plan represents a network of threads and milestones – as well as their logical interlinkages – which are regarded by team members as important or critical for the project. When creating the PERT plan, a holistic view should be applied.

> 'PERT is the preferred scheduling procedure for a large-scale system in which there are large numbers of events and activities that must be identified and tracked. This technique, in distinction to Gantt charting, deals explicitly with dependencies between the various tasks and activities. A PERT network is normally devised by starting with known 'end' events and milestones and asking the question: What activities need to be accomplished before this event or milestone can be achieved? By working backward in this fashion, eventually an entire network is developed. ... The PERT procedure leads to a network of serial and parallel paths of events and activities.'[3]

Very often, it is not necessary to start creating a new PERT-chart from scratch: as the way how commercial aircraft are developed does not fundamentally change from one programme to another, the team planning exercise can be performed on the basis of PERT-plans – or equivalent planning experience captured otherwise – from previous programmes.

As good planning tools are capable of converting PERT-charts into GANTT-charts automatically, PERT-charts should be set up in a way that they can be used as GANTT-charts on the appropriate GANTT-level (typically Medium level). These plans will then reflect a top-down approach using the total know-how of the team in charge of, say, a Product & Assembly Tree (PAT)-pyramid. In fact, at this point they represent 'Target Plans', which set the stage for more detailed, bottom-up planning on lowest GANTT-level. Target plans at highest PAT-levels are often called master schedules, representing the top level of the schedule hierarchy. They may consist of key milestones only but certainly cover all three phases, that is, Pre-Assembly, Assembly and Post-Assembly, respectively.

Project team planning workshops should take place shortly after the key functional representatives – including the team leader – have been appointed. If this is done by all multi-functional teams than target schedule plans on Medium GANTT-level should come into existence relatively quickly for virtually all those important aspects of the aircraft which are already known at the early stages of the Development sub-phase.

Following the team planning workshops, target plans have to be populated using more detailed plans on the lowest GANTT-level. Because the target plans for individual PAT-phases are linked together via the logics embedded in the Project architecture, it will quickly become obvious whether the resulting overall target schedule plan meets the final Must Finish Date (MFD) (for example the Entry into Service (EIS) date). If not, then schedule mismatches must be highlighted in the GANTT-charts (for example by using different colours) and teams need to work on removing them.

> 'The ... [final Must Finish Date] should be given to the team as the *primary constraint on the project*. The team should be explicitly told to propose whatever modification in project scope and resourcing that might be necessary to meet this date. Then the team can build a detailed schedule to meet the mandated introduction date. There is always a certain scope of work that can be fit into any allowed amount of time. If we are willing to be flexible on scope and

3 Eisner, H. (2002), *Essentials of Project and Systems Engineering Management*, 2nd Edition, (New York: John Wiley & Sons), p. 92.

resources we can get any schedule we need. If we attempt to tightly constrain schedule, scope, and resources simultaneously we will create insoluble problems for development teams who will respond by refusing to buy-in to these objectives.'[4]

A GENERIC WAY TO PRAGMATIC SCHEDULE PLANNING

In Chapter 13 a top-down approach to integrated planning was introduced. Based on the described Project architecture and its PAT sub-structures it is now possible to actually perform the detailed bottom-up schedule planning for all those elements where the Aggregated Threads (A-threads) (or dummy A-threads) have been defined. A pragmatic recipe for the schedule planning to be performed by a multi-functional team could be as follows:

1. Start at the lowest GANTT-level with the setting-up of all tasks/activities and associated milestones. Using an optimistic duration (D_O), most likely duration (D_L) and a pessimistic duration (D_P), which all result from a discussion between team members and other stakeholders, calculate the estimated duration D_E to be used in the schedule plan according to:

$$D_E = \frac{D_O + 4D_L + D_P}{6}$$

    Make sure that tasks/activities are logically linked with each other and milestones are not 'hanging in the air' but are linked to tasks/activities.
    The start date of the earliest task/activity and the finish date of the last one should correspond with the Must Start Date (MSD) and MFD, respectively, of the corresponding A-thread.
    The resulting plan should be truly cross-functional reflecting the planning needs of all Functions represented in a multi-functional team as well as of other stakeholders, if any.
    The available (or expected) resources within each multi-functional team in charge of PAT-elements' or uncommon modules' planning should be capable of performing regular updates of all tasks/activities. In other words, the availability of planning resources sets a natural limit to the degree of schedule detailing, thus limiting the number of tasks/activities. So only detail plans to a degree commensurate with the authorised level of resources in charge of planning.
    In order to avoid unmanageable detail in the plans, it is worth noting that for many tasks/activities or work packages simple bar charts, production rate charts with trend analysis or checklists are all that is needed, provided that the key task/activity interface events (such as 'Start' or 'Finish') are included in the project network plan. In addition, for some or even many PAT-elements it may not be possible to establish full-scale planning as the PAT-structure is not yet developed enough, so that planning for lower level PAT-elements cannot be performed.

4   Smith, P.G. and Reinertsen, D.G. (1998), *Developing Products in Half the Time* 2nd Edition, (New York: John Wiley & Sons), p. 193.

In dynamic projects such as commercial aircraft development, the quantity of change resulting from 'unknown unknowns' makes too detailed plans very difficult to maintain. Plans with excessive details are therefore often found to be misleading and are abandoned in favour of a higher-level rolling wave approach. Thus, the planning detail is progressively developed as the project progresses and as more is learned. Planning details may then only be required for, say, the next two years. This, too, will contribute to a reduction of schedule complexities.

2. Where appropriate, group tasks/activities to threads, which are represented on Medium GANTT-level. This is a major architectural job and requires careful consideration:
    –   The workload of each thread (for example in terms of Full Time Equivalents (FTEs) required to do the job) is aggregated from the workload of all lower level tasks/activities which are covered by the thread. Ideally it should show a more or less even distribution across the duration of the thread. This is an important prerequisite when it comes to a pragmatic determination of project status using the Earned Value concept, to be explained in Chapter 15. If threads tend to have uneven workload distributions, it may help to approximate the workload distribution by a sufficient number of smaller threads with even workload distributions, see Figure 14.1.

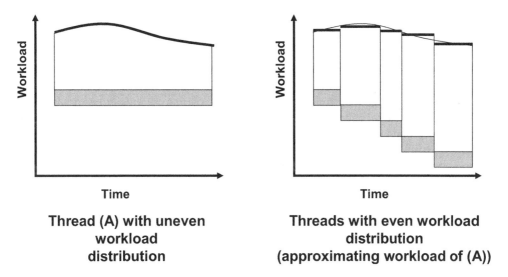

Thread (A) with uneven workload distribution

Threads with even workload distribution (approximating workload of (A))

Figure 14.1   Approximating threads with uneven workload distributions

    –   The link between threads and booking codes needs to be considered. One booking code, that is, one elementary cost unit, may cover one or more threads (see again Figure 13.15) – but no thread should ever be linked to more than one booking code. Wherever possible, there should be individual booking codes for different planned costs per unit of work. If, for example, work is performed by different departments featuring different hourly rates, there should be booking codes for each department. This is to later enable easier monitoring of Earned Value. Check the resulting number of booking codes: a reasonable balance should be achieved between the needs of cost control on the one hand and the added cost and administrative burden resulting from a higher number of accounts on the other.

– In commercial aircraft development there are very many basic components for which individual GANTT-chart planning would be a waste of time; instead, a faster standard database approach is preferred. Thus, where threads represent tasks/activities, which are better and in more detail monitored in a database rather than in a GANTT-chart, provide threads *directly* to schedule plans on Medium GANTT-level and ensure that each thread is linked to such a database.

If more than one (critical) milestone is regarded necessary for a given database representing a thread, split the thread up into as many threads as milestones required but otherwise keep the same booking code. For example, if a thread represents the period of drawing generation and release, than a sub-thread representing the period until completion of, say, 25 per cent of the drawings, another for 50 per cent completion and so on may be helpful to monitor progress. Note that the phasing of the planned cost described in the Work Package Description remains unaffected from this breakdown.

3. Ensure each thread is finished by a milestone.[5] By definition, the latter is always regarded as critical, see Chapter 13. All milestones identified as critical are also to be made visible on High and Medium GANTT-levels.

4. There could be additional milestones associated with tasks/activities on the lowest GANTT-level, which might be requested to be identifiable at Medium or even High GANTT-levels, respectively. Progress-, payment-, interdependency- or Critical Path-milestones might be examples for this. Any of these milestones should be categorised according to the criticality for the overall plan. Of course, all milestones on the Critical Path become critical milestones automatically.[6] Apart from that, the criticality of all other milestones is to be revisited on a regular basis. Do not replicate the detailed plans of suppliers. Instead, ensure that suppliers deliver on a regular basis the list and achievement status of critical milestones in Medium GANTT-level format, so that this information can be easily added to one's own plans.

5. Ensure grouping of threads into A-threads and work packages, respectively. Schedule data contained in the Work Package Descriptions (such as start date, finish date, inputs, outputs) should be compatible with the threads.

6. Ensure that for each work package there is a single booking code.

7. Ensure that booking codes can only be released on the basis of signed Work Package Descriptions.

---

5 This can, for example, be done by attaching a milestone to the finish date of the last tasks/activity which is to be grouped into the thread. However, in the case where an uneven workload distribution of a thread is to be approximated by splitting the thread into various ones with an even distribution (as shown in Figure 14.1), it may need some iteration before reasonable milestones can be identified while achieving an appropriate approximation at the same time.

6 As modern schedule planning tools calculate the critical path automatically, it is possible to identify the critical milestones too.

8. Grouping tasks/activities to a number of threads in balance with an acceptable number of booking codes should result in not more than ten critical milestones per month per multi-functional team to be monitored.

9. As higher-level PAT-elements aggregate or summate the data attached to lower-level elements, additional aggregation of data contained in the GANTT-charts (such as durations of threads) or of data attached to threads (such as milestone data, costs) can be achieved by moving upwards in the PAT-structure. It can then be checked whether the integration or aggregation of all schedule plans attached to the PAT-structures is in the end consistent with the delivery requirements. This applies to major interim dates as well as the final project completion date. If consistency cannot be achieved the schedule plans have to be continually reworked until it is met. If this is not possible, then the plan is not viable and there is an impasse. The schedule situation must be escalated to higher Management levels to find a solution. Until then, work should not begin.

10. Ensure plan is bought-off by the team leader and the key functional representatives (for its content and as a manifestation of their agreement and commitment) as well as by higher-level Management (to approve and to state that their requests for schedule visibility and transparency are met). Signing-off the detailed zero mismatch target schedule plan converts it into an agreed baseline plan for the teams, against which progress will be monitored.

Engineers, Manufacturing Engineers, Procurement representatives and so on provide all the detailed inputs for the planning: tasks/activities, their belongings to threads, the milestones and critical milestones. If the plan is then constructed in the way just described, a team can view the complete plan for all the PAT-elements or uncommon modules, for which it is accountable.

CONCURRENT OVERLAPPING

Introduction

The Concurrent Engineering approach, which represents a flow of interdependency information, continuously exchanged between individuals belonging to multi-functional teams, is a cardinal tool for faster aircraft development. However, the intense and continuous exchange of detailed data cannot be depicted any more in a schedule plan in any meaningful way. Attempting to do so would blow the manageable size of any plan. As a result, many threads will tend to overlap with others. Concurrent Engineering leads to overlapping threads in GANTT-charts. However, overlapping bars can represent both intentional concurrency as well as undesired but real mismatches between activities. One should be able to distinguish the one from the other.

With Concurrent Engineering, overlapping in a schedule plan means that one activity is not only kicked-off after a predecessor activity has been completely finished, but rather as soon as possible, that is, as soon as a useful piece of information can be exported from the predecessor activity to the successor activity, see Figure 14.2a. At first glance, for example, requirements are usually expected to be completed before design can start and parts can only be manufactured once they are completely designed. But by looking inside these activities and dissecting them, one usually finds that part of the succeeding activity can be started before the predecessor activity is complete, see Figure 14.2b. The corresponding bars in a GANTT-chart would be *planned* to overlap, thus indicating where concurrent ways of working are applied.

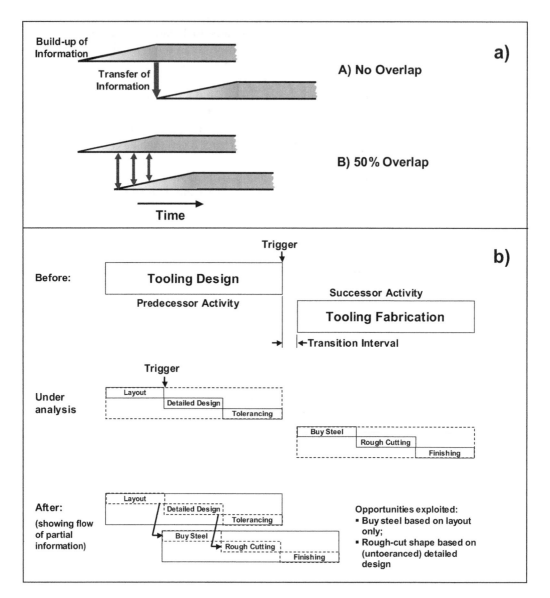

Figure 14.2   The concept of concurrent overlapping

Figure 14.2a adopted from: Smith, P.G. and Reinertsen, D.G. (1998), *Developing Products in Half the Time*,
2nd Edition, (New York: John Wiley & Sons), p. 168, with permission; Figure 14.2b adopted from:
Smith, P.G. and Reinertsen, D.G. (1998), *Developing Products in Half the Time*, 2nd Edition, (New York:
John Wiley & Sons), p. 174, with permission. Reproduced with permission of John Wiley & Sons Inc.

'Because the first activity is still incomplete when the second one starts, the information available
to it is by necessity incomplete. Working with this partial information requires a completely
different style. ... In the former case [i.e. without concurrent engineering] the information
transfers in a single large piece, in just one direction. In contrast, in the overlapped case the
information is transferred in small batches as it evolves. Because the information is incomplete,
communication must go both ways as recipients ask questions about the data to find out

what it means and provide feedback as to how well it meets their current needs.'[7] (see again Figure 14.2a)

Mismatches, on the other hand, result from the fact that the time left between 'Time Now' in a plan and a future milestone is no longer sufficient to perform all the planned activities in the planned sequence (that is, with the planned pattern of interdependent activities). Good planning tools automatically highlight identified mismatches. However, what is important to note here is that the magnitude of mismatches depend on two aspects:

- the remaining time; as well as
- the way the plan is designed and resourced.

Unless a future milestone is shifted further to the right, potentially causing delays, not much can be done about the remaining time. But even if no additional resources can be added to the plan, a lot can still be done about the working philosophy imbedded in a plan. Concurrent overlapping is an important enabler to remove mismatch overlapping. Let us therefore look at some examples where concurrent overlapping can be applied in aircraft development.

Examples of Concurrent Overlapping

Generation and Release of Drawings   A key area for application of concurrent overlapping is where drawings are generated and released by Engineering. The associated leadtimes for drawing generation and release depend largely on the complexity of the component or assembly described by the drawing as well as of the evolution stage of the aircraft, see Figure 14.3. New developments consume the longest leadtimes. With thousands of drawings to be released during the Development sub-phase, even little savings in process time per drawing accumulate to significant overall leadtime reductions. This is why concurrent overlapping is so powerful in this field.

Traditionally, drawings were generated and released by the Design department, to be checked by, say, the Stress and Weights departments, respectively, see Figure 14.4 part a). If the design did not withstand the loads, Stress sent the drawings back to Design. If the designed part was too heavy, Weights sent the drawings back too. The drawing release loop started again until all departments gave their 'OK'. Only then were drawings approved for release to Manufacture. This more or less has been the process for many decades and has often resulted in separate Engineering departments for Design, Stress, Weights and others.

Compared to this rather sequential and lengthy process, it is today within reach of existing design and stress Information Technology (IT)-tools to perform stress and weights analyses almost in parallel with the design job, see Figure 14.4 part b); and even by the same person. If adequately qualified, a modern engineer can perform the different jobs (design, stress analyses as well as weights calculation) concurrently without having to specialise in one of them as in the past. To cash the potential process time reductions, one therefore not only has to invest in modern tools but is equally required to invest in the qualification of people too.

---

7   Smith, P.G. and Reinertsen, D.G. (1998), *Developing Products in Half the Time*, 2nd Edition, (New York: John Wiley & Sons), p. 168.

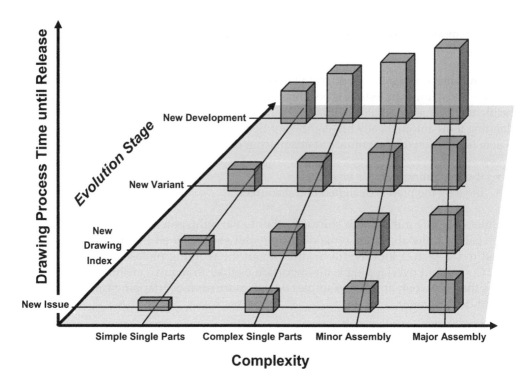

Figure 14.3   Process time drivers for drawing generation until release

It is this latter investment which is much more difficult to achieve than the former. In fact, it might take a generation of engineers to achieve the required level of qualification. And it is to be expected that the possibility for concurrency of the design generation and release process will result in massive organisational changes within the Engineering departments too, completely blowing away the walls which have built up over the years between Design, Stress and other departments. Until then, Design, Stress and Weights engineers working on the same components should be co-located wherever possible and whenever a new commercial aircraft is to be developed.

The drawing generation and release process shown in Figure 14.4 part b) may be part of a bigger meta-process, which sees the process of Figure 14.4 part b) repeated each time design intends to release data at a new level of maturity. Maturities are 'marks' given to design products and are used to describe how confident one can be in the design products' content and how far detailed they are. Those maturities, validated by all appropriate stakeholders, aim to:

- *describe what has been achieved so far;*
- *share relevant data between all actors; and*
- *engage further design work.*

So each time a maturity level is to be achieved, the process of Figure 14.4 part b) would have to be applied for the design data available at that point in time.

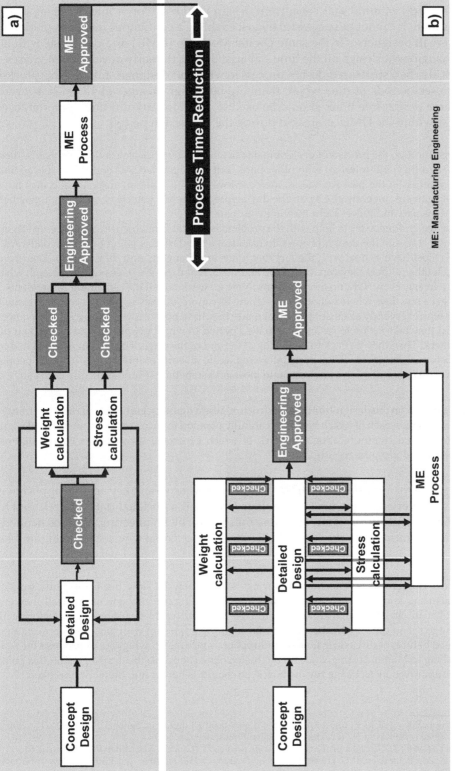

Figure 14.4   Overlapping opportunities in drawing generation and release process
(a) classical approach; (b) concurrent approach

One must bear in mind that concurrent design requires a lot of interpersonal, direct communication. It cannot be successful by just enabling a continuous design evolution in 3D. Because all designers use the same Digital Mock-Up (DMU) as a data base, with all of them changing the design all the time – at least during the early development phases – there must also be a structured design process to which all designers adhere. This process should foresee periods of time where there are no design changes at all, such as every Friday. These design-free times should be used to check for any unwanted interferences and clashes within the DMU as created during the last design period.

'As we go through the process of designing day in and day out, people continue to change their designs and they will interfere with other parts. But they have a [CAD and PDM] tool to find it now, whereas in the past we had no idea until we tried to build an airplane. You may have checked against somebody else's part the day before, and when you check against it today he's changed his part and it goes right through your part.

What we've done to try to help with the problem is to establish what we call a series of stages. There's six stages to the design process. In each stage the design is going to change daily. And designers just have to deal with the fact that plans are changing, and they have to coordinate with each other as best they can. But at the end of each of these six stages we go through what we call a freeze. I say, "OK, no more designs. Now go work out all the remaining fit problems." So you get a few days where nobody is designing any more: all they're doing is comparing their designs with everybody else's design to make sure they have no more interferences. Interferences that exist they take action to go and fix. So it's a period in which people stop designing and do the fit check. Then they go back to designing at the end of the stage. The negotiating that had to go on between designers whose parts interfered led to all sorts of interactions between people who wouldn't normally have any reason to meet, and sometimes found difficulty doing so.'[8]

'Superimposed on the design-build-team structure was a top-secret schedule of drawing freezes and design stages, each of which had been carefully planned to ensure smooth, error-free design, consultation and communication, and each of which generated its own flow charts, sign-off documents, and progress reports.'[9]

Release of Data for Manufacture (DfM)   Another very important key area for application of concurrent overlapping is where drawings (and their associated data) are released to Manufacturing. Historically, drawings were finished-off by Engineering and then handed-over to Manufacturing Engineering (ME) in a strictly sequential way, see again Figure 14.4 part a).

'The "toss is over the wall to Manufacturing" system may be slow but is elegantly simple. Manufacturing knows that if the drawings are on the Engineering side of the wall they are susceptible to change and therefore should be regarded with suspicion. Manufacturing is reluctant to spend time on the drawings or do anything important with time, such as designing associated fixtures or discussing them with suppliers. Once the drawings are tossed over the wall they belong to Manufacturing, and can be trusted. Any change in the drawings after this point must be approved by a change review board, so change is harder and therefore less likely.'[10]

---

8   Dick Johnson, Boeing's Chief Project Engineer for digital product design in: Sabbagh, K. (1996), *21st Century Jet. The Making of the Boeing 777* (London: MacMillan Publishers), pp. 56–7.
9   Sabbagh, K. (1996), *21st Century Jet. The Making of the Boeing 777* (London: MacMillan Publishers), p. 64.
10 Smith, P.G. and Reinertsen, D.G. (1998), *Developing Products in Half the Time*, 2nd Edition, (New York: John Wiley & Sons), p. 253.

But in commercial aircraft companies with computerised design processes, where ME can receive digital Engineering data, there is potential for significant concurrency. Many ME processes can already start while the design is still on-going, thus saving a considerable amount of time, see Figure 14.4 part b).

'As people implement concurrent engineering they add ... levels of release for different purposes. For example, a drawing can be released for purposes of material procurement, and then be released at a different level to start fixture design, and finally released at another level to finalise machining dimensions. For each release level different information will be frozen. For material procurement, we need to know the type of material and the rough size of the part. For fixture design we need to know the dimensions at the points at which the fixtures will come in contact with the part. Only in the final machining drawing do we need all dimensional information on the part. Such a system with multiple release levels is common in highly concurrent processes. For example, on the Boeing 777 the drawing release procedures often had six to eight distinct levels of drawing release depending on the type of part.'[11]

In order to cash overlapping opportunities at the interface between Engineering and ME, it is therefore important that Engineering and ME jointly agree on the number and content of data drops. This agreement should be reached as early as possible, for example, when the Project architecture is starting to be jointly set up. The data drop dates need to be pulled by Manufacturing. Subsequently, data drops have to be carefully schedule planned and imbedded on lowest GANTT-level or in databases attached to threads.[12]

Because of the highly concurrent way of working possible in the area of Engineering data release as well as ME processes generation, some companies are considering bringing these detailed activities under one organisational roof all together, for example, under the overall responsibility of ME. This would leave the Engineering department of a commercial aircraft integrator company with the conceptual and aircraft-level design work. It would also allow more value added work to be subcontracted to outside integrators of individual parts of the aircraft, thus allowing for levering more sources of funding for an aircraft to be developed. Boeing seems to have chosen this path with the development of its 787.

Note that the concept of stepped drawing releases is by no means limited to parts produced in-house. It equally applies to items procured by Manufacturing, and is particularly applied to so-called longlead items: in aircraft development there are many components, such as complex forgings, which need much longer leadtimes to be produced than other parts. In order to protect the overall development schedule, they need to be ordered from suppliers much earlier during the Development sub-phase compared to other parts. The key question as expressed by P.G. Smith and

---

11 Smith, P.G. and Reinertsen, D.G. (1998), *Developing Products in Half the Time,* 2nd Edition, (New York: John Wiley & Sons), p. 254.
12 An example comprising four data drops and describing a converging degree of accuracy for tolerances of, say, a spoiler hinge position might be as follows:

+/- 50 mm: March 5
+/- 5 mm: May 10
+/- 0,5 mm: August 12
final release of data: September 4

D.G. Reinertsen is then: 'When is enough information available to begin the procurement process?' Note the difference to the traditional approach which was: 'When do we have *all the information* that we need to *complete* the procurement process?'[13] To order forgings, initially only the rough size and the type of material are necessary inputs. At later stages, more design details can be delivered to the supplier as and when available until the final drawings are released.[14]

Early Supplier Involvement   Another aspect of concurrent overlapping is the early participation of supplier staff in the multi-functional teams. This should be done early during the Development phase until start of detailed design. Detailed design can subsequently be performed at the suppliers' premises, perhaps even using the suppliers' own methods and processes, rather than methods and processes used by the contractor. In this way, the contractor benefits directly from the enormous experience of suppliers, accepting them to make own suggestions for improvements of the design. Notwithstanding this, the contractor can also positively influence the production costs by sharing its own know-how with the supplier in an open attitude. This, however, should only be done with a small number of highly capable, reliable suppliers, which are regarded as strategic partners.

Concurrent Use of Aircraft and Digital Test Facilities   W. Spitz et al. describe how concurrent way of working in flight testing can save development cycle time:

> 'It is important to note that significant time can be expended in designing and installing the flight test instrumentation, as opposed to the flight testing itself. For example, the [Boeing] 757 and 767 had approximately the same number of test flight days but the 757's non-flying days were much higher due to additional ground support and documentation expenditures. One suggested rule of thumb was that a single test airplane could provide somewhere between 30 and 50 flight hours per month. Thus, if an entirely new design ... were to require, say 1,800 to 2,000 flight hours to reach certification, this would entail approximately 40 aircraft months of test flying. With three or four aircraft included into the test program, the actual calendar time spent is approximately 1 year. The flight test span for derivative designs is currently on the order of 10 months for a typical program. Current goals are to reduce the time spent on flight testing from 12 months to 8 to 10 months for new designs, and from 10 months to 4 to 5 months for derivative designs. This could be accomplished by relying more on the observed performance of ground test rigs and the use of simulations in lieu of actual flight tests.'[15]

Purchase of Equipment before Required   Another example for concurrent overlapping is described by P.G. Smith and D.G. Reinertsen:

---

13 Smith, P.G. and Reinertsen, D.G. (1998), *Developing Products in Half the Time*, 2nd Edition, (New York: John Wiley & Sons), pp. 255–6.
14 However, when first ordering forgings they should be dimensioned big enough to protect potential future design changes too. The potential magnitude of these changes needs to be judged on the basis of experience resulting from previous aircraft programmes. Also, if a particular aircraft under development is supposed to be the first member of a larger aircraft family, longlead items are often designed to protect *all* members of that family.
15 Spitz, W., Golaszewski, R., Berardino, F. and Johnson, J. (2001), *Development Cycle Time Simulation for Civil Aircraft*, NASA/CR-2001-210658, pp. 3–11.

A company had obtained new machinery before actually required. It then immediately started making parts of the best precision possible, using normal products.

'The company was in fact giving away precision because these parts [required by a new customer] cost more to make than those normally used, but its staff was learning, preparing for the new product's arrival. When the product design was initially ready to manufacture there was a stable, high-precision manufacturing process ready for it, with no start-up time need for the new machinery and very few scrapped parts. Note that in order to use this approach, you must justify the acquisition of the new equipment before it is required by the new product – a substantial challenge at many companies.'[16]

Just in Time (JIT) Deliveries   Once the start dates of assembly stations are known they need to be communicated to all internal and external suppliers. In fact, what should be communicated to the suppliers are individual on-dock delivery dates per assembly station which are common to all components to be delivered to that station. The on-dock delivery dates should in fact be sufficiently earlier dates than the start dates of the assembly station (that is, the MSDs, see again Figure 13.11). It is wise to add some margins here so that the company's exposure to delays resulting from delayed receivables is reduced. Give someone a delivery date and he or she will try to achieve it. Try! In reality, most parts planned to be delivered at a certain date will be late, with some parts being delivered early and a few others exactly on time. Thus, sufficient margin has to be accounted for.

For initial industrial ramp-up, suppliers should be given on-dock delivery dates which are considerably earlier than the assembly start dates identified in the internal schedule plans, such that the later, real delivery dates can be expected by the goods receiver to vary around these adjusted dates. It is on the basis of these adjusted dates that suppliers should establish their own schedule plans as well as their Business Case.[17]

As time approaches the individual components' on-dock delivery dates, it will not come as a surprise that many components are still far from being delivered, in particular for the first new aircraft. However, only once the contractor is convinced that cashing all recovery possibilities (such as concurrent overlapping opportunities) at the supplier would still result in a delay to the agreed on-dock delivery date, it would be time to negotiate with the supplier a date closer to the planned assembly start date.

However, usually not all parts are needed precisely at planned start of assembly. Some could be delivered later because they are in fact needed later in the assembly process. It may then be possible to agree with the supplier on detailed Just in Time (JIT) delivery dates on a part-by-part basis.

Finally, as a last resort, it could be required to rearrange the assembly operation sequence. In other words: a planned assembly operation defines theoretical JIT dates for each component, but further lateness of components could drive a rescheduling of the assembly operations. The principle of concurrent overlapping needs then also to be applied to the assembly operations: start as much as possible with the available components and be prepared to continuously amend the sequence of assembly operations depending on the parts' availabilities.

---

16 Smith, P.G. and Reinertsen, D.G. (1998), *Developing Products in Half the Time*, 2nd Edition, (New York: John Wiley & Sons), p. 251.
17 Note that suppliers are accountable to plan the entire Assembly phase of their PAT-elements, even if task/ activities like 'Goods Receive' or 'Logistics' are taking place at the premises of the customer. The customer needs to provide typical durations for such tasks/activities to the supplier, including the desired margins.

Obviously, continuously changing the sequence of assembly operations and, thus, reshuffling of the corresponding PAT-elements within the affected PAT-sub-structures is a tremendous change control task and should be avoided wherever possible. As a prerequisite, negotiations with the suppliers about JIT dates should take place as late as possible. But if delays still persist, JIT dates have to be negotiated with the suppliers.

This last example demonstrates well how much both concurrent overlapping in particular as well as Concurrent Engineering in general depend on communication and negotiation tactics.

What it Takes to do Overlapping

The pressure for on-time delivery of data or parts must be kept high at all times. Only if mismatches against the plan are non-recoverable, downstream opportunities like:

- JIT-delivery;
- elimination of planning margins; as well as
- further concurrent overlapping;

should be cashed – and not all at once but on a step-by-step basis.

This is because at a given point in time only partial information is available and used, which can be based on incorrect assumptions. There is the risk that some work has to be redone as a result of overlapping. Overlapping therefore places a burden on the developers: the safe use of partial information despite its inherent fragility can only be assured by even stronger communication among the aircraft developing teams. Here we have another very strong argument in favour of co-locating multi-functional teams. In fact, they are a necessity to make concurrent overlapping possible at acceptable risk.

To exploit the advantages of concurrent overlapping developers must in addition change their mindset. The new way of thinking requires an ability to move ahead proactively with much less information than most developers would like to have. They rather must constantly assess, 'whether the muck ahead is firm enough to support a few more steps'.[18] Every member of a team should always scout for concurrent overlapping potentials as these represent opportunities to reduce leadtime. If team members are not scouting enough for overlapping opportunities, then the team leader should encourage them to do so.

> 'The overlapping mindset is reflected in the types of questions we ask when seeking overlapping opportunities:
> - What is the bare minimum of information I need to get the next step started?
> - When is the earliest I can produce this information in the preceeding step?
> - Is there anything I can do to make the requirement for having this information unnecessary?
> - Are there assumptions I can make about this information that will provide a high likelihood of it being accurate enough to being work?

18 Smith, P.G. and Reinertsen, D.G. (1998), *Developing Products in Half the Time*, 2nd Edition, (New York: John Wiley & Sons), p. 176.

- How great are the consequences of being wrong, and how long will it take me to get back to where I am now if I am wrong?
- Can I save enough time by starting early to allow for making a mistake and still finish early?
- What information would allow me to take another step?
- Who could use the information I have to enable them to take another step?'[19]

The Future of Phase Project Planning

In Chapter 3 the concept of Phased Project Planning (PPP) used by many aircraft companies was introduced. In particular it was mentioned that the Detailed Design period is distinct from the manufacturing and assembly processes.

With this PPP concept, a project passes sequentially through checkpoints at the intersections of phases to ensure that all the items required by that checkpoint are in good order. However, if a project does not pass the checkpoint review because not all checkpoint criteria have been met, does it really come to a standstill? Usually, this is not Management's intention, as stopping the 'marching army' of thousands of aircraft developers would generate all sorts of other problems. What is usually done is to record all of the checkpoint criteria discrepencies, and to work on their resolutions with more effort and accelerated speed. This is to be done until all discrepencies can be declared 'closed' at a new, specifically organised checkpoint date.

As a result of overlapping, the PPP approach is somewhat not applicable, at least when looking at its original intent. It is, for example, much more important to manage as one single phase and in a concurrent manner the difficult period from start of release of Data for Manufacture (DfM) until the point in time where a component is completely manufactured and ready to be assembled. This corresponds to a concurrent way of working much better than waiting until *all* drawings are released, jointly availability- and quality-checked at a PPP checkpoint prior to start of manufacture. This is why the Pre-Assembly phase was introduced in Chapter 11, with the basic components' manufacture being part of it.[20, 21]

Checkpoints should still take place but most of them should have more the character of 'Are we sure we have not forgotten anything?' rather than a gate, at which the project could come to a stop. However, top Management should be present at these checkpoints to provide the right level of Management attention which these checkpoints clearly deserve.

---

19 Smith, P.G. and Reinertsen, D.G. (1998), *Developing Products in Half the Time*, 2nd Edition, (New York: John Wiley & Sons), pp. 176–7.

20 There is no concurrent overlapping possible at the intersection between Pre-assembly and Assembly phases: the physical parts need to be available fully and completely manufactured before they can be used for assembly.

21 Sometimes dummy parts are used to overcome shortfalls associated with non-availability of live components. But even dummy parts need to be fully available before any assembly can start. However, in this case the PAT-structure would need adjustment to account for the fact that the live component will be used later. The dummy would then need to be de-assembled. This could be represented in the PAT by what is called a 'minus' PAT-element.

PROJECT TEAM PLANNING PART II

The team leader should encourage the person in charge of schedule planning to optimise the team's schedule with respect to both time and utilisation of resources, and to make the corresponding suggestions during team meetings. 'Most projects require juggling and rescheduling as the differences between reality and plan begin to surface. For this reason, a good project controller who is able to reschedule quickly and efficiently can have a major influence on the success, or lack of it, of a complicated project.'[22] Not only for reasons of good project control, but also for good communication and reporting, changes to durations of tasks/activities and threads as well as shifts in dates of milestone should be highlighted in the GANTT-charts on a regular basis.

For regular updating, the joint team planning approach proves useful again. However, no more workshops are required. Instead, regular team meetings are sufficient to identify those elements of the plan which need updating. If team leaders pull for planning discussions during normal team meetings, they can be a lot more confident that the plans correctly reflect the actual situation of the teams' projects.

During team meetings the Critical Path of the schedule plans should be jointly analysed. The team should then concentrate on any possibilities for leadtime reduction on the Critical Path. If the team is able to shorten the Critical Path, then it may be possible to reduce the overall project schedule.

With so many teams involved in aircraft development, it is the need for proper management of complex interdependencies which drives the regular planning cycle. In small projects, with the schedule plan kept in one and the same IT-tool, scheduling can be done at any time once the work progress has been updated. Work progress is measured against the baseline plan which has been signed-off by all involved Functions. If – as part of the update – the delivery of, say, an interdependency milestone is delayed, the associated milestone in the GANTT-chart moves to the right and with it all subsequent tasks/activities which are linked to it.[23]

For large projects all of this is still true for individual plans but in addition there is the need to touch base with other teams on a regular basis to update the interdependency data *between* plans. A pragmatic approach for the regular (for example monthly) planning cycle, which takes the needs of cross-team Interdependency Management into account, could be as follows:

- Based on the scheduling of his own plan, a supplier A ('Mike') of an interdependency provides an update of the forecast delivery date for his deliverable. This new forecast date is to be manually fed by the supplier into the common database, see again Figure 13.19.
- The database tool automatically forwards a notification of the change to the customer, for example by e-mail.
- When customer B ('John') incorporates progress updates in his plan, the interdependency update is taken into account.
- As soon as all the updates are fed into the plan, customer B runs the scheduling activities eventually leading to a recalculation of his Critical Path.

---

22 Eisner, H. (2002), *Essentials of Project and Systems Engineering Management*, 2nd Edition, (New York: John Wiley & Sons), p. 95.

23 Note that the baseline plan must be subject to stringent change control, otherwise progress will get reported as too optimistic.

- B probably also has interdependency milestones in his plan where he is a supplier. The corresponding delivery dates may have adversely changed as a result of the scheduling. The schedule planning tool will provide to him a list of all interdependencies where the delivery date has changed.
- B would now do the same what A did before: in the database he would update manually the forecast date for his deliverables.
- The cycle begins again.

However, before doing so both, A and B should try to recover new delays wherever possible, for example by rearrangement of the plan or cashing concurrent overlapping opportunities. One must bear in mind that scheduling is a very arithmetic task following clear logics. Thus, in companies where there is no established delivery-oriented corporate culture there is the danger that repeating the iterative planning cycle on a regular basis could result in exaggerated delays to the overall schedule. Caution needs to be applied.

To remove a forecast delay from the schedule there are two principal possibilities:

- either all required recovery activities are schedule planned in the GANTT-charts to demonstrate how late milestones will be brought back to their baseline achievement dates. For example, the forecast delay to the delivery of a sub-assembly to the next Assembly Stage (AS) may be brought back by planning outstanding work during the next AS;[24]
- or the forecast dates for milestones and start/finish dates of tasks/activities/threads calculated by the automatic scheduling are manually re-edited. This solution can be applied if known recovery possibilities or any additional margins were intentionally not captured in the original baseline plan.

Schedule planning of recovery activities is the cleaner approach. But it requires improved communication between those who know about the recovery possibilities and the project planners. Again, there is a higher chance of this happening in a co-located multi-functional environment. In any case, teams should discuss the issues during regular team meetings. They must provide their planners with guidelines of how to change the plan and how to recover as well as with information about which dates the team is committed to achieve.

---

24 The price of this approach is higher costs associated with the work in the next AS. This cost is usually higher by a factor of 3 to 6 compared to the planned costs. This is because outstanding work can only be performed during identified free slots in the schedule plan for that AS, where there is in addition much less flexibility to perform the outstanding work.

MAIN CONCLUSIONS FROM CHAPTER 14

- Involving all team members simultaneously in the planning process can bring about individual team members' commitment to and motivation for the Project Plan. This is a much better approach compared to a series of bi-lateral meetings between functional representatives in the multi-functional team on the one hand and the person in charge of schedule planning on the other.
- A good way of living through an initial team planning workshop is by jointly creating a Program Evaluation and Review Technique plan. This plan represents a network of tasks/activities and milestones – as well as their logical interlinkages. Later, regular team meetings should be used to jointly analyse the team's schedule plan.
- A pragmatic way of how to perform schedule planning is presented. The level of planning granularity plays a major role in view of the limited resources authorised to perform schedule planning tasks. In other words, the availability of planning resources sets a natural limit to the degree of schedule detailing, thus limiting the number of tasks/activities.
- In dynamic and complex projects, such as commercial aircraft development, the quantity of change resulting from 'unknown unknowns' makes too detailed plans very difficult to maintain. Plans with excessive details are therefore often found to be misleading and are abandoned in favour of a higher-level rolling wave approach. Thus, the planning detail is progressively developed as the project progresses and as more is learned. Planning details may then only be required for, say, the next two years. This, too, will contribute to a reduction of schedule complexities.
- Grouping tasks/activities to threads is a major architectural job and requires careful multi-functional consideration and discussion. It is at the same time a very important enabler to link Project Management to the needs of Finance and Engineering, in particular.
- Certain planning rules must be respected to ensure a simplified Earned Value approach, to be described in the next chapter.
- There are many possibilities to shorten schedule leadtimes. Every member of a team should continuously scout for concurrent overlapping potentials as these represent opportunities to reduce leadtime.

15

Integrated Project Monitoring and Control

INTRODUCTION

Complex projects like the development of a commercial aircraft require tight monitoring and control, using many tools and methods developed by classical project control. The basis of classical project control consists of a recursive feedback loop comprising the following steps:

- initial planning of activities;
- performing the planned activities;
- monitoring Key Performance Indicators (KPIs), in particular with regard to
  - variances to baseline plans; as well as
  - gaps to targets and objectives;
- identification of corrective actions, if any; leading to the
- establishment of a recovery plan with a new and improved forecast trend;
- implementation of recovery actions;
- back to monitoring of (updated) KPI status;
- and so on.

This sequence is also called the PDCA cycle ('plan, do, check, act'). However, what gets often forgotten with this rather mechanistic approach is the fact that good project control is also based on important soft skills like:

- trust and respect required to encourage full transparency of issues;
- creativity and pragmatism in problem solving; as well as the
- preparedness on behalf of managers and project leaders to take a certain level of personal risk.

Within a multi-functional environment where teams work concurrently using an integrated architecture and a common DMU, classical project control will not suffice. It needs to go beyond the classical methods and tools. For example, new KPIs adapted to the application of 3D data, such as 'number of DMU clashes', need to be introduced. But above all, it needs to become integrated. Integrated project control can only be exerted if it is based on integrated Product, Process and Resources architectures. In the following, it will be described how this can work in practice.

In addition, it must be understood that the needs of project monitoring and control influence the way how the interlinkage of Product, Process and Resources data is structured and set up. Project control has its own requirements towards an integrated Project architecture. They come on top of the overall design and development needs for an integrated architecture, which have been outlined in the previous chapters. This requires a lot of upfront discussions and agreement within Management as well as within the teams, about which KPIs to monitor and how. In fact, it requires a lot of experience in the company's aircraft development processes in order to anticipate the needs of project control over the many years of the Development phase. It must be avoided that project control cannot be performed in a professional, efficient and pragmatic way just because of inadequate architectural integration.

ALLOCATION OF OBJECTIVES AND THEIR LOCATION WITHIN THE INTEGRATED ARCHITECTURE

Before being able to monitor and control, reference values need to be created against which relevant KPIs are to be measured. If targets or budgets or milestones – in short: objectives – are known in advance, well-selected elements of the Project architecture should be 'populated' with such objectives prior to start of monitoring. It is therefore important to explain how these architectural elements should be selected and at which level(s) of the architecture elements should be populated with objectives. The distinction between:

- time dependent KPIs;[1] and
- KPIs which are not (or almost not) a function of time

is helpful in this context.

Location of Budgets for Resources and Cost

The way the elements of the Project architecture with their threads and booking codes were set up in Chapter 13 implies that *actual* bookings (such as of hours worked) are captured on the level of threads. However, the same may not apply for *budgets* allocated to individual teams. Project budgets are most important control tools for project and functional managers. It is therefore necessary to understand how they are generated and where they are located in the architecture. What is the right level (for example in terms of Product & Assembly Tree (PAT)-level or GANTT-level) to allocate budgets? And how does this link to the planned cost of individual work packages, which are – after all – located on thread level or higher (see again Figure 13.18)?

The minimum requirement for the selection of potential 'budget-carrying' elements should be that they are sufficiently stable so that budget allocation can be performed on a stable architectural basis. 'Stable' means that changes to the aircraft design will not –

---

1   Such as:
    – resources, cost and schedule performance KPIs; as well as
    – KPIs, which need to be monitored during a certain time period only (such as deliveries of a certain category of drawings).

or no more – change the position of those elements within the integrated Project architecture with its PAT sub-structures, see Figure 15.1. This requirement would lead to budgets being allocated reasonably high up in the Project architecture, particularly for the early periods of the Development sub-phase.

However, it has also been said that budgets are allocated to multi-functional teams because teams are not only made responsible for achievement of time and quality objectives but also for cost objectives, see Chapter 5. But teams may well be in charge of different elements or different PAT-pyramids at different locations of the architecture. Thus, allocating budgets to teams means in fact allocating budgets to all the different heads of Project architecture sub-structures (such as PAT-pyramids or view-specific modules), for which teams are made responsible, see example of Team A in Figure 15.1.

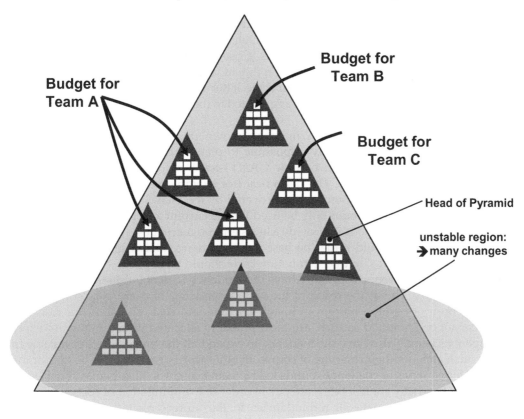

Figure 15.1   Location of budgets for resources and costs within the Project architecture

It is therefore reasonable to conclude that the lowest levels to which budgets should be allocated should be the levels where one finds heads of pyramids, for which individual multi-functional teams are responsible.[2] It is suggested to allocate budgets directly to

---

2   Note that allocation is different from aggregation. Clearly, budgets can be aggregated to any higher level compared to the level to which they have been allocated.

Summary Threads (S-threads) of individual pyramids for which multi-functional teams are in charge and not to do this on any lower level. This can be seen in Figures 15.2 and 13.18.[3] Note, that in Chapter 13 it had been shown for PATs how S-threads represent the envelope of all threads which belong to the same phase of all PAT-elements of a selected PAT-pyramid. The same can be done for view-specific elements external to PATs too. Thus, in terms of budget allocation to S-threads, there is no difference between PATs and other view-specific modules of the Project architecture.

Allocation of Budgets for Resources and Cost

Budgets to be allocated must be derived from initial budget *requests* based on integrated schedule, resource and financial plans. More specifically, budget requests should be based on the work described in the various Work Package Descriptions. For each work package, the workload required must first be individually estimated by elaborating individual tasks/activities on a period-by-period basis. Costs encountered during a financial or accounting period can then be estimated on thread level by converting the work-loaded thread data into cost data and aggregating the latter to work package level, see Figure 13.18. Finally, these estimations must be aggregated to the corresponding S-thread to yield the budget to be requested.

However, in large projects with many unknown unknowns such as commercial aircraft development, it is virtually impossible to plan all detailed tasks/activities and associated resources on Medium or Low GANTT-level, respectively, for all the years of the Development sub-phase. The periodical (for example yearly) planning accuracy will degrade with every additional period covered by the plan. Thus, detailed plans (on the lowest GANTT-level) should only be used for the current as well as for a reasonable number of subsequent periods to provide a basis for budget request estimations. For later periods, estimations should rather be based on experience and assumptions. In other words, a rolling periodical budget estimation cycle should be applied.

Bottom-up budget requests usually have the tendency to demand more funding than actually necessary. This is the result of the natural tendency of every person to allow a certain amount of cushion margin in his or her time and cost estimates. But reserves, which cannot be identified as such, are most probably allocated wrongly, and are likely to be spent as people also have the tendency to expend all the time and cost available to them, which includes their reserves, of course. In other words: reserves will be lost for the project. Thus, a series of challenge rounds led by team leaders on all organisational levels is necessary to strictly control this tendency in order to avoid inflation of costs.

The goal is to eliminate all redundancies in the budget requests and to remove contingencies from all teams (and corresponding Project architecture elements) where they are not needed. Contingencies should rather be collected and held on higher Management levels compared to the level of multi-functional teams, on the assumption that only a few project areas should encounter major unforeseen difficulties.[4] Management should be given the freedom to allocate funds from the overall project's contingencies to overcome them. This approach minimises budgets for all lower-level teams. If in the end a contingency of about 10–15 per cent of Non-Recurring Costs (NRCs) is used,

---

3   As PAT-solutions represent configured PAT-pyramids, the rule suggested here implies that budgets should be allocated to S-threads of the heads of the highest-level PAT-solutions for which teams are in charge.

4   Contingencies are on top of the provisions foreseen in the Business Case to fund risk mitigation actions, see Chapter 9.

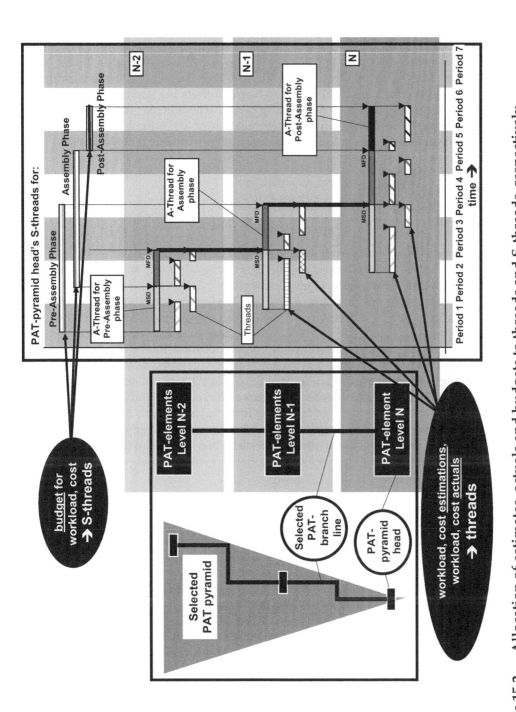

Figure 15.2   Allocation of estimates, actuals and budgets to threads and S-threads, respectively

then this figure is in line with contingency levels commonly encountered on complex projects. However, certain areas of the projects with many unknowns may require much higher levels of contingencies, say 30–50 per cent. This may, for example, be the case in areas where new or as yet immature technologies are applied.

Eventually, the budget requests are to be compared with the period budgets, which are derived from the strict cost limits expressed in the Business Case as Target Must Cost (TMC) or Target Cost at Completion (TCAC). They therefore represent an upper limit to the budgets which can be allocated to teams. If there is a gap between top-down Business Case costs and bottom-up work package cost, schedule plans and Work Package Descriptions must be reworked until the aggregation of all individual planned work package costs equals or comes at least much closer to the available budget. Besides:

- taking out reserves;
- correcting assumptions;
- clarifying the scope of work packages; as well as
- eliminating redundancies;

this might for example be achieved by omission of all the nice-to-have work, which is not really required to achieve project objectives. For the latter it is suggested to always remember that 'better is the enemy of good enough'.[5]

This convergence process may take a few challenging iterations and possibly many months. But there must be a milestone in the overall development schedule where the convergence process has to come to a halt. The attempts to reduce the gap between top-down and bottom-up costs cannot continue forever. Beyond this milestone, which all project leaders should know well in advance, no further attempts are made to resolve the total remaining gap but instead it is allocated to work packages on, say, a pro rata basis. After all, project leaders are accountable managers of the company and as such must share the burden of achieving the strict cost targets set forth by the project's Business Case. This additional challenge released to project leaders formally closes the gap.

Note that only the convergence process as described above, involving an iterative challenge process with many inconvenient negotiations with the budget receiving teams, will develop:

- the necessary confidence that a realistic but challenging budget for the work to be performed has indeed been identified; as well as
- the knowledge about where contingencies are still held (that is, to which Project architecture element contingencies are allocated) and what their magnitude is.

Once the convergence process has come to a final conclusion, budgets can at last be allocated to S-threads. Allocated cost budgets now represent what is called Budgeted Cost of Work Scheduled (BCWS), see Figure 13.18. BCWS are budgets allocated to S-threads of pyramid heads, for which individual multi-functional teams are in charge. More precisely, BCWS(t) is the cumulated BCWS until time t. It may show a non-linear increase over the span of the S-thread.

---

5   This quotation is an adaptation found in many sources which goes back to 'Le mieux est l'ennemi du bien' from Voltaire's *Dictionnaire Philosophique* (1764), literally translated as 'The best is the enemy of good'. Voltaire wanted to emphasise that, when trying to improve things, there comes a point where further improvements will turn out to be negligible in comparison to the effort required.

The BCWS(t) of each S-thread can then be cascaded down to the corresponding work packages, A-threads and threads, of which many have been updated during the convergence process. This cascade also includes the additional challenge released to the project leaders, if any, as described above. Let the resulting budgets on thread-level be called Planned Cost of Work Scheduled (PCWS), see again Figure 13.18.

Each thread now has a defined value. More precisely, for a given thread, PCWS(t) is the cumulated planned cost of work scheduled until time t. It may also show a non-linear increase over the span of the thread. However, the distribution of PCWS(t) may be linearised for reasons of simplicity as no further budget cascading is required. In summary, therefore:

- the total value of the threads is a result of the cascade of the converged budget on S-thread level; while
- the distribution of the BCWS(t) over the span of the S-thread is a result of the aggregation of the final and possibly linearised distributions of the many PCWS(t) at thread level.

Release of Budget

Once the budgets have been allocated, costs/resources/workload can be allowed to run against them by opening and closing the individual booking codes associated with the work packages. There are two major rules of how booking codes should be used:

1. they are opened at the start date of a work package, for which a Work Package Description has been signed-off, and are closed either on its finish date or when the planned cost/resources/workload have been used up, whatever comes first;[6] and – while open –
2. booking codes should be used for the associated scope of work only.

These essential rules of Project Management should always be obeyed.

Baselines for Data Deliveries and Milestones

Commercial aircraft development requires the management of a huge number of data, among which one can find:

- Digital Mock-Up (DMU) models;
- drawings, such as conceptual design drawings, detailed parts' drawings, interchangeability drawings, rigging drawings as well as Principle Diagrams (PDs), Wiring Diagrams (WDs) and so on;
- Data for Manufacture (DfM) for parts, assemblies, interfaces;
- material orderings;
- assembly plans;
- parts' demand data;
- purchase orders;

---

6 In practice, the time accounting system may issue a warning to the booking code owner at a certain date before the finish date has been reached or when a certain amount of the budget has been consumed.

- parts delivery and goods receipts data;
- job cards and work orders; as well as
- test results.

In Chapter 13 it was said that this kind of data would typically be planned in databases attached to threads, and not individually or directly in GANTT-charts. The common theme among this data is that it represents countable units of deliverables between a supplier and a customer, appearing in large quantities created during a dedicated and limited period of time. As such, these deliverables can and must be planned in advance.

Thus, a baseline plan for data release and delivery can be created and actual release performance can be monitored against that baseline. In fact, each successful data release and delivery marks a milestone in the database. Because of the large quantities of milestones in aircraft development, it is common to use S-curves depicting cumulated data over time. They can be generated directly from the information provided in the databases attached to PAT-elements or view-specific modules. A generic example can be seen in Figure 15.3.

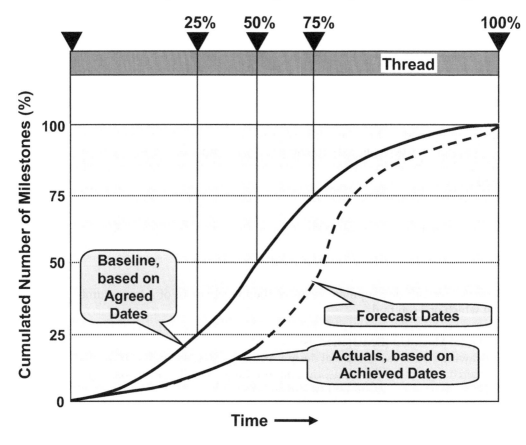

Figure 15.3   Generic S-curve describing progress on, for example, drawing
              release

However, it is wise to still have some additional milestones related to this type of databases visible in the GANTT-charts, not only the one milestone marking the finish date of the thread to which a database is attached. Additional milestones could for example represent 25 per cent, 50 per cent and 75 per cent, respectively, of all data releases and deliveries. This can also be seen in Figure 15.3.

Please note that S-curves can, of course, be used to graphically represent progress of *all* milestones of a PAT-element or view-specific module, not only the milestones contained in databases. In fact, because of the links within the Project architecture it is principally possible to depict in S-curves all milestones of any selected architectural sub-structure. This eases greatly the generation of reports for different customers.

Targets Independent of Time

Targets independent (or almost independent) of time which can be allocated to the elements of the integrated architecture include:

- requirements;
- mass and weight targets;
- maturity targets;
- targets for Non-Recurring Costs (NRC) for the total Development phase; as well as
- targets for Recurring Costs (RC) (using for example aircraft #100 as a reference);

to name a few. These requirements and targets usually remain constant during the whole Development sub-phase (subject to some minor adaptations and corrections) and therefore are virtually independent of time. Again it is necessary to identify adequate Project architecture element to which these become allocated. Only then can actual KPI performance of individual elements be aggregated and monitored against these targets. Wherever possible it is suggested to use the same elements for target allocation than for the allocation of budgets, that is, it is suggested to use S-threads too.

KEY PERFORMANCE INDICATOR MONITORING: VARIANCES AGAINST BASELINES

Resources and Costs

It is, of course, an essential element of effective Project Management to maintain continuous monitoring of project schedule, cost and performance. But how to monitor schedule, cost and performance in an integrated, multi-functional environment?

As far as workload and costs are concerned, their *actual* figures are always linked to threads. This is a direct result from linking booking codes to threads, as was explained in Chapter 13. However, if BCWS-budgets are allocated to the level of S-threads of individual heads of, say, PAT-pyramids, then an analysis of variances between actual cost and *allocated budgets* is of course only possible on this level. Consequently, workload and cost actuals captured at thread-level need first to be aggregated to the corresponding S-thread level before any analysis of variances can be performed.[7]

For this to work, actuals should be captured with the same frequency applied for the phasing of allocated budgets. If, for example, the budgets for a particular year are broken down to a monthly phasing, the actuals should be captured and aggregated on at least a monthly basis as well and not, say, on a quarterly basis.

While a synchronous phasing between budgets and actuals can usually be achieved relatively easily for costs resulting from worked hours booked against the projects' booking codes, it can be difficult to be achieved for *procured* products and services. This is because costs are not incurred continuously but at certain times only. In this case, only the careful outline of Purchase Orders helps. As it is usually the goods receiving date which is cost-relevant and not any upfront, mid-course or final payment date(s), every Purchase Order should list the planned individual receiving date of each item, which it comprises. This will ensure a quasi-synchronised KPI monitoring. Where this is not possible, such as for fixed price packages covering sub-contracted Engineering work, periodic accruals should be applied.[8]

Work Progress

As far as work progress is concerned, there are various methods to measure the variances of actually performed work against the work planned:

- Percent Complete Method: Work package owners (for example the leaders of the multi-functional teams) are asked on a regular basis about the percentage of completed work in relation to the total planned work.
- The 50/50 (or other ratio) Technique: Instead of asking a work package owner on a frequent and regular basis about work progress, it is for the work package owner to report to the control system if and when 50 per cent and 100 per cent, respectively, of the work has been achieved.
- Milestone Method: By definition, milestones represent events with measurable deliverables (see Chapter 13). Work progress can therefore be measured by output milestones, if and when declared as achieved by the customers in accepting the outputs.

One has to bear in mind that percentages provided for the Percent Complete Method are often based on subjective judgements, which tend to be over-optimistic. As a rule of thumb – which only too often holds true – it takes as much time to complete the last 10 per cent of a work package as it took to complete the first 90 per cent.

'The use of subjective, unsubstantiated estimates of "percentage complete" of long-duration tasks, or of the total project, is generally unreliable, since these normally show that everything

---

7   Having said this, a comparison between actual and planned costs (or resources or workload) could, of course, in any case be made on thread, A-thread and work package level, too, see Figure 13.18. But, again, it is only suggested here to do it on S-thread level.

8   Accruals are adjustments made at the end of an accounting period to recognise expenses that have been incurred during the period but for which no invoice has yet been received and no cash has been paid out.

is on schedule and budget – until the last 5 or 10% of the task or project, which will then require an inordinate amount of time and money to complete.'[9]

Therefore, for the Percent Complete Method to work, firm guidelines must exist and must be applied in a disciplined manner. In a multi-functional team environment this is a difficulty on its own. But even then, with so many work packages in place, doing regular checks on work progress is very resource intensive and therefore not sufficiently pragmatic for commercial aircraft development. The 50/50 Technique is less resource intensive compared to the Percent Complete one. However, it still relies on a subjective judgement. The Milestone Method is a far more objective method than the other ones. But it is also the most pessimistic one as it only credits work achievement once all the work towards a given milestone has been completed.[10]

It is nevertheless suggested here to use the Milestone Method for a pragmatic work progress measurement, albeit not using the finishing milestones of tasks/activities but of threads.[11] Applied in this way, the Milestone Method represents a cheaper method to monitor work progress as it requires fewer resources. However, it still yields a sufficiently accurate work progress picture at higher levels in the Project architecture as sufficient thread-finishing milestones are available.[12] Once the work progress variances have been determined, they can be represented by S-curves, see again Figure 15.3.

Earned Value

R.D. Archibald states that, 'Experience over the years on many projects in many industries, as reported in the extensive Project Management literature available today, has shown that trying to control physical progress and schedules separately from costs usually results in ineffective project control.'[13, 14] For example, Cost Variances (CVs) can be very misleading if they alone are taken as performance indicators separate from schedule variances.

Imagine, for example, that at a given point in time t the difference between the Actual Cost of Work Performed (ACWP(t)) on the one hand and the PCWS(t) on the other is positive. In this case, there is an overspend compared to plan. It could be a result of higher cost for each unit of work (such as higher hourly rates compared to plan) or simply a result of earlier achievement of some additional work originally planned for later. From the CV calculated in this way there would be no unambiguous conclusion. But no Project Manager can accept not to know what the root cause of an overspend is.

---

9   Archibald, R.D. (1992), *Managing High-Technology Programs and Projects*, 2nd Edition, (New York: John Wiley & Sons), p. 278.
10 Note that the Percent Complete method as well as the 50/50 Technique are based on tasks/activities while the Milestone Method is based on milestones. If every task/activity would be finished by a single milestone (and there would be no other milestones to be achieved while performing each task/activity), the 0/100 Technique would be identical to the Milestone Method. With a ratio of 0/100, no value is earned at start, none while it is in progress, and 100 per cent is only earned once a task/activity is fully completed.
11 Having said this, it can be considered to apply the Milestone Method in one area on below-thread level: it can be used for measuring work progress with regard to the milestone data contained in databases attached to threads (see explanations in Chapter 13). The other methods would not work for this at all.
12 As suggested, if ten thread-finishing milestones per team and month are taken as the guideline, times hundred teams times twelve months ≈ 10,000 milestones to be monitored per year.
13 Archibald, R.D. (1992), *Managing High-Technology Programs and Projects*, 2nd Edition, (New York: John Wiley & Sons), p. 277
14 Among other, this is an important argument for keeping schedules and finances monitored and controlled under the common umbrella of a Project Management Office (PMO).

Therefore, for an improved project control, the Earned Value (EV) methodology has been developed. EV is the value of the work which has been earned until a given date, expressed in units of cost (such as $ or €). The EV at a given time t is therefore equal to the Planned Cost of all the Work Performed until t (PCWP(t)), independent from when the work had originally been scheduled.[15] It is a cumulated value. Thus, in order to calculate the EV at time t, one must be able to measure the work performed until t and to identify what its planned cost was.

As it has been argued many times by now, for commercial aircraft development projects there is the necessity to limit the resources for project control. It is therefore not suggested to follow the classical approach when monitoring the EV evolution during the Development sub-phase. Classical EV is based on a bottom-up approach where planned costs are established for *all* lowest level tasks/activities of a project (that is, in the 'language' of this book: on the lowest GANTT-level).

> '[However,] breaking down a project into a large number of tasks may create such an administrative workload that it is too cumbersome and costly to maintain the information up to date.'[16]

It is rather suggested here to collect ACWP(t), PCWS(t) as well as EV(t) on thread level only and then to aggregate them individually on S-thread level or any higher level in the Project architecture.

The aggregated values of ACWP(t), PCWS(t) as well as EV(t) (that is, Σ ACWP(t), Σ PCWS(t), Σ EV(t)) can subsequently be used to calculate some important KPIs on S-thread or higher level. These KPIs are very powerful and there is evidence that fewer milestones are delayed if rigorously monitored.[17]

- Cost Variance: $CV_C(t) = \Sigma\ EV(t) - \Sigma\ ACWP(t)$
- Schedule Variance: $SV_C(t) = \Sigma\ EV(t) - \Sigma\ PCWS(t)$

or, alternatively, the

- Cost Performance Index: $CPI_C(t) = \Sigma\ EV(t)\ /\ \Sigma\ ACWP(t)$
- Schedule Performance Index: $SPI_C(t) = \Sigma\ EV(t)\ /\ \Sigma\ PCWS(t)$

(The subscript 'C' denotes that these KPIs are based on cost information). Note, that on S-thread level Σ PCWS(t) = BCWS(t).

With a code-based booking system in place, it is standard practice to capture ACWP(t). However, as two or more threads may share the same booking code (see explanations in Chapter 13), it is not possible to break down the ACWP(t) for the different threads. In the Project architecture suggested here, it is therefore impossible to calculate the $CPI_C(t)$ on thread level. However, on S-thread or higher level it is absolutely possible, see again Figure 13.18.

With the above described budget allocation process in place, which leads to valued threads, the identification of thread-level PCWS(t) is also possible. And as it was suggested

---

15 In literature, PCWP is usually called Budgeted Cost of Work Performed (BCWP). However, for the purpose of this book 'budget' is a synonym for *allocated* funds not for planned costs. Budgets are only allocated to S-thread level, where BCWP = Σ PCWP.

16 Archibald, R.D. (1992), *Managing High-Technology Programs and Projects*, 2nd Edition, (New York: John Wiley & Sons), p. 277.

17 See for example: Chase, J.P. (2001), *Value Creation in the Product Development Process* (Boston: Massachusetts Institute of Technology, Department of Aeronautics and Astronautics, Master of Science Thesis), pp. 99–101.

above to use the Milestone Method, it is sufficient to concentrate the planned cost value of the threads in the thread-finishing output milestones. Whenever, therefore, such a milestone has been achieved, the corresponding thread-value is earned.

One may wish to call the output milestones on thread level which are valued in this way 'EV-milestones' as they are relevant for the calculation of the EV. This distinction is important because there may well be other milestones associated with threads, even critical ones, which are not EV relevant. For example, a critical milestone in a plan of a multi-functional team A may be represented by an important delivery from another team B. This milestone is not EV relevant for team A as no work and no budget of team A are associated with it.

Another example could be the work progress at a Risk-Sharing Supplier (RSS): although there may be milestones included in the schedule plan of a contractor's team A to monitor work progress at the RSS, they are not EV-milestones for team A. By definition, all NRCs to be covered by a RSS are amortised over the number of shipsets to be delivered in the future. Therefore, there is no NRC budget provided by any multi-functional team to a RSS for its development work. Seen from the perspective of team A, the EV-methodology does not apply to the work performed by the RSS. Only the monitoring of work progress does.

The definition of what threads are all about can now be formulated more concretely compared to what was said in Chapter 13: each thread is an aggregation of tasks/activities which finishes by an EV-milestone. Please note that this definition requires an a priori modelling of the schedule plans to enable easier EV measurements: tasks/activities have to be aggregated to threads in a way that the latter can be finished by EV-milestones, which are valued according to the budget allocation process described above, and where the value is earned once the customers of the EV-milestones have accepted the associated deliverables. The EV methodology therefore strongly influences how schedule plans should be set up.

For individual threads, therefore, the PCWS(t) is either 0 per cent or 100 per cent of the total Planned Cost of Work Complete (PCWC) for this thread, depending on whether the thread-finishing EV-milestone was *originally scheduled* before (100 per cent) or after (0 per cent) time t. Similarly for EV(t): it amounts to either 0 per cent or 100 per cent of the total PCWC for this thread, depending on whether the thread-finishing EV-milestone was *de facto achieved* before or at time t (100 per cent) or not (0 per cent).

Figure 15.4 illustrates this. It compares a 'real' situation with the approximation using the Milestone Method. Clearly, the 'reality' in the example shown is already a simplification as the distribution of PCWS(t) over the span of the thread would usually not be of a linear nature.[18] In addition, the example is assuming that the work progress can be measured in the first place. As we saw above, it is impossible to do this in an objective way except if the Milestone Method is used. However, the important point of Figure 15.4 is that in 'reality' an EV of 20 per cent of the PCWC has been achieved, while the approximation using the

---

18 In Figure 15.4 the Earned Value is based on a linearised approach: it is assumed that work progresses with time in a linear way and that the planned cost per unit of work remains constant.

   EV(t) is therefore simply expressed as the percentage of work complete at time t (WP%(t)), times the total PCWC.

   WP%(t) is calculated as the actually performed number of work units (for example: manhours) at time t (WP(t)) over the total number of work units forecast to complete the work.

   The forecast total number of work units is equal to WP(t) + RTD(t), with RTD(t) being the estimated number of work units required for the remain-to-do work until full work completion. To avoid misunderstandings it is at this point important to note that the forecast total number of work units is neither calculated as the total work scheduled until time t (WS(t)) nor as the total number of work units needed to represent the Planned Work Complete (PWC) at the Planned Time of Work Complete (PTWC).

   Therefore, in Figure 15.4, WP% = 20 per cent, and not 33 per cent or 25 per cent, respectively, and EV(t) = 20 per cent of PCWC.

Milestone Method yields 0 per cent. This is because the thread-finishing milestone is not achieved before time t.

One has to bear in mind that this approximation of EV has the tendency of showing a pessimistic picture of project status as work progress is only credited once a thread has been completely finished. This is a direct result of having selected the Milestone Method. However, there is also an advantage of using this method: the knowledge about the precise distribution of the PCWS(t) over the span of each thread becomes irrelevant, which otherwise would have been a necessary (but difficult to achieve) prerequisite for the application of the Percent Complete Method.

EV(t) on S-thread level is now equal to the sum of the corresponding values of all achieved finishing milestones, which belong to threads, which in turn belong to the S-thread. This is shown in Figure 15.5, where an S-thread is composed of five threads. The threads are at different achievement status. Their relative location is according to the original baseline plan. As in Figure 15.4 it is also assumed here that work progresses with time in a linear way. For reasons of simplicity all threads have the same PCWC.

For the same reason, the total forecast work required to complete a thread is in each case assumed to be equal to the Planned Work Complete (PWC). This contrasts with the example provided in Figure 15.4, where the total forecast work required to complete a thread was bigger than PWC.

Note that Thread (E) shows out-of-sequence work, that is, work which is performed earlier than planned. All work- and cost-related values are at time t. Also for simplicity reasons, planned thread-level costs increase again on a linear basis with the work. This, of course, does not apply any more on S-thread level because of the relative position of the threads along the time axis. The BCWS and EV at S-thread level at time t are calculated using a graphical approach and are shown on the top of Figure 15.5.

While on thread level the PCWS(t) and EV(t) is either 0 per cent or 100 per cent of the total PCWC, for an S-thread it could be any number between 0 per cent and 100 per cent, see Figure 15.5. One can easily see that the Milestone Method leads to numbers for BCWS and EV on S-thread level, respectively, which are somewhat away from the 'reality'. While in Figure 15.5 the exact value for the 'real' BCWS(t) is 55 per cent, the approximated value is 40 per cent, thus corresponding to a 27 per cent deviation. For the EV, the 'real' figure is 45 per cent as opposed to the approximated figure of 40 per cent. The corresponding deviation amounts to 11 per cent. The accuracy of the approximations would be further decreased if long duration threads would be applied. If, for example, Thread B in Figure 15.5 would last longer, the above mentioned EV-based KPIs would not yield any reasonable information for a long time.

The only way to improve the accuracy is to ensure a sufficiently high number of threads to be aggregated to S-threads, while these threads must be of reasonably short duration compared to the reporting periods. Thus, in order to save project control resources, it is worth considering *a priori* how many threads should on average aggregate to one S-thread and what their average duration should be. This is another good example of how requirements for project monitoring and control influence a project's architecting and structural build-up. The latter must indeed be a truly multi-functional endeavour!

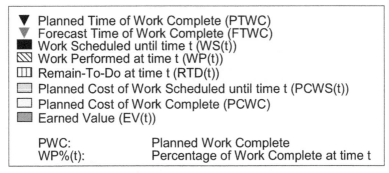

- ▼ Planned Time of Work Complete (PTWC)
- ▼ Forecast Time of Work Complete (FTWC)
- ■ Work Scheduled until time t (WS(t))
- ▨ Work Performed at time t (WP(t))
- ▥ Remain-To-Do at time t (RTD(t))
- ▢ Planned Cost of Work Scheduled until time t (PCWS(t))
- ▢ Planned Cost of Work Complete (PCWC)
- ▩ Earned Value (EV(t))

PWC: Planned Work Complete
WP%(t):   Percentage of Work Complete at time t

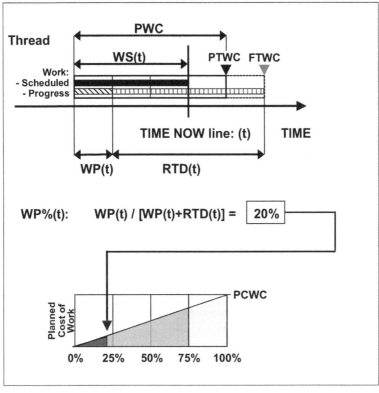

"Reality" Example:

PCWS(t): 75% of PCWC
EV(t):   20% of PCWC

Using Milestone Method:

PCWS(t):   0% of PCWC
EV(t):   0% of PCWC

Figure 15.4   Calculation of Planned Cost of Work Scheduled and Earned Value

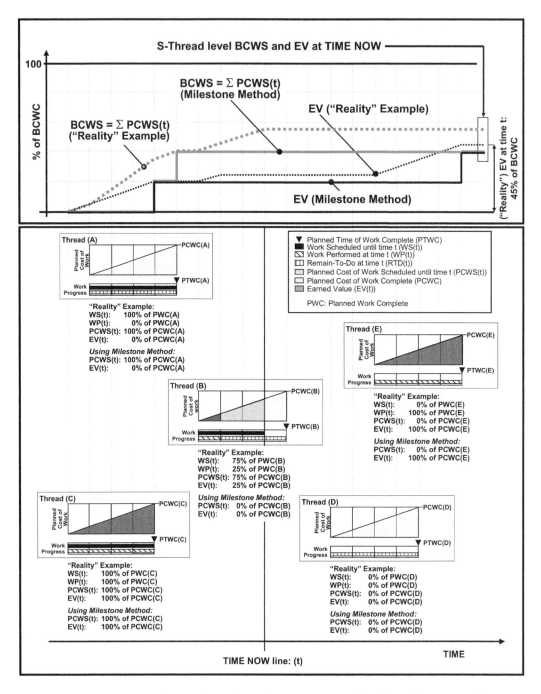

Figure 15.5   Comparison of 'real' and approximated values for Earned
Value and Budgeted Cost of Work Scheduled on S-thread level.
Simplified example

However, there are two problems with the above mentioned EV-based KPIs:

1. The times, at which ACWP(t) on the one hand and PCWS(t) and EV(t) on the other are measured, are not identical. ACWP(t) is usually captured at month end, but by far not all thread-finishing milestones are planned to be at ends of months. This gap in timing is a direct consequence of having preferred the Milestone Method. But by using the other methods one would not be better off: as they represent subjective estimations of work progress, an element of inaccuracy would also in this case be introduced. Consequently, one definitely has to live with some sort of compromise and approximation when using EV.

2. Schedule Variance (SV(t)) and Schedule Performance Index (SPI(t)) are broadly recognised for failing when projects continue execution past the planned end date. For late finish projects, SV(t) converges and concludes at the value 0 while SPI(t) behaves similarly, converging and concluding at 1. With this flaw, forecasts on schedule variances cannot be performed reliably.[19] However, more reliable schedule performance information can be obtained by applying what is called Earned Schedule (ES(t)).[20] The latter identifies when the amount of EV(t) should have occurred. On S-thread level, this is the point in time ES(t) where the BCWS is equal to the Σ EV(t): BCWS(ES(t)) = Σ EV(t), see Figure 15.6. Note that ES(t) is denoted in units of time, while EV(t) is in units of money.

    Using ES(t) on S-thread level, the schedule-relevant KPIs can be calculated as follows:

- Schedule Variance (SV$_S$(t)) = ES(t) – t
- Schedule Performance Index (SPI$_S$(t))= ES(t) / t

(The subscript 'S' denotes that these KPIs are based on schedule information).

Although the EV/Earned Schedule method will always be an approximation of reality, there are now various KPIs available which provide the Project Manager with reasonably unambiguous status information about the project. In particular CPI$_C$ and SPI$_S$ are very powerful KPIs in monitoring projects. They are dimensionless relations which can be easily depicted in charts such as the one used in Figure 15.7.

---

19 This flaw does not occur if PCWP(t) is calculated as the number of performed work units times the planned cost per unit of work (example: hourly rate). However, this method of calculating Earned Value is not the classical and normally used one.
20 See for example: Lipke, W. et al. (2009), 'Prediction of Project Outcome. The Application of Statistical Methods to Earned Valued Management and Earned Schedule Performance Indexes', *International Journal of Project Management* Vol. 27, pp. 400–407.

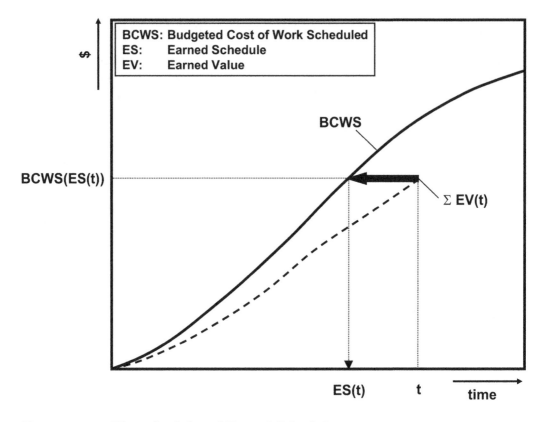

Figure 15.6   The principles of Earned Schedule

Requirements and Their Verification and Validation

In Chapter 8 the overall Requirements Management process was described. Apart from ensuring traceability, it is all about an agreement between the issuer and the receiver of a requirement in terms of its content, attributes and means of verification. It should therefore be the actual number of fully agreed requirements compared to the total number of requirements which should be monitored over time as a KPI for each multi-functional team.

When analysing a requirement, engineers usually already have an initial design solution in their mind which could meet the requirement. Experienced engineers will therefore have no problem agreeing to the majority of the requirements quickly. Thus, the initial achievement rate on the above KPI will be high. However, subsequently the rate will slow down as more effort needs to be taken to properly understand the remaining requirements and their repercussions. The rate on the final 5–10 per cent of requirements, which are not yet agreed, needs to be monitored very carefully as a slow agreement rate may well indicate a problem of the team not knowing how to cope with these requirements. Unfortunately, with the overall project proceeding quickly, it is only too often at this point that further work on

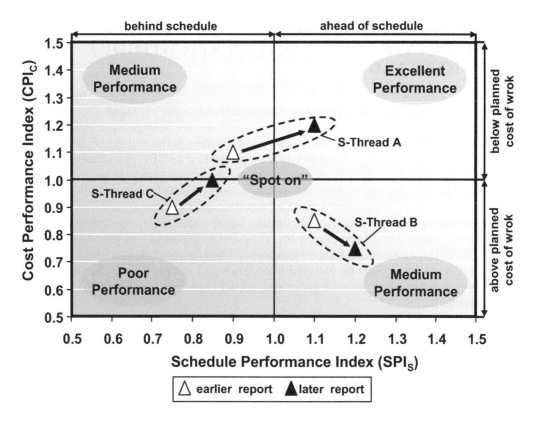

Figure 15.7   Schedule Performance and Cost Performance Indices used for project control

agreeing requirements gets delayed *ad infinitum* because other development aspects receive higher Management attention. It is usually the new, critical requirements – unexpected by the Engineering community – which get pushed aside. Deferring the agreement process into the future and not addressing this issue immediately may cause significant delays and cost overruns at a later stage.

One encounters a similar situation when looking at the verification and validation status of requirements. The number of verified and validated requirements, respectively, compared to the total number of agreed requirements are two important KPIs to monitor. Again, the rate of verification/validation will be high initially as many requirements can be verified/validated easily. But then the rate will slow down as more verification and validation effort is required and the last 5–10 per cent will be very difficult to achieve too. Project managers should carefully investigate the criticality of not having verified or validated the last few requirements. If design cannot meet requirements (verification) or the produced parts or assemblies cannot meet the requirements (validation) then the certification process or customer satisfaction or both could be in jeopardy.

KEY PERFORMANCE INDICATOR MONITORING: RUN-RATE EXTRAPOLATION FOR KPIs WITHOUT BASELINES/TARGETS

Wherever the actual status of a KPI has been monitored over a period of time, a trend can be extracted and extrapolated into the future based on the run-rate averaged over a number of recent reporting periods. In particular where no baselines or targets are available, the only way to gain visibility about potentially arising issues is the run-rate extrapolation. This run-rate KPI is a trend KPI and not a KPI representing a status of variances.

It can be applied, for example, to events like modifications, concessions, outstanding work and so on, the amount of which cannot be foreseen. Modifications can be regarded as mini-projects with their own schedule, resource and cost repercussions once the rationale for a change to an aircraft configuration has matured and the process for gaining modification approval has been kicked-off. They can be planned and monitored as any other project. The same applies to concessions. However, new design changes or the raising of concessions cannot be planned in advance and can only rarely be anticipated. As no baseline S-curve can, thus, be generated, the KPI can only measure actual numbers of new events.

Run-rate KPIs are very important as, for example, the accumulation of new design changes or more and more concessions sends a strong message into the organisation with regard to the corresponding downstream work: As long as the run-rate increases (that is, more and more changes or concessions are generated over time) the downstream workload for change implementation or concessions approval will increase too – often in a non-linear way.

CORRECTIVE ACTIONS

Team Meetings

Once the team has visibility on the status of its project, corrective actions may have to be taken to reduce variances and to bring the project back on track. Actions should be launched and commitments to fulfil them should be provided during team meetings. It is these multi-functional team meetings which really lie at the heart of the whole project control system. All debates to find the best way forward need to take place there.

During team meetings, the team leader and key functional representatives in the team are confronted with both the current, integrated analysis of progress against the plan as well as with the forecasts for the project's future status. It is important to assess progress and future status jointly so that everyone has the same level of understanding and status awareness, and agreement can be obtained that a problem really exists and does indeed require action.

'An important footnote to the formulation of the plan for solutions … is the question of who it is that devises such solutions. Overall responsibility rests with the P[roject] M[anager], P[roject] C[ontroller], and … [key functional representative], but it is suggested that a team approach to problem solving be undertaken. In other words, information and proposals for solutions should be consciously elicited from members of the project team.'[21]

'Once a project is in trouble, it is very often the case that there is disagreement on how to fix it. It is very important to achieve a common understanding of the problem and to jointly search for a solution.'[22]

However, as the atmosphere at project team meetings can be crisp and tense, some guiding principles should be communicated and agreed to ease discussion and keep time discipline.

'[During the 777 development, Alan Mulally had] ... a set of "Principles and Practice" that ... meant to govern ... programme reviews. They ... [were] put up on a slide at the beginning to remind people of the ground rules. They ... [included]:

- use facts and data;
- no secrets;
- whining is OK, occasionally;
- prepare a plan, find a way;
- listen to each other;
- help each other, include everyone;
- enjoy each other and the journey;
- emotional resilience.'[23]

Sometimes guidelines need to take into account cultural differences of team members too.

'According to the German culture, the efficiency of a meeting is judged by the degree to which it has adhered to the foreseen time and agenda. This is why a meeting facilitator's task is to bring discussions back to the agenda whenever necessary. In contrary to this, in France and in other Mediterranean countries it is not so important to stick to the agenda provided one gets a better overview on the project status.'[24]

After all, if an environment of open communication is desired among team members, the tone at team meetings is important. Team leaders play a key role in setting the right tone and in getting the team meeting to focus on solving problems rather than discussing the blame for them. It is in these meetings where the leadership skills of the team leader are most important for the project success.

The frequency of team meetings should be chosen adequately. There will be times during the Development sub-phase where weekly meetings will suffice while at other times teams need to hold daily 'drumbeats' to ensure progress and no falling back behind the plan. Every day counts. When large amounts of drawings need to be released or many modifications

21 Eisner, H. (2002), *Essentials of Project and Systems Engineering Management*, 2nd Edition, (New York: John Wiley & Sons), p. 117.
22 Hoffmann, H.-E., Schoper, Y.-G. and Fitzsimons, C.J. (2004), *Internationales Projektmanagement* (München: Beck-Wirtschaftsberater im Deutschen Taschenbuch Verlag), p. 54, free translation by the author.
23 Sabbagh, K. (1996), *21st Century Jet. The Making of the Boeing 777* (London: MacMillan Publishers), pp. 253–4.
24 Hoffmann, H.-E., Schoper, Y.-G. and Fitzsimons, C.J. (2004), *Internationales Projektmanagement* (München: Beck-Wirtschaftsberater im Deutschen Taschenbuch Verlag), p. 259, free translation by the author.

need to be digested, losing any one day hurts. Reviews at lower frequency (weekly, monthly) would simply not do the job during these crucial periods. In addition, daily meetings generate a discipline and mindset among people to think about making daily progress.

Apart from technical, contractual and commercial problems, there should also be Project Management-related topics on the agenda of team meetings in order to properly understand KPI variances and their repercussions. The latter may include any of the following:

- As engineers often tend to make designs more sophisticated than necessary, there is the danger that design work extends beyond the agreed scope of work. To keep the scope of work under control the team leader not only has to insist on signed Work Package Descriptions (WPDs) based on agreed requirements but also has to constantly monitor that the actual work performed is within the scope of the WPDs. In Concurrent Engineering, this can be very challenging as the level of risk taking is high (see Chapter 9) and work package owners may be forced to divert from the originally intended work content.

- A typical temptation of teams is to bring forward milestones which can be achieved easily and quickly although they were originally planned to be achieved at a later stage only. As a result, a team's performance may initially look in line with the baseline (or even better) leading to over-optimistic forecasts. In reality, bringing forward the easy milestones usually implies that other deliverables, which should have been delivered earlier, have been pushed back, thus causing delays in other areas of the project.

- Team members performing the work often do not appreciate the importance of correctly recording their time against open booking codes. Many of them regard this as a bureaucratic activity which distracts them from their work. Consequently, time sheets get filled out in a careless fashion and costs are allocated incorrectly. Team leaders should remind all members of the team about the correct recording of time. If necessary, additional training may be required.

- Also, Functions' managers often are tempted to charge costs for work done on a project, which is over budget, to another project that still has an unexpended budget balance. This behaviour damages company performance in the longer term as real problems remain hidden. They therefore distort the records for retrospective cost analyses, which represent the basis for future projects' cost estimations. Team leaders need to identify such misbookings and ensure corrective actions.

- Bookings should happen at frequencies in line with the reporting frequency of work progress. If, for example, work progress gets reported on a monthly basis, all bookings for the work performed should be completed on at least a monthly basis too. Large, singular bookings at lower frequencies, such as quarterly or yearly, should be avoided as they tend to create surprises, such as unexpected cost overruns, with limited possibilities for countermeasures.

- Companies usually have established procedures for work authorisation, sub-contracting, purchasing, project control and many others. The team leader as well as the key functional representatives must understand and know how to use them. If they don't, team meetings should be used to share knowledge.

- Sub-contracting requires special attention, in particular where it represents a major part of the project, or where the Critical Path depends on the delivery of a sub-contract product or service. Some unique steps may need to be taken in this case for the team leader to stay in control of the project. H. Eisner suggests the following:
  – 'placing project personnel at the subcontractor's facility to monitor status and progress;
  – establishing interface control and documentation as a more prominent aspect of the Systems Engineering effort;
  – holding more frequent status review sessions for subcontractors;
  – meeting with the Management of the subcontractors to obtain commitment to cost, schedule, and performance requirements;
  – providing parallel developments and backup sources as insurance, if they can be afforded;
  – using incentive award contracts for on-time, high-quality deliveries.'[25]

- For complex projects it is particularly important to not only manage the interfaces with the direct suppliers (that is, the first tier suppliers), but in fact as well with critical lower-level tier suppliers in the supply chain. First tier suppliers may themselves not have the experience or capacity to perform supply chain management professionally. In this case, teams need to discuss whether support to first tier suppliers is required.

- Developers must also be in direct contact with the customers in order to receive constant feedback whether the aircraft development proceeds into the right direction. However, the customer should not be bombarded by questions and comments from different organisations in the company. Direct contact should not result in uncontrolled contact. Contacts should rather be structured in customer focus groups, which act as the sole focal point between the company and the customer. Team meetings should be used to coordinate customer contacts.

- Leaders and facilitators of team meetings must be aware of the different multi-cultural backgrounds of team members when trying to identify and solve problems. Some cultures, for example, tend to have more difficulties in communicating problems openly than others. US Americans, for example, will not communicate problems as openly as Germans do as this could rapidly have adverse consequences for their jobs and careers.[26]

Wherever unfavourable variances exist, teams should identify a series of corrective actions and should nominate individuals to be in charge of their implementation. This should be done according to a priority list separating the urgent and the important issues to be resolved from the not so urgent and not so important ones.

---

25 Eisner, H. (2002), *Essentials of Project and Systems Engineering Management*, 2nd Edition, (New York: John Wiley & Sons), p. 105.
26 Hoffmann, H.-E., Schoper, Y.-G. and Fitzsimons, C.J. (2004), *Internationales Projektmanagement* (München: Beck-Wirtschaftsberater im Deutschen Taschenbuch Verlag), p. 86, free translation by the author.

'A priority list is intended to force a discipline that assures that key problems cannot be ignored or placed on the back burner. Without this discipline, a project manager might be otherwise inclined to tackle more tractible issues that are of little or no real importance and avoid handling critical problems that might be difficult to confront.'[27]

Corrective actions should get recorded in an actions' tracker, which is visible in the 'war room' so that progress on the actions' achievement can be discussed and recorded at each meeting. Resulting recovery plans should always be regarded as what they are: (important) guidelines for taking decisions and measuring progress based on regular updates reflecting the decisions taken by a team of how to progress. The validity of the plan is therefore by definition limited until the next update.

Project Evaluation Meetings

Frequent multi-functional team meetings should be complemented by regular project evaluation meetings on, say, a monthly or quarterly basis. These latter meetings should be chaired by team leaders, which are one or two levels higher up in the organisation compared to the team in charge of the project to be evaluated. The major objectives of these evaluation meetings is to check whether:

- the lower-level team is in fact able to exert proper project control, demonstrated by its success in achieving milestones and in implementing recovery actions; and whether
- there are any obstacles, which higher Management might be able to remove.

There is an obvious question with regard to the communication of issues and problems to higher-level Management during an evaluation meeting (or at any other time): 'If new problems arise, at what point does the team give an alert to the higher Management?' First of all, with proper Risk Management in place (see Chapter 9), the communication of risks to higher-level Management should work. Management should not be surprised if mitigation plans sometimes fail and risks materialise as issues.

However, other issues might come as a surprise. In this case, the recommended approach is the following:

- inform the Management about the issues while telling it that a recovery plan will be set up until a given date;
- present up to that date the recovery plan to the Management which gives the latter a last opportunity to provide input into the plan, or to modify the plan if necessary;
- in any case, do not 'hide' problems from Management as initiating a recovery activity without consulting it may be risky, in particular if the issues are severe.

There is a risk with project evaluation meetings, though, in particular for mega-projects such as aircraft development. Mega-projects:

---

27 Eisner, H. (2002), *Essentials of Project and Systems Engineering Management*, 2nd Edition, (New York: John Wiley & Sons), p. 117.

- have a long planning horizon, thus making it easier to plan them, but potentially offer opportunities for redirection, which is attractive to Top Management, but which could cause disturbances;
- involve huge budgets. Managers give huge budgets a lot of attention, in particular if budget overruns have catastrophic consequences for the company as a whole; and they
- increase business risks, in particular if concurrent ways of working are applied.

All of this drives Management to seek methods and tools which are perceived to increase predictability. Management therefore has the tendency for micro-management, requesting more detail in the planning as well as in the reviews. Project evaluation meetings can become a nightmare. More reporting, more redirection, more detailed planning all make a review of a mega-project even more 'mega'. Mega-projects are of a self-reinforcing 'mega-' nature.

'The sad truth about the mega-project is that giant steps require us to take even bigger giant steps.'[28]

Unfortunately, this approach is misusing scarce resources for too detailed reporting without offering the benefit of drastically improved predictability.

Top Management must be aware of the counter-productiveness of micro-management. Instead, the principle of subsidiarity should be applied on all levels. Only if lower level teams have all the levers themselves to take appropriate decisions can they manage Rapid Developments. They need the authority and capability to react quickly to move on. Decisions necessary for Rapid Development projects can only be made quickly if they are located at a level in the multi-functional matrix organisation that has sufficient power to take them quickly. Not to provide the right capabilities to the teams is one of the major stumbling blocks in many companies.

The attitude Top Management has to adopt is therefore quite different from the one frequently applied in the past. It has to change its mindset from making teams dependent on their wisdom and experience to a mindset where imperfect decisions taken quickly are more valuable than perfect decisions taken too slowly. Decisions should be taken at the lowest adequate level. They will usually turn out to be better and faster decisions.

RECOVERY-BASED FORECASTING

Based on corrective actions committed during the multi-functional team meetings, improved forecasts can be established. This requires, however, that recovery activities are added to the GANTT-schedule plans and the corresponding threads are amended (for example by updating resources profile, changing delivery sequences and so on). If scheduling[29] of the plans then delivers the desired leadtime or cost improvements, the recovery activities have been properly implemented in the schedule plans. Forecasts can subsequently be generated with a high level of precision.

---

28 Smith, P.G. and Reinertsen, D.G. (1998), *Developing Products in Half the Time*, 2nd Edition, (New York: John Wiley & Sons), p. 75.
29 Remember that 'scheduling' is a planning activity, which calculates the correct location of task/activity bars along the time axis in a GANTT-chart according to the interdependency logics (such as Start-to-Finish, Start-to-Start, and so on) as well as the dates or durations associated with those tasks/activities.

Note that there is a strong link between Risk Management and cost control too. Most risks have potential cost impacts.[30] Thus, whenever future CVs have to be forecasted (such as the cost situation at year end or the Forecast Cost at Completion (FCAC) of the entire project) the potential cost (and opportunity) impacts of risks should also be taken into account.

Establishing forecasts – in addition to baselines and performance actuals – demonstrates control, if based on recovery plans and not on simple extrapolation of the run-rate averaged over recent periods. Management should always ask for the detailed recovery activities which are behind the forecasts. However, as this represents a lot of work for the teams, it may not be possible to generate more than a very small number of forecasts per year. Asking for detailed recovery activities is not so much a question of confidence but more a pull from Management to ensure that in fact the teams in charge develop recovery scenarios in the first place. Only if delays result in a situation where the original baseline is unrecoverable, controlled re-baselining needs to take place. This must always result in a new delivery agreement for the data, service or product, to be approved by the affected stakeholders.

However, as far as schedule and cost forecast are concerned, one may wish to use the EV-based KPIs introduced earlier. Unfortunately, EV-KPI-based forecasts need to be taken cautiously as it remains principally impossible to predict a project's future, in particular if there are many unknowns. Under this condition and based on findings by W. Lipke et al.,[31] the following formulae are suggested here for forecasting purposes:

- Forecast Duration at Completion (FDAC):
  $FDAC(t)_{>50\%} = PDAC / [SPI_S(t)_{>50\%} \pm 0.10]$

- FCAC:
  $FCAC(t)_{>50\%} = BCWC / [CPI_C(t)_{>50\%} \times (1-\alpha)]$

with PDAC the Planned Duration at Completion and BCWC the Budgeted Cost of Work Complete. The correction factor α amounts to 20 per cent at 50 per cent of PDAC and linearly decreases thereafter until it is 0 per cent at 100 per cent PDAC. Different analyses have shown that FDAC and FCAC can only be interpreted with some confidence once the project is at least at 50 per cent of PDAC (hence the index '>50 per cent').

REPORTING

With all the different data (drawings, documents, data entries and so on) linked in the described way to the integrated Project architecture, there exists now a single source from where data for reporting purposes are to be generated. Project data can be reported in many different user-specific formats, such as S-curves, schedule mismatch charts, diagrams, as well as actual-versus-plan variances for costs, resources, milestone achievements and so on. They can all be generated to monitor the progress of the project. An example for reporting of weight evolution is depicted in Figure 15.8. In addition, it is possible to

---

30 Even those risks which primarily result in potential schedule impacts will have cost repercussions as a secondary effect.
31 Lipke, W. et al. (2009), 'Prediction of Project Outcome. The Application of Statistical Methods to Earned Valued Management and Earned Schedule Performance Indexes', *International Journal of Project Management* Vol. 27, pp. 400–407.

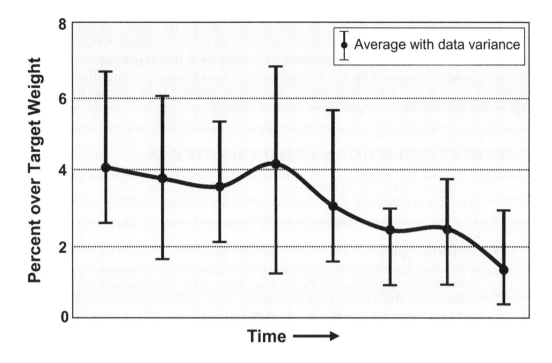

Figure 15.8   Example of how to report on the evolution of weight

generate reports according to a whole range of breakdown possibilities as defined by the different views.

However, good KPI reporting is not just a representation of measured statistics. By combining different KPIs it should rather be possible to generate a status 'picture' in the mind and imagination of stakeholders. Only by looking at a problem from different angles can an in-depth comprehension of the real status be obtained. Each page of a report should therefore contain different KPIs, which nevertheless deal with the same subject. It should cover the KPIs' individual:

- status of variances (against baselines, requirements, targets and so on); as well as
- trends (historic, forecasts);

as these are the essential ingredients of good KPI reporting. Quite importantly, explanatory text needs to be provided too. It generates the required level of situation awareness beyond statistics and recovery plans. Forecasts should always be enriched with information on risks and opportunities to demonstrate the reliability of the forecast. These risks and opportunities should be managed using the Risk Management process described in Chapter 9.

Reporting is useless if there is no customer for the reports. Thus, every team generating reports should identify who the customers for the different reports are and what their needs are. Of course, one of the customers for KPI reports should always be the team leaders and key functional representatives themselves. They need to pull for the reports and use them in their periodic team meetings to take decisions. However, on a higher level of the project's organization, not only the aggregated performance of the many multi-functional teams with their individual outputs should be monitored but also the entire Business Case of the programme in terms of Net Present Value (NPV), Internal Rate of Return (IRR) and so on, as introduced in Chapter 4. It is very important for the company to know in case of budget overruns to what extent and in which direction the Business Case evolves, identifying perhaps, how many additional aircraft may have to be sold before the Business Case objectives can be achieved.

MAIN CONCLUSIONS FROM CHAPTER 15

- It is suggested to allocate budgets directly to S-threads of individual pyramids of Product & Assembly Tree-elements or view-specific modules, for which multi-functional teams are in charge, and not to do this on any lower level.
- A challenging, iterative convergence process is required to match the top-down cost targets as derived from the Business Case with the bottom-up estimations originating from the work packages. Only then, budgets can be allocated to S-threads and further cascaded down to the corresponding work packages, A-threads and threads. Each thread then has a defined value.
- Workload and cost actuals captured at thread level need first to be aggregated to the corresponding S-thread level before any analysis of variances can be performed. For this to work, actuals should be captured with the same frequency applied for the phasing of allocated budgets.
- It is suggested to use the Milestone Method on thread level for work progress measurement. The value of each thread is to be concentrated in a thread-finishing milestone. Whenever a thread-finishing milestone has been achieved, a value is earned.
- Measuring Earned Value in this way is an approximation of real value earned. It is showing a pessimistic picture of project status as work progress is only credited once a thread has been completely finished. But it is simple to use.
- For a reasonable accuracy of the approximation, there must be a sufficiently high number of threads, while these threads must be of a shorter duration compared to the reporting periods.
- The Cost Performance Index CPI_C, based on Earned Value, as well as the Schedule Performance Index SPI_S, based on Earned Schedule, are very powerful Key Performance Indicators, which provide the Project Manager with reasonably unambiguous status information about the project.
- With all the different data (drawings, documents, data entries, actions and so on) linked to the integrated Project architecture, there exists a single source from where data for reporting purposes can be generated.
- Once a team has visibility on the status of its project, corrective actions may have to be taken to bring the project back on track. Actions should be launched and commitments to fulfil them should be provided during team meetings. It is these team meetings which really lie at the heart of the whole project control system and where the leadership skills of the team leader are most important for project success.

PART VI

Summary

16

Summarising Remarks

There is a fundamental dilemma typical for all commercially funded projects of very high complexity: on the one hand they require highest levels of sophistication in various management disciplines. On the other, there is a natural tendency to limit the costs associated with the management aspects of such projects. The theme of this book was that the dilemma can be resolved by concentrating on some essential elements of the various management disciplines and integrating them in an intelligent and pragmatic, yet holistic way. In other words: complex commercial projects require a dedicated management approach.

It is the project's complexity which drives the need for an integrated, holistic management approach, while the strong need to save management costs in commercial projects drives the tendency to concentrate on some essentials of the available management disciplines.

Commercial aircraft developments represent an extreme example of such demanding projects. They are uniquely complex undertakings from a technical, managerial, processes and business point of view. They are multi-cultural and top-of-the-learning curve mega-projects exposed to high financial risks and strong public interest. This book therefore concentrated on the example of commercial aircraft developments to demonstrate what a holistic management approach using essential management ingredients could look like.

The views elaborated in this book intentionally concentrated on the *management* challenges related to complex, commercial projects. Major *technical* challenges were not tackled although, for example, technology readiness, technical integration capabilities and many other Engineering disciplines are of course also crucial contributors to the success of a complex development project. Equally, other important disciplines were not addressed in depth, such as Manufacturing Engineering, Logistics and Product Support.

This book identified a series of really essential (project) management ingredients, the most important of which are:

- full exploitation of modern Information Technology (IT) capabilities;
- application of Concurrent Engineering techniques and its major enablers;
- significantly improved streams of communication;
- matrix organisations supporting multi-functional, multi-cultural and highly productive project teams led by enlightened, courageous leaders;
- change controlled project data based on a fully integrated, equally change-controlled Project architecture;
- the right balance between creativity on the one hand and process adherence discipline on the other;

- simplified Earned Value Management during project execution;
- a robust Business Case; as well as
- a Project Plan at project set-up which details what needs to be generated at which point in time.

The secret of successful management of complex, commercial projects lies in the integration of the Product, Process and People aspects related to the above-mentioned ingredients.

To this end, the starting point for an integrated approach is represented by the capabilities offered by modern IT solutions. This is because the full exploitation of today's IT capabilities is about the only way to satisfactorily manage the complexities embedded in such projects.

With modern IT becoming so important to master project complexities, the relevant question then is: How should complex product developments be managed and organised to fully exploit the advantages offered by IT solutions? Or asked in a different way: Provided adequate, robust IT solutions are made available and people are sufficiently well trained to use them, is anything else needed to fully exploit the potentials of modern IT capabilities? Analysing this question, it should have become clear in this book that there are cultural, organisational and process-related changes that must come along. After all, this is the real challenge of applying computing tools.

For example, thanks to modern IT tools, Rapid Development techniques such as Concurrent Engineering became a reality, offering the potential for drastically reduced development leadtimes. But, and this may look like a paradox, they do require drastically improved levels of interpersonal, direct communication between people working on the project in order to ensure faster communication.

This is because digitally managed design changes can be implemented so much quicker compared to the use of classical drawing boards some 20 years ago. In highly complex projects with many thousands of interfaces it becomes increasingly difficult to manage information and to ensure that everyone who needs to know about, say, design changes is informed in due course. Significantly improved ways of communication are therefore a mandatory prerequisite for complex projects seeking rapid product development. The problem is not so much the speed of sending information but the rate at which people can absorb it.

As it has been shown in this book, the demand for much improved communication drives the need for multi-functional, co-located teams. Ideally, these need to be set up in such a way that the most difficult or critical aspects of the product to be developed, such as complex technical interfaces, can always be managed *within* these teams, with the relatively easier aspects to be dealt with *between* teams. The knowledge related to interface complexities must come from experience with previous projects. Not to have this experience introduces a significant amount of additional risk into a project.

This book advocated the design and establishment of office space architectures, which pro-actively support the product development process in general and the communication streams within and between teams in particular. If designed well, they represent very suitable platforms for the co-location of multi-functional teams – rather than forcing people into an office space which happens to be available but is otherwise not suitable to support the product development process.

It has been demonstrated in this book that the multi-functional approach requires complex projects to be organised within a matrix structure. There are pros and cons of such an organisation, with the cons certainly not to be overlooked. In particular for large-scale projects, where a single multi-functional team is not sufficient to develop the product, there is a need for an organisational hierarchy to manage the multiple teams. This introduces a third dimension to the matrix structure, thus adding further complexity to an already complex project. It tends to raise many concerns among people affected by it, which need to be addressed with sensitivity and persuasion.

This is why complex projects require significantly more advanced leadership of the project teams. This is even more necessary where large-scale projects, such as aircraft development, are managed in an international environment, requiring a significant multi-cultural experience of project leaders and team members.

Leadership in a multi-functional and multi-cultural environment is all about generating teamwork and efficiency within a team as well as promoting the interaction with other teams. One of a team leader's most important functions is therefore to develop the project team and to help it move through the various stages of its evolution. But leadership in the context of matrix organisations also has the additional role of counter-balancing their deficiencies.

As professional project leaders are a scarce resource in any company, there is the risk for companies to be limited in their application of IT tools by the shortage of sufficiently skilled project leaders. It must be strongly underlined that the lack of skilled leaders is the single biggest threat to the whole concept of multi-functional teams supporting concurrent ways of working. The not so obvious conclusion therefore is that the application of IT tools in complex product development not only requires well-trained and experienced engineers as members of project teams but also, and perhaps primarily, well trained and experienced project team leaders at all levels of the project's organisation.

More IT tool capabilities also mean more (design) possibilities, but less personal freedom in how things can be done. As IT tools are based on programmed and therefore fixed algorithms and codes they can only work properly and deliver the expected results if people adhere to the processes required by them.

Process adherence is necessary for the organised collection and sharing of the huge amount of business and technical information related to a complex project or product. This is even more important as in complex projects with many unknowns one usually finds a high rate of change.

Strict process adherence is, however, difficult to accept for many engineers. Quite normally, technical departments make use of engineering ingenuity to solve product development problems and they, thus, encourage engineers in their ingenious creativity. But applying high levels of creativity freedom is somewhat in contradiction to process adherence.

Unfortunately, complex projects are not manageable to time, cost and quality without adherence to proven processes. This is where Project Managers who are used to process adherence can add significant value. They should lead their project teams in a way which encourages creativity while staying within the boundaries of process adherence. Only in cases where extreme situations require it may they authorise limited diversions from processes' steps to speed up overall process times. However, personally keeping such diversions under control is then a mandatory leadership requirement. Finding the right balance between process adherence discipline and being flexible in exceptional cases must be regarded as a major leadership challenge.

Process adherence is also necessary to develop a single, integrated Project architecture. As advocated in this book, it really represents a new way of breaking down a project into smaller, manageable parts, to be used by all the company Functions involved in the project. What has been proposed in this book is a common Project architecture that all Functions can understand and agree to, and to which all the traditional Breakdown Structures used in Project Management and Systems Engineering can be linked to. Of course, this integrated architecture and the project data attached to it must be subject to change control. Ways of how to do this have also been explained in this book. Using this Project architecture, information related to the Product, to Processes and Resources can be managed concurrently.

It is in fact a project's tasks/activities which turn out to be the key elements to enable interlinkage of the various classical frameworks and, thus, of the Product-, Process- and Resources-related project information. Tasks/activities first of all form the building blocks of processes. But they also describe the steps required to produce an output (drawing, document, infrastructure, part, assembly and so on) from an input. They, thus, link Product data with Process data. They also can be clustered to belong to work packages, to which Resources (that is, teams, funds, infrastructure and so on) are allocated (thus linking Process data with Resources data). Architecting a complex large-scale project in a way, which allows for integrative management, must therefore regard tasks/activities as architectural centrepieces.

Any integrated Project architecture has to satisfy the needs of Product Data Management (PDM), Systems Engineering, change control, Project Management as well as communication management, among others. It has been demonstrated in this book that fully integrated Project architectures lead to new interpretations of classical items such as a 'work package'. To implement new ways of structuring complex projects is therefore also a Change Management challenge in its own right, as new interpretations need to be widely communicated and people trained. It should have become clear in this book that architectural decisions are not simply technical decisions that can be left to one Function only. As they impact on the very essence of business, they should be taken by multi-functional teams.

Process adherence is, of course, also required in order to have a structured and controlled approach, for example, for:

- the establishment and cascading of requirements for all the teams involved in a complex project;
- risk- and opportunities management;
- schedule and Interdependency Management; as well as for
- correct booking of workloads spent on a project to enable meaningful Earned Value measurements.

The pros and cons of Earned Value Management (EVM) have been discussed in this book. One conclusion is that the commercial nature of the projects discussed here does not allow a full bottom-up EVM as this would be too onerous. Instead, a simplified EVM approach has been presented offering various dimensionless Key Performance Indicators, which provide the Project Manager with reasonably unambiguous status information about the project at acceptable levels of accuracy.

Last, but not least, one must secure robust Business Cases as an essential aspect to the management of complex projects. A programme's Business Case should be challenging but realistic. To this end, it should not only include a contingency to cover issues and problems, which inevitably will materialise sooner or later. It should also include a provision to cover all perceived costs of potential actions required to mitigate risks.

In this context, the difference between provision and contingency is the following: one knows from the onset of a project that risks will need to be taken to achieve project objectives. At the beginning, one may not know precisely which risks there might be and therefore which mitigation actions, at which cost and leadtimes, one would have to launch. But that some sort of mitigation actions will have to be launched is as certain as experience in managing projects can be. It is therefore important to introduce a provisional lump-sum into the Business Case, based on risk mitigation experience from previous projects. A contingency is needed in addition for cases where risks have turned into issues because mitigation actions failed or in cases where unforeseen issues arise without any pre-warning.

In most companies there is still much to do to have all the aforementioned essential ingredients in place to successfully manage complex projects. The most severe challenges are still related to the lack of skilled project leaders, as well as to the availability of tools, which offer Project architectures as the backbone to the integrated management of projects. With the advancements of Product Life-cycle Management (PLM) major steps have been taken in this direction. However, the link to Processes and Resources is still missing, and a merging of PLM with management tools covering Processes- and Resources-related data is still pending.

But one day there will be *Programme* Life-cycle Management tools available which will offer this capability. Despite the globally observed trend that companies are less and less capable of managing complex projects to time, cost and quality, there is no fundamental reason why this cannot be possible: managing complex projects is not hopeless! It does require significant efforts on behalf of a company, though, to get the right management essentials in place.

PART VII

Appendices

Appendix A1:
The Business Case

As was stated in Chapter 4, a Business Case defines the expected cost and revenue streams of a programme as well as its associated funding strategy. Every aspect affecting Life-Cycle Costs (LCC) – such as application of new technologies, material maturity and prices, sub-contracting opportunities, design capabilities, skills, capacities, certification requirements – to name but a few, should be analysed and considered in the Business Case. The launch of an aircraft programme should always be based upon a well-prepared Business Case in order to minimise the financial risks.

On the *revenue side* of the Business Case, market demand projections are taken into account as well as a unit price, which is deemed achievable on the basis of the aircraft's competitive situation. The unit price assumed by the Business Case is established – among others – by:

1. looking into the aircraft's market segment (Are there any competing new aircraft in the market? Is there a risk that second-hand aircraft will depress prices? And so on...);
2. the attractiveness of the aircraft to airlines based on the projected life-cycle operating costs and its residual value,[1] among other; as well as
3. the margins which the company believes can be realised given the life-cycle production costs.

As far as the operating costs of the aircraft are concerned, they play a significant role for the competitiveness of an airline. Airlines are therefore always interested in new aircraft developments which promise significantly reduced operating costs for their fleet compared to the fleet of their competitors. Major factors of the total operating costs can be directly influenced and defined while designing the aircraft, such as costs related to fuel consumption and maintenance, see Figure A1.1. Others are indirectly influenced. An example for this is user charges, like airport fees, which are based on aircraft weight. Operating costs need to be carefully analysed in order to be able to establish reliable sales forecasts.

On its *cost side*, the Business Case accounts for the estimated LCC for development and production, that is, Non-Recurring Costs (NRCs) and Recurring Costs (RCs), respectively, including the cost of financing.

The top-down estimating of NRCs is an art on its own. It is a prerequisite for defining development funding needs as well as for preparing later negotiations with the internal and external suppliers of work packages. NRC estimations are usually based on models using costs encountered during past experiences, which are scaled to a reference unit wherever possible, such as $/kg. The cost per reference unit is then used to calculate

---

1 The residual value is the value of an aircraft (or engine or other item) at a future date. It is particularly important in connection with the conclusion of a lease term.

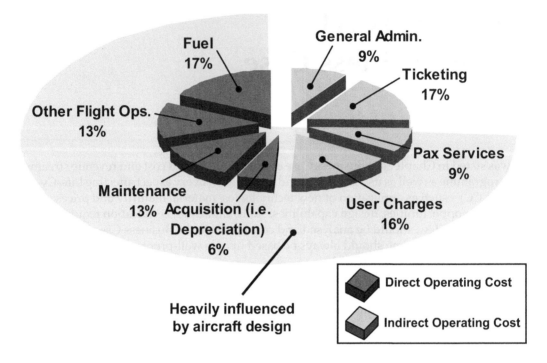

Figure A1.1   Typical breakdown of total operating cost for a commercial aircraft operator

the costs for the development of the equivalent equipment on the new project. More sophisticated models take further aspects into account, such as:

- complexity evolution;
- challenges associated with new technologies;
- different supply chain architectures;
- degree of available workforce experience;
- and many more.

Of much higher importance for the long-term well-being of the aircraft manufacturing company, however, are the RCs. As an input to the Business Case, they need to be challenging but should be judged as realistically achievable. Historically, for the portion of the total RCs, which can be influenced by the company itself, human learning curves have played an important role in defining production costs.

'It is a fundamental human characteristic that a person engaged in a repetitive task will improve his performance over time. If data are gathered on the phenomenon, a [learning] curve representing a decrease in effort per unit for repetitive operations can be developed.'[2]

'The concept of the learning curve was introduced to the aircraft industry in 1936 when T.P.Wright published an article in the February 1936 *Journal of the Aeronautical Science*. Wright

---

2   From: United States Department of Energy, http://www.directives.doe.gov/pdfs/doe/doetext/neword/430/ g4301-1chp21.pdf, accessed 10 October 2009.

described a basic theory for obtaining cost estimates based on repetitive production of airplane assemblies.'[3]

'In Wright's model, the learning curve function is defined as follows:

$E_N=E_1 \times N^s$

with E = effort per unit of production (e.g. manhours, time, etc.) to produce the Nth unit, and s = slope constant when plotted on log-log paper. The equation implies a constant fractional or percentage reduction in effort for doubled (or tripled, etc.) production. Every time production is doubled, the effort per unit required is a constant of 2^s of what it was. It is common practice to express the learning-curve function in terms of the gain for double production. Thus, a 90 percent learning-curve function requires only 90 percent of the effort per unit every time production is doubled.'[4]

However, as a large portion of the total RCs usually result from outsourced activities, and as production processes have become more and more automated, the relevance of human learning curves for the aircraft manufacturer's Business Case has become rather limited over the years. It is nevertheless still used in Business Cases to estimate costs of assembly and integration activities with significant manual involvement.

Every effort has to be made to reduce production costs as quickly as possible. Too slow a RC decrease during Series Production has devastating effects for achieving the Business Case. For example, the 300th unit on a 90 per cent curve requires almost three times the production costs as the 300th unit on an 80 per cent curve, see Figure A1.2. Achieving RC reductions is a must throughout the life-cycle of an aircraft programme. This is why aircraft companies launch major cost-saving projects every few years or so and also have continuous improvement measures in place in Manufacturing.

However, as Figure A1.2 shows, the most significant cost savings will only be achieved up to, say, aircraft No. 100. Absolute prerequisites for achieving this are robust design processes, high-quality drawings (or their 3D equivalents) released on time and a design which is highly production-oriented. This is why companies apply Design to Cost (DtC) methodologies, among others.

DtC is intended to introduce the mindset of low production- or procurement costs into the design during the early phases of the development. Its principal intention is that every time a design solution has to be selected out of a range of alternative possibilities, manufacturing easiness and, thus, lower unit costs, are taken as an important decision criteria.

However, in reality this approach is often not possible because of lack of time or skills or budget or a mixture of all of it. In addition, a cost-optimised design for one type of aircraft may not allow for operational advantages for airline customers, which may also use other types of aircraft. What counts for them is more a cost-optimised fleet rather than different aircraft types which are individually cost-optimised.

What is important to note is that for the Business Case both revenues and costs are converted in cash-in and cash-out streams, respectively. Combining cash-in and cash-out streams yields the stream of cash flow, see Figure A1.3.[5] During the early phases of

---

3   From: http://cost.jsc.nasa.gov/learn.html, accessed 20 November 2009.
4   From: United States Department of Energy, http://www.directives.doe.gov/pdfs/doe/doetext/neword/430/ g4301-1chp21.pdf, accessed 10 October 2009.
5   From an accounting point of view, cash flow is essentially defined as cash resulting from revenues plus depreciations minus costs.

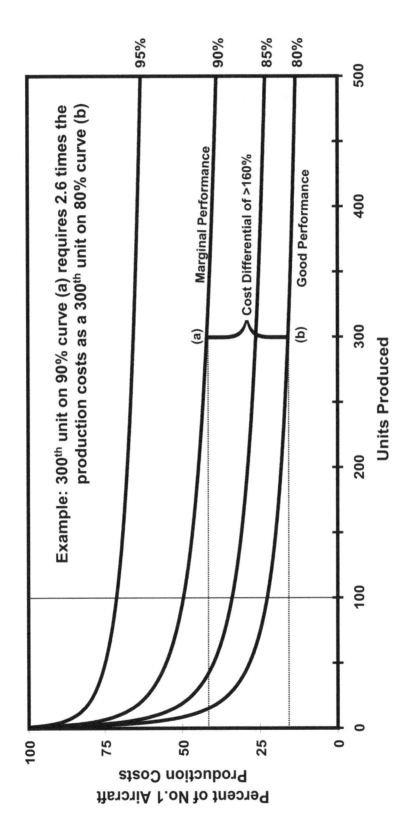

Figure A1.2   The mathematics of learning curves

development there is no (or almost no) cash-in as no aircraft has been delivered yet. Hence, the cash flow is dominated by the cash-out (based on the NRCs) and is negative. Only when deliveries have commenced does the cash flow curve have a chance to gradually increase: the aircraft programme eventually generates cash for the company. The point in time where the cash flow changes from negative to positive cash is called the Break Even Point.

Cash flow curves are calculated using cash-in and cash-out streams expressed in terms of *present* values. Net Present Value (NPV) is a capital budgeting technique in which the present value of all future cash inflows is netted against the present value of all future cash outflows for a pre-determined period of time. In commercial aeronautics, 20 years or more are usually chosen. NPV is based on the intuitive concept that cash in hand today is more valuable than cash in the future.

To arrive at present values, future cash flows associated with a new aircraft programme have to be discounted. This is done using a dedicated interest rate called Opportunity Cost of Capital (OCC) (expressed in per cent of interest over borrowing). The latter is equal to the sum of the Cost of Capital[6] for the company plus a programme-specific risk premium to take into account business risks or financial risks associated with the programme. Thus, the aircraft company decides upon an OCC specific to the Business Case of the aircraft programme. The theoretical OCC where NPV = 0 – when looking over a time-span of, say, 20 years – is called the Internal Rate of Return (IRR). In the example depicted in Figure A1.4, the IRR amounts to 26 per cent.

The aircraft programme clearly creates shareholder value if NPV > 0 for the selected OCC. In this case, the Business Case is definitely viable. However, in cases where NPV ≤ 0 at selected OCC the aircraft programme could nevertheless be regarded viable if for NPV = 0 the programme still earns returns sufficiently above its pre-fixed Cost of Capital (IRR > Cost of Capital). It will depend on the shareholders' judgement whether it is acceptable or not to lower the programme-specific risk premium to arrive at a viable Business Case.

To launch a programme, the Business Case must pass two tests:[7] First, the product to be developed must have enough value to enough customers to support prices and volumes that exceed the costs of supplying it – including the OCC. Second, the developing company must have enough sources of sustainable competitive advantage (such as skilled people, innovative technologies, robust processes, sufficient cash, and so on) to exploit, develop and defend the opportunity. When launching a programme the trick is to encourage a

---

6   A company's Cost of Capital is the minimum rate of return its business can earn on existing assets and still meet the expectations of its capital providers. It is also the return a firm must earn on existing assets to keep its stock price constant. The usual way of arriving at an approximation of the Cost of Capital is to take a simple weighted average for the cost of equity and dept. The Weighted Average Cost of Capital (WACC) is the cost of the individual sources of capital (that is, equity, debt, pensions), weighted according to their importance in the firm's capital structure. The cost of debt can, for example, be taken as the interest rate applying to each debt item net of tax. The cost of equity can be approximated as the sum of an expected riskless interest rate plus the expected rate of inflation (together usually represented by long-term government bonds) plus a company (that is, not programme-) specific risk premium. The latter is taken as a standard industry risk premium (typically 6 per cent) multiplied by what is called the Beta. The Beta measures the volatility of a particular share relative to the overall market. A Beta of 1 suggests that the company's share price moves precisely in line with the market. A Beta of 1.5 suggests that the company's share price will on average move 1.5 per cent with every 1 per cent move in the market. With a sound economy, high Beta-companies can be expected to generate very good returns, but in a recession they will be poor performers. Aerospace companies typically have Betas in the area of 1.1.

7   Barwise, P., Marsh, P. and Wensley, R. (1987), 'Strategic Investment Decisions', *Research in Marketing* Vol. 9, pp. 1–57.

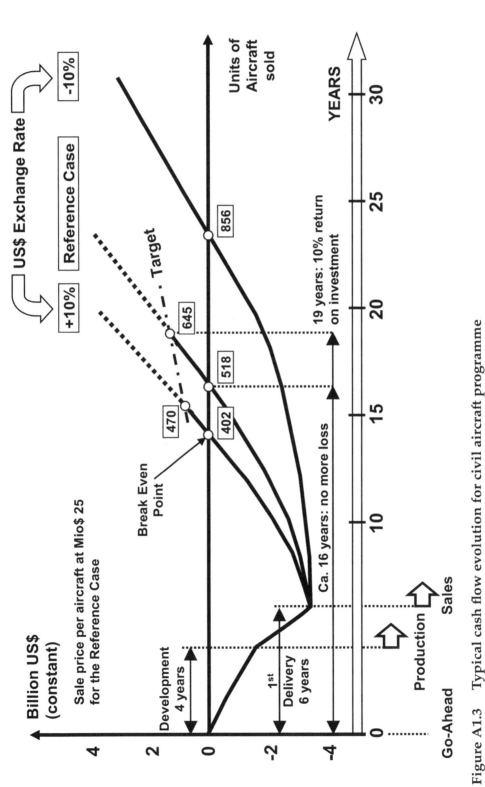

Figure A1.3   Typical cash flow evolution for civil aircraft programme

Based on example calculations performed and published by: Zabka, W. (1987), 'The Judgment and Evaluation of Long-Term Investments Demonstrated by Means of a Civil Aircraft Program', NATO Advisory Group for Aerospace Research & Development (AGARD) Conference Proceedings AGARD-CP-424, pp. 7–9, © NATO RTO.

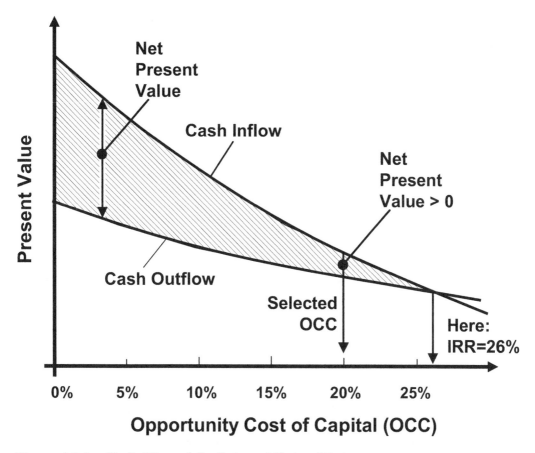

Figure A1.4   Definition of the Internal Rate of Return

decision-making process in which the financial analysis highlights rather than masks these two fundamental conditions.

However, before these tests can be passed, the Cost Estimation or Finance department will run many (literally thousands of) scenarios where the values of variables in the equations, such as the expected cash flow, investments, cost of capital and so on, are varied within given limits. A Monte Carlo analysis will then reveal what the theoretical probability for programme success is, that is, how many of the total number of calculated scenarios yield viable Business Cases.[8, 9]

However, the scenario analyses will also reveal what the drivers are for achieving programme success. As a result, quantifiable objectives for (1) product performance (such as quality, maturity, operating costs), as well as (2) development time, (3) NRCs and

---

8   Modern IT-tools can calculate thousands of scenarios as well as a complete Monte Carlo analysis within seconds.

9   It is interesting to note that the average industry standard for launching major projects is only a mere 25 per cent theoretical success chance. In contrast, most people would express a probability of 25 per cent as a 'remote chance' and would rather like to see a probability of 75 per cent or thereabout to have confidence in a programme's success. But one should not forget that in a theoretical Monte Carlo analysis all scenarios have equal probability of occurrence. Management must take a stand on whether or not the probability of occurrence is sufficiently high for those scenarios which are theoretically viable.

(4) RCs will emerge. It is these objectives the programme must achieve. There are six major interdependencies between them:

- a late Entry into Service (EIS) resulting from a slower than planned Development phase means keeping the development teams together for longer than expected ('marching army' effect), thus leading to higher NRCs;[10]
- more DtC activities during the Development phase can lead to lower RCs at the expense of higher NRCs, lack of these or similar activities may have the opposite effect;
- the necessity to improve projected low product performances (such as too high operating costs) can delay the market introduction date;
- it also can lead to higher NRCs; and
- to higher RCs,
- late EIS can mean that the overall accessible market is becoming smaller, for example because the competitor makes use of the sudden market opportunity and sells more of its own products; the production costs can therefore not come down the learning curve as quickly as was anticipated, resulting in higher overall RCs.

It is important that project leaders understand the necessity of keeping NRCs and RCs within acceptable limits as well as to get a good feeling for the degree of susceptibility to the various interdependencies. According to analyses performed by, for example, W. Zabka, it is in particular the delays in development leadtimes which have a tremendous effect on the success of the programme.[11]

Where, after all the analyses, a programme's overall financial risk is perceived to be too high, risk mitigation strategies must be imbedded in the Business Case. Standard mitigation plans include outsourcing of work packages to risk-sharing partners (with no or almost no transfer of cash for development activities) or to low-cost non-Risk-Sharing Suppliers (RSS).

As the former CEO of Boeing, Phil Condit, once said:

'When you look at the life cycle of the revenue generated, I see a case for the suppliers paying to be on the program and not just getting that revenue stream for nothing. Then we all share the risk and the reward.'[12]

In the case of risks resulting from the unpredictabilities of exchange rate fluctuations (as described in Chapter 1 and shown in Figure A1.3) to which, for example, Airbus is greatly exposed,[13] mitigation actions include the following:

---

10 The magnitude of cost overruns can be grasped by a statement made by Boeing's Chief Financial Officer James Bell who said that the delays encountered with the 787 'are growing $800 million a quarter'. From: Mecham, M., Anselmo, J.C. and Norris, G. (2009), 'A Wing and a Prayer', *Aviation Week & Space Technology* 27 July 2009, p. 22.
11 Zabka, W. (1987), 'The Judgment and Evaluation of Long-Term Investments Demonstrated by Means of a Civil Aircraft Program', *NATO Advisory Group for Aerospace Research & Development (AGARD) Conference Proceedings* AGARD-CP-424, p. 7.
12 Condit, P. (2004), *Flight Supplement* 17–23 June 2004.
13 'A great part of the revenues of EADS, the 80% owner of Airbus, are in US$, with only about 50% of this currency exposure "naturally hedged" by US$ dominated costs. A €0.1 change in the average 2003 €/US$ spot rate (€/US$=1.13) would have an impact of approximately 2% on EADS' consolidated revenues assuming 300 deliveries at Airbus.' From: *EADS Annual Report 2003*, p. 28.

- ensuring revenues in the home-currency wherever possible;
- denominating more purchases of goods and services in US$;[14] as well as
- hedging.[15]

---

14 As Noel Forgeard, CEO of Airbus outlined: 'A continuous improvement process has allowed … [Airbus] to have more than 70% of $ content in the unit cost of every model in production. It will be even higher for the A380. We buy in dollar all around the world and not only in the US even if we procure there more than 5 billion a year.' In: Forgeard, N. (2005), *Airbus Annual Press Conference* (Paris), 12 January 2005.
15 For the hedging policy of EADS see: http://www.finance.eads.net/GIF2004/gif2004_2uk.pdf, accessed 2005, or the *EADS Annual Report 2003*.

Appendix A2:
Desirable Behaviours for Leaders of Multi-Functional Project Teams

CRITERIA FOR A SUCCESSFUL TEAM

The complex matrix organisation outlined in Chapter 5 will inevitably lead to some conflicts regarding responsibilities, levels of authority, goals, priorities and other related subjects, in particular during its set-up phase. On the other hand, this type of organisation suits best the needs for an integrated, concurrent way of working. It is leadership behaviours which must compensate for the organisational deficiencies to ensure project success. The latter depends so much on having well-performing teams that leadership in a multi-functional environment means primarily to do everything necessary to generate teamwork and efficiency within a team as well as in the interaction with other teams.

Multi-functional team organisations require a different management style compared to one within a single Function. Leaders of multi-functional teams are also more exposed than bosses or managers of uni-functional teams. They are under intense scrutiny and are watched by more than one Function all the time. Fortunately, being more exposed also offers the opportunity for faster communication: if a leader recognisably demonstrates positive leadership behaviours then people will not only know more quickly about this but more people from different Functions will be willing to follow him or her too. Positive leadership behaviours have a multiplier effect which can make the life of a leader much easier while having a successful team at the same time.

What are positive examples of respected behaviours a leader can demonstrate to be a role model for the other members of the team? Let us now look at some desirable behaviours towards individual team members as well as towards the team as a whole.

ENABLING PEOPLE TO TAKE OWNERSHIP OF THEIR WORK

Introduction

For an effective team to come into existence, leaders must encourage team members to take ownership of their work by contributing to planning, problem solving, implementing solutions and solving issues. In order to achieve that, leaders need to remember that individuals:

- want to know what their roles and responsibilities are;
- need to be adequately empowered;
- need to know what is expected from them;

- need to know all applicable rules and procedures;
- must be given the work they need to do;
- need to be open to change; and
- may not be courageous enough to take risks.

Let us therefore start by having a look at the needs of team members.

Clarify Roles and Responsibilities

For multi-functional teams to work effectively, it is mandatory to initially define the roles and responsibilities for each team member and to reinforce them throughout the project. This already is quite difficult to achieve for individuals working in a complex matrix environment. Even the use of the much favoured RASCI methodology, identifying Responsibilities, Accountabilities and Supporting roles for any given process as well as areas where individuals must be Consulted or Informed, will leave ample room for confusion and misunderstanding.[1] This is particularly true for processes which are not defined in enough detail, which most often is the case. The main purpose for such methodologies like RASCI is therefore to use them as a tool to initiate and lead discussions about roles and responsibilities. They will in most cases not provide the full answer.

In reality, roles and responsibilities can only be described on a sheet of paper to a limited extent, for example in the form of a job description. Thus, there is a conflict between the need for clarity on roles and responsibilities on the one hand and the almost impossible task to define precisely what everyone has to do on the other. Fortunately, over time, the interaction between different people will gradually lead to a common and in-depth understanding of the roles and responsibilities within a working organisation, provided an initial job description was available.

Delegate Authority

In order for team members or lower-level team leaders to be able to fulfil their roles and responsibilities, higher-level team leaders must empower them through delegation of authority. The former must be given the levers they need to do the job. In many companies the principle of subsidiarity is applied. This means that the organisational level, which could deal best with a decision, should have the empowerment to take the decision. The problem with this principle lies in the different views people have about which level would be 'best'. Should project team leaders, for example, be given the authority to sign purchasing contracts with their suppliers? Or is this a question of the wider company's procurement strategy? After all, other departments may also procure goods from the same suppliers. In which case the contract would need to be negotiated and signed by someone higher up in the organisation in order to benefit from larger economies of scale.

Individuals should therefore be encouraged to request details of the authority they are supposed to have before deciding to take on the leadership role. Leaders need guidance

---

1   RASCI describes the participation by various roles in completing tasks or deliverables for a project or business process. It is especially useful in clarifying roles and responsibilities in cross-functional projects and processes.

to determine which decisions they can take on their own, which they need to make as a team and which they need to escalate to Upper Management.

As a visual sign of authority transfer, appointments should ideally be done during a ceremony in front of all team members. This should also be communicated to a wider audience via the company's magazines, intranet or e-mail system.

However, trying to specify in a job description or an exposition document all details of authority and empowerment for all imaginable cases would, again, not add much value. This is because surely there will be unforeseeable situations which cannot be described in such documents and it would be rather bureaucratic to try to do so. Transferring authority within a complex organisation can therefore usually not be done to the entire satisfaction of all individuals involved. Accepting a leading role in a multi-functional environment therefore entails accepting a certain level of uncertainty and, perhaps, frustration about not exactly knowing where the limits of the authority are. Team leaders must try and test their authority limits by means of trial and error. The solution to the problem of empowerment will in the end depend on the skills of individuals, as well as on the specific circumstances of a project and a company.

Set Visions

A leader should trigger the inspiration of people by creating visions which everyone can share. It is particularly important for international projects, such as commercial aircraft developments, to share a common vision as it provides a means to bring multi-cultural, multi-functional team members closer together.

> '[J.P. Lewis says:] ... I have many times asked people to tell me the mission of their company. Most are unable to do so. However, I say, 'Well, I can tell you what it is.' They always look surprised. Then I say, "More than likely, your mission statement says something like 'Our mission is to produce world-class stuff that will make a lot of money for our stockholders.'" They all laugh and agree that this is true. ... [However,] such statements do not inspire the people who must make it happen.'[2]

So, what is needed is a compelling vision, a vision of the end result that the project is supposed to achieve and that inspires people. But, as J.P. Lewis writes, 'It is not so easy to achieve a *shared* vision throughout the project team, and yet this is necessary if that vision is to be achieved.'[3]

Here is an example of how it was done during the development of the Boeing 777:

> 'The vision ... was that ... [the 777] would be intermediate in size between the 747 and 767, a plane that would be the most technologically advanced in the world, and one that would take [Boeing] into the twenty-first century well positioned in the marketplace. [Boeing's VP for the 777 project Alan] Mulally simplified that vision by drawing a little cartoon that said, "Denver to Honolulu on a hot day."'[4]

---

2   Lewis, J.P. (2002), *Working Together. 12 Principles for Achieving Excellence in Managing Projects, Teams and Organizations* (New York: McGraw-Hill), p. 53.
3   Lewis, J.P. (2002), *Working Together. 12 Principles for Achieving Excellence in Managing Projects, Teams and Organizations* (New York: McGraw-Hill), p. 46.
4   Lewis, J.P. (2002), *Working Together. 12 Principles for Achieving Excellence in Managing Projects, Teams and Organizations* (New York: McGraw-Hill), p. 46.

Figure A2.1   Alan Mulally's vision for the 777

Source: Lewis, J.P. (2002), *Working Together. 12 Principles for Achieving Excellence in Managing Projects, Teams and Organizations* (New York: McGraw-Hill), p. 47, with permission.

The cartoon is shown in Figure A2.1.

'To illustrate that Mulally was able to create this excitement with what seems to be a simple slogan, Walt Gillette, who was chief project engineer for airplane performance, safety, and reliability on the 777 program, wrote: The Statement, "Denver to Honolulu on a hot day" meant all the following:

It was very visual – in our mind's eye, we could imagine ourselves as the captain and first officer of this flight, seeing the heat waves rising off the concrete runway in the thin air of Denver, and having full confidence that our silver machine would take us safely into the air; "Denver" meant that the airplane had the high-altitude capability from the onset to do this difficult mission;

"Hot day" meant that the airplane had gone into revenue service in the summer, as promised five years earlier;

"Honolulu" meant that the airplane had ETOPS ability at the start of revenue service.

These images in the minds of the 777 creation team evoked by "Denver to Honolulu on a hot day" spoke to the heart in a way that facts and data could not. Each of us was able to internalize what our share of the assignment meant to achieving this vision.'[5]

---

5   Lewis, J.P. (2002), *Working Together. 12 Principles for Achieving Excellence in Managing Projects, Teams and Organizations* (New York: McGraw-Hill), pp. 48–9.

Set Objectives

Objectives for the entire team are usually known prior to setting objectives to individual team members. However, to convert the one into the other requires a dedicated team effort (such as during a workshop) so that every team member understands and is committed to his or her objectives. Leaders must ensure that individual objectives which are to be achieved are set. They also must seek agreement and buy-in to those objectives. This must be done on a regular, say annual, basis. In addition, leaders need to ensure that objectives of individual team members are defined in a way consistent with the objectives set for the whole team.

Ensure Availability and Common Knowledge of Company Rules and Procedures

Teams cannot work efficiently if company rules and procedures are not available or not known to their members.

> 'Trying to achieve good teamwork on a complex project without having established reasonable rules, procedures and practices for how the project will be planned, the work authorized, progress reported and evaluated, conflicts resolved, and so on, is like collecting the best athletes from six different sports and turning them loose on an open, unmarked field with instructions to "play the game as hard as you can".'[6]

The team leader must ensure that rules and procedures are in place and team members are briefed accordingly.

Delegate Work while Ensuring Appropriate Balance Between Freedom and Control

Leaders must be able to delegate as delegation is a visible sign of trust into the capabilities of the team members. If a leader communicates a sense of confidence in their own abilities, people are more willing to commit to give their best for the benefit of the project. The result is a higher level of motivation among people working for this leader. Delegation teaches them to become self-assured and independent, stimulates their creativity and makes them happy.

Perfectionist leaders will find delegation difficult: if team members cannot deliver the perfection deemed necessary, then perfectionists tend to do the job themselves. Consequently, there is a big danger that the motivation level among team members declines rapidly – and for the leader there will not be sufficient time in a day to do all the jobs which need to be done. Aggressiveness and stress are common symptoms of such a situation.

---

6 Archibald, R.D. (1992), *Managing High-Technology Programs and Projects*, 2nd Edition, (New York: John Wiley & Sons), pp. 96–7.

Of course, there will always be individuals who need guidance, coaching and instructions. And a leader should certainly dedicate time for coaching others. But in order to be able to coach and provide advice well, one often has to dig into the details and complexities of a given problem situation. And this can be very time consuming.

Nevertheless, there are situations where a leader must dig into the details, for example, to test the carefulness applied at work or to find out whether there is sufficient substance behind statements made by others. Again this can be very time consuming. A leader therefore must be capable of quickly grasping the essentials of what is going on while resisting micro-management at the same time. Leaders must be able to keep an appropriate balance between freedom and control, between delegation and micro-management.

Find the Right Balance Between Creativity and Process Adherence

Complex projects are not manageable to time, cost and quality without adherence to proven processes. This is where project leaders, who are used to process adherence, can add significant value if they lead their project teams in a way which is encouraging creativity while staying within the boundaries of process adherence.[7] Only in cases where extreme situations require it may they authorise limited diversions from processes' steps to speed up overall process times. However, personally keeping such diversions under control is then a mandatory leadership requirement. Finding the right balance between process adherence discipline and being flexible in exceptional cases is a major leadership challenge.

Encourage Openness for Change and Innovation

Not only must leaders themselves be open-minded to any changes which might come along, but should also encourage the members of their teams to be open to change too. People often feel an aversion to any change, even if it would improve results.

For example, during a commercial aircraft development, there could be (design) changes which are intended to be for the benefit of the overall aircraft but have adverse repercussions to some Design-Build Teams (DBTs). As a consequence, these teams may no longer be able to meet their deadlines. Teams which are unfavourably affected will therefore be reluctant to accept the change. Yet the change is regarded necessary by higher-level Programme managers.

In this situation how can leaders generate sufficient openness for change? How can they encourage the search for innovative solutions to solve the problems their DBT may now facing? How can they find access to their people to harness their naturally curious, adaptable and immensely resourceful attitudes? They must provide emotional security! Leaders should therefore never lead by threatening people ('management by fear'). People under threat will tend to become defensive and, thus, unimaginative, uncreative and certainly not open to change.

---

7   Note that creativity is not only required for design work but also to continuously improve processes.

Encourage Risk Taking

Encouraging change implies leaving known paths in order to try out new ones. This is not possible without accepting some level of risk. A good leader should therefore also encourage individuals to take risks. However, risks must remain acceptable. Consequently, the leader must provide clear guidelines to individuals about what an acceptable degree of risk taking would be. In addition, the leader must make it clear that risk taking needs to be accompanied by additional actions to keep the risk under control. In Chapter 9 it is described how risks can be contained.

> 'The ultimate test of how effectively a ... leader is able to encourage and support risk is measured by the way he or she deals with failure. If you want risk-taking behaviours, you must also be willing to accept a certain (but limited) number of failures.'[8]

The leader needs to remember that if people are not making any mistakes, they may well be too complacent. The goal is not 'no mistakes'; instead, it is continual learning by the individuals. This requires some failure tolerance on the side of the leader.

This concludes the list of some important aspects required for individuals to take ownership of their work. But more is required to achieve objectives. Motivation is certainly one essential ingredient if teams want to achieve objectives. In the following we will look at some means to motivate people.

MOTIVATING PEOPLE

An Interesting Product is not Enough as a Motivator

The single biggest motivator one often encounters in projects is the product to be developed itself. If people see the beauty or understand the competitive advantage of a product or comprehend the rationale why a product makes sense, they will be motivated. Aircraft are fascinating products as they satisfy the old dream of mankind to be able to fly. Thus, on a global scale, there usually is no problem in commercial aircraft companies to find motivated employees.

In principle. However, on the level of individual roles and responsibilities this may not be the case. This is where local work aspects become more predominant. As on this level the most critical factor in keeping team members motivated is interesting and challenging work, team leaders need to have an adequate 'antenna' to be able to receive signals of potentially unfulfilled needs residing in the team. They need to be able to appraise motivational requirements and – if necessary – adjust the roles and responsibilities to meet those needs. People, for example, will be motivated if the tasks assigned to them enable them to use a broader range of their skills, or to develop new skills. This requires a deep sense of what is challenging to each person, which can only be obtained through sufficient communication with team members. Only through direct contact can one obtain the necessary feedback about the level of motivational satisfaction.

---

8   Archibald, R.D. (1992), *Managing High-Technology Programs and Projects*, 2nd Edition, (New York: John Wiley & Sons), p. 120.

Provide Positive Feedback and Constructive Criticism

Providing positive feedback is another very powerful motivator. This seems to be a universal rule and therefore can be applied to individuals of whatever cultural and ethnical background. If someone has done a good job he or she should be told so. Even if performance of the individual is not exactly according to the expectations of the leader, positive feedback may still be justified because of its longer-term motivational effects.

However, if performance is repeatedly below expectations, this needs to be discussed with the individual. This should be done in a way as to encourage change without destroying the ego and motivation of the person. As leaders should always seek to raise the level of performance of individuals, they need to be capable to provide this kind of constructive criticism.

The leader must be careful, though, that the criticism is based on fair judgement. Almost by definition, team leaders in their very role as project leaders tend to concentrate on negative effects and results (such as on diversions from the schedule plan) rather than on good contributions made otherwise. Obviously, while the achievement of the project's time, cost, performance and quality objectives is paramount, in a multi-functional team and matrix organisation other contributions also deserve recognition, such as for showing:

- Effort: it is not as good as result, but it demonstrates commitment. In any case, in a matrix organisation it is often difficult to achieve expected results as an individual alone, for example, because no consensus can be reached between the functional and project needs. So, effort counts a lot, even in relation to results.
- Team spirit: individuals, which collaborate well with other individuals and show active contributions to keep the number of conflicts down are vital for strong team-building. They are directly raising the performance of a group of people as a team. They actively promote a climate where cooperation among individuals with different skills, values and perspectives is supported, and this should be recognised.
- Openness to communicate: communicative individuals not only support team spirit within a team but directly support the interrelations a team has with other teams too. Through their communications they will feed back to other team members what is happening in the 'world' outside of the team.

Constructive criticism will help people develop. If individuals recognise that constructive criticism allows them to broaden their skills and know-how, they will feel motivated and will give more.

Evaluation and Rewarding

Achievement of objectives should also be revisited regularly, but at a higher frequency than that of setting objectives. When three-quarters of a term are over, the leader should encourage his team members to look again at their objectives. At this point there is still some amount of time left for them to improve their individual rates of achievement. Sometimes it is also necessary to realign the objectives.

Evaluation of performance of any key team member in the multi-functional team – but in particular the performance of the team leader – is often more difficult than evaluating other managers in uni-functional teams.

'This is because so many factors affect the project over which the project manager has little direct … authority and control. A highly successful project may or may not be the direct result of the [key team members] efforts, just as an unsuccessful project may have had the best [key team members] in the organization in charge of it. At this stage in the evolution of the project management approach, no formalized or systematic method of evaluating performance has been developed, but obviously the overall project results do count heavily in this evaluation.'[9]

In principle people will also feel more motivated if a generous supplementary financial reward – in addition to the standard company salaries, benefits, and bonuses – is granted to them in recognition of their individual strong contributions. But one must bear in mind that financial rewards only have a short-term motivational effect. In addition, financial rewards may contradict the overall company or project policy to save costs. One could, of course, limit the rewards to smaller financial amounts, but not below a certain standard. Sub-standard rewards will be counterproductive. In other words: one should grant rewards either at standard or higher than standard levels – or better not look into additional financial rewards at all.

Rewarding a complete team is an even more delicate subject, even at standard level. Other teams would start to get jealous and so would other employees in the company who are not working for the project but who are still doing an excellent job. As a result, it usually turns out to be almost impossible to identify where to draw the line.

Leaders might therefore decide to refrain from providing financial rewards to individuals or entire teams. Fortunately, monetary rewards are not so important as one might think. Non-financial rewards such as public recognition are often much more effective and at much less of a cost. Public recognition can be used whenever a milestone has been achieved successfully or achievements beyond the call of duty have been identified.

This could be done, for example, by showing pictures of the successful team in the company's newspaper, producing a video as a souvenir for the team members, by writing an article in a respected newspaper, and so on. An example for this can be seen in Figure A2.2 where Airbus employees have been acknowledged in an internal Airbus magazine to have done extra work over Christmas to protect the 'Roll-out' date of the first A380.

'The bottom line is that a relatively small number of non-monetary incentives and actions can be extremely effective in supporting team motivation. On the other hand, failure to pay attention to these non-monetary actions will be a de-motivator that cannot be made up by cash awards and raises. People who are not appreciated ultimately will seek an environment where they are able to get the psychic rewards they feel they deserve. Finally, and inevitably, the best performers will normally be the first ones to leave.'[10]

'Elmer [a Boeing employee to be rewarded] was given a bright-green 777 ski jacket from a supply of goodies that cannot be bought in the Boeing shops – there are some things that you can only earn, not buy.'[11]

---

9   Archibald, R.D. (1992), *Managing High-Technology Programs and Projects*, 2nd Edition, (New York: John Wiley & Sons), pp. 89–90.
10 Eisner, H. (2002), *Essentials of Project and Systems Engineering Management*, 2nd Edition, (New York: John Wiley & Sons), p. 178.
11 Sabbagh, K. (1996), *21st Century Jet. The Making of the Boeing 777* (London: MacMillan Publishers), p. 81.

**Figure A2.2   Airbus employees rewarded by having their pictures in a
company newspaper**

Source: Airbus (2005), '10,000 Squaremeter Teamwork', *One-Airbus News For Airbus People,* 9 January 2005,
p. 2, with permission.

As in commercial aircraft development projects the total number of team members
is sufficiently large, it is worth having one or more full-time employee(s) taking care of
public recognitions. However, it is team leaders who need to proactively propose reward
events and ensure they are happening.

Pro-actively Manage Careers

Another motivator for people is the feeling of being looked after in terms of career
development. If a leader develops potential career paths for an individual or proactively
explores paths suggested by the individual, the latter will certainly react in an appreciative
and motivated way. For leaders to be able to do that, though, they need to have a network
of contacts within the company and they need to find support from the Human Resources
department.

Avoid Demotivation

During the intense and tough Development sub-phase there are many traps which can lead to demotivated team members. If, for example, a team has to redesign most of its components because of late changes the level of frustration will be high. Leaders must demonstrate that they protect the team as much as possible from changes originating from outside the team, such as change of loads, requirements, interfaces or any other kind of unfavourable change resulting in rework, longer leadtimes and added cost. Protective measures could, for example, include the consequent challenging of inputs before allowing them to pass into the team's remit. However, while protecting the team raises the motivation of team members, this is at the expense of causing some friction with people outside the team. Leaders need to find the right balance in order not to overstretch protective attitudes.

Another cause for demotivation emerges if obstacles, which team members have identified and flagged up as a problem, are not removed by team leaders or Upper Management. Thus, to keep teams motivated, leaders and Management must ensure that disturbing influences from outside are rejected and obstacles are escalated rapidly to a level where they can get removed quickly.

Understand the Social Needs of Others

To have an understanding for the social needs of others is what social competence is about. The individual with sickness problems, the colleague whose car has just broken down, the young mother working, the man who dedicates much of his private time to his club: they all face the same tension between their very demanding project activities on the one hand and their social needs and desires on the other. In a project environment, the latter often are not sufficiently respected or even not respected at all.

> 'An organization conceived as a machine has no purpose of its own, but exists solely to serve the purposes of its owners, and that usually means making a profit. An organization seen as a social system, however, is an entirely different matter. Not only does the enterprise have a purpose but so do the constituent parts. ... What is important to understand is that every employee in the Company has his or her own ideals, interests, and purposes, and in the view of an organization as a machine, these were ignored. People were viewed as replaceable parts. As managers began to realize that an enterprise is a social system, however, they also realized that they could not ignore the concerns of their employees.'[12]

Social needs often represent an important contribution to the society in general. Leaders should therefore exhibit social competence by respecting the social needs of team members. They should support them in pursuing their social interests wherever possible, for example by discretely offering more time flexibility. Social competence of a leader supports the motivation of individuals. If they realise that their social issues and needs are considered – which often lead to a reduced performance at a given moment in time –

---

12 Lewis, J.P. (2002), *Working Together. 12 Principles for Achieving Excellence in Managing Projects, Teams and Organizations* (New York: McGraw-Hill), p. 32.

they will give extra effort at other times. In addition, it will help the leader to better understand the company and the people working for it.[13]

> 'When the organization was viewed as a machine, the belief was that it could be understood through the methods of scientific reductionist thinking. That is, by understanding the parts, one could understand the whole. That view is totally incorrect when the enterprise is a social system, rather than a machine. You cannot understand how a company works as a whole by understanding the functioning of the maintenance department alone. The subsystems of a larger system *interact*, and it is these interactions that give it the unique properties that differentiate it from other organizations.'[14]

DEMONSTRATING THE ABILITY TO RESOLVE CONFLICTS

Potential Conflicts in a Multi-Functional Team

Relationships between individuals are always somewhat complex and often not without friction. In a multi-functional team they are even more complex. Multi-functional teams are designed by purpose to foster multi-functional debates about the right technical and business solutions. Numerous confrontations and conflicts are inevitable. A leader of a multi-functional team needs to be able to resolve them. Typical conflicts in multi-functional teams are about:

- project objectives and priorities;
- multi-functional trade-offs involving schedule, cost, technical alternatives and performances;
- budget allocations to teams (for resources, cost, and so on);
- applicable methods, tools and processes;
- roles and responsibilities, reporting lines, ways-of-working rules; as well as
- personal issues resulting from ego-centered mindsets, disrespect, team-busters, and so on.

If teams are not only multi-functional but also multi-cultural, additional conflicts may arise from misinterpretations of culturally dominated behaviours.

How to Solve Conflicts

On the contrary to constructive debates, which leaders should always foster, every effort should be made to mitigate erupting conflicts. With multi-functional teams, this is usually done by trying to establish a strong team spirit. This includes:

---

13 However, while willing to support the social needs of others, leaders could face a conflict situation themselves. For example, when offering more time flexibility to an individual they may have chosen not to adhere to the company rules applicable to everyone else. This is a conflict situation they must be capable of enduring.

14 Lewis, J.P. (2002), *Working Together. 12 Principles for Achieving Excellence in Managing Projects, Teams and Organizations* (New York: McGraw-Hill), p. 82

- developing an acceptance among team members that conflicts are unavoidable in projects;
- establishing team relationships, which are stable enough to endure differences in opinions, interests, values and so on;
- accepting that all parties potentially involved in a conflict are being entitled to equal rights, with reciprocal respect and the reciprocal insight that anyone can fail; as well as
- fostering, developing and keeping alive the desire among all team members that conflicts of any kind can be solved.

Even larger problems will then be perceived as to be manageable. Common adventure tours, barbeques, uniforms T-shirts and so on can all contribute to raise team spirit.

However, where there are conflicts, they need to be addressed. The five commonly used modes for handling conflict, as reported by H.J. Thamhaim and D.L. Wilemon, are:[15]

- *'Confrontation Problem Solving.* Facing the conflict directly, which involves a problem-solving approach whereby affected parties work through their disagreements. Using informal team sessions can be a useful resolution strategy.
- *Compromising.* Bargaining and searching for solutions that bring some degree of satisfaction to the parties in a dispute. Characterized by a "give-and-take" attitude.[16]
- *Smoothing.* De-emphasizing or avoiding areas of difference and emphasizing areas of agreement.
- *Forcing.* Exerting one's viewpoint at the potential expense of another. Often characterized by competitiveness and a win/lose situation.
- *Withdrawal.* Retreating or withdrawing from an actual or potential disagreement.'

Confrontation Problem Solving, Compromising, Smoothing and Withdrawal could all be applied by a strong facilitator to solve conflicts between team members. Leaders need to understand that most conflicts, in fact, contribute enormously to the experience and well-performance of a team – and for that reason must be regarded as constructive elements of multi-functional teamwork – as long as the conflict is managed properly and channelled by a facilitator. Facilitation helps others to voice their view comfortably and avoids power and personality conflicts. It leads to participative decision making which meets the needs of individual team members and contributes toward effective decisions and team unity.

Ideally, therefore, the team leader can lead the parties in conflict to agreement and consensus by means of facilitation. However, to be a successful facilitator one needs to have earned the respect of the team members. Apart from some degree of technical skills, this requires exerting all the other leadership behaviours mentioned in this chapter. Typical facilitation methods include:

---

15 Thamhain, H.J. and Wilemon, D.L. (1975), 'Conflict Management in Project Life Cycles', *Sloan Management Review* Summer 1975.

16 Negotiations are a primary means of compromising. But people from different cultures exhibit very different approaches of how to negotiate. In some countries the relationship between the negotiators is initially more important while in others the matter of conflict stands in the foreground. To stay on the safe side, negotiations should generally sort out relationship aspects first, then concentrate on the subject the conflict is all about, only to finish with the detailed wording for the achieved compromise.

- ensuring all involved parties contribute to the conflict resolution, with every party making its individual contributions visible to the other parties;
- reminding involved parties of their common objectives and goals;
- describing the conflict and making it accepted as such by everyone involved ('Yes, we have a conflict');
- ensuring everyone understands the importance (or unimportance) of the problem;
- breaking the problem down into smaller problems, which are easier to solve;
- addressing the conflict to the right people with the right delegated authority to remove them;
- identifying practices, with which solutions can be found, which are acceptable to everyone involved;
- ensuring that all conflict parties have the same data or information or at least gain an in-depth understanding why the data is different;
- analysing the causes of the problem in a systematic manner by agreeing on criteria for a solution evaluation before selecting a solution; as well as
- ensuring that the found solutions are tested in practice (which may include to alter them if necessary);

and all of this while clearly differentiating between the actual subject of the conflict on the one hand and the personal relationships between those in conflict on the other.

The leader should always look for an integrated solution, which – as a compromise – satisfies the needs of all functional representatives in the team. In order to do so, he or she needs to pay attention to the interactions between the team members while at the same time taking into account the organisational constraints, the available processes, methods and tools as well as the needs of the overall aircraft development project. The leader basically applies a holistic or 'helicopter' view: he or she keeps a certain inner distance to a conflict or person involved in a conflict in order not to lose oversight and to be able to intervene as a neutral person.

In trying to solve conflicts in *multi-cultural* teams, facilitators need to be aware of the fact that native speakers have a definite advantage compared to others. The latter often tend to use this advantage to push their own interests and arguments. Facilitators must understand that there is a strong relation between mind and language. And they must find means of compensation for the other parties involved in the conflict which are at a disadvantage. For example, they should remind the native speaker to speak more slowly and to explain unknown terminologies.

'The strong relation between mind and language is obvious to people who can speak different languages: they represent different styles of thinking. English or German, for example, use a linear thinking process. Roman languages use the same basic structure than English. But they enable linguistic excursions or the introduction of irrelevant subjects into a discussion while focussing on emotions and behavioural expectations. Oriental languages (Japanese, Chinese) are based on an indirect approach and a cyclical movement of ideas. The latter are developed while speaking, and are described by explaining what they are not rather than what they are. The Russian language uses a deductive thinking process, which relates to a series of possible parallel constructions. It requires particular patience on the side of the listener.'[17]

---

17 Hoffmann, H.-E., Schoper, Y.-G. and Fitzsimons, C.J. (2004), *Internationales Projektmanagement* (München: Beck-Wirtschaftsberater im Deutschen Taschenbuch Verlag), p. 87, free translation by the author.

However, in cases where consensus and agreement cannot be achieved through facilitation, the leader must be capable to decide for himself. He or she then leaves the facilitator role and becomes a deciding boss. In this situation it definitely helps to have some good arguments in favour of the decision at hand. Once a decision has been taken by the team leader 'all team members must be aware of and accept the prerogative of management in terms of the final decision, and once it is made, must use their best efforts to implement that decision. There is no room for the team member who undermines the team leader's decision, whether or not he or she agrees with it. If the team member cannot ultimately support a ... [team leader's] decision, the next step is to leave the project team.'[18]

Apart from conflicts *between* team members, there may also be conflicts between the team leader and other individuals of the team. However, even in this case some of the conflict resolution strategies mentioned earlier can work. If the conflict is about an individual blowing the team spirit then the strategy of 'three warnings and you are out' should apply. Good team spirit is fundamentally important to project success, so team busters should not be given any chance to live their damaging behaviours.

Looking again at the multi-cultural aspect of conflicts, it is important to try not to gauge the seen and heard according to one's own cultural standards but to take into account what might be the cultural background of the person(s) involved in the conflict. Knowing about the models developed by multi-cultural researchers, such as the model by Hofstede and Hofstede, helps understanding the cultural backgrounds of conflicts and how to resolve them. Lacking such knowledge could mean not knowing how to counteract possible conflicts. In this case, a downward spiral of mistrust would probably develop. For example, one needs to know that the methods for conflict resolution facilitation outlined above can only be applied in individualistic cultures and those with little Power Distance. For other cultures, different facilitation methods have to be applied. Sometimes facilitation is not even the right approach in the first place.

LEADING BY EXAMPLE

Introduction

The key to good leadership is to be an example for all good behaviours. This is the best way to positively influence staff in multi-functional teams in a sustainable way. In particular the team leaders must recognise that their actions and attitudes speak far louder than their words and that they are role models in and for the organisation. Leaders need to understand that the behaviours they want to see in the teams depend to a large extent on what they do themselves, how they do it and where they set the priorities.

---

18 Eisner, H. (2002), *Essentials of Project and Systems Engineering Management*, 2nd Edition, (New York: John Wiley & Sons), p. 158.

Practise What You Preach

When leaders preach the benefits of sandwiches, they cannot eat caviar themselves. Preaching the benefits of sandwiches means eating sandwiches. The following examples may illustrate this principle:

- unless leaders are truly interested in controlling cost – and show it – little can be done by lower-level team members or other individuals that will have a sustained effect on cost control;
- if leaders request their teams to plan and schedule the project, and then never look at the schedules produced or make key decisions without reference to these schedules, it will be clear to the team members that planning and scheduling are not so important after all;
- if leaders preach integrative teamwork, they should demonstrate how they integrate themselves into a (higher-level) team.

Leaders must always practise what they preach. Words and actions must correspond. If that is not the case, people will not show the intended behaviours because leaders don't either.

Be a Respectful Communicator

One of the key behaviours of a leader is his ability to communicate with team members and other individuals. Leaders also have more confidence to employees with whom they deal on a daily basis. A good leader therefore keeps in touch with individuals – of his own or other teams – on a daily basis, mostly by walking around and thereby informally exploring progress, issues, problems and needs. This 'Management by Walking About' (MBWA) provides leaders with unfiltered information resulting in detailed background knowledge on the project's progress, as well as with information which could be important to take decisions or to solve conflicts. The direct observation and contact with contributors through MBWA is invaluable to determine physical progress at all stages of the project.

MBWA is particularly helpful with people from cultures with a high context relation, where the importance of informal channels is high. In these cultures it is expected that everyone tries to get information proactively. Messages, which are received personally from individuals, are regarded as more reliable compared to official news. Constant communication is therefore the norm, while in cultures with weak context relation MBWA is regarded more as an annoying duty.[19]

> 'The first secret of effectiveness is to understand the people you work with and depend on so that you can make use of their strengths, their ways of working, and their values. Working relationships are as much based on the people as they are on the work. The second part of relationship responsibility is taking responsibility for communication. … Personality conflicts … arise from the fact that people do not know what other people are doing and how they do their work, or what contribution the other people are concentrating on and what results they expect.

---

19 However, there is a problem with unofficial messages communicated during MBWA: they could leak to destinations outside of the project and if wrong, could do much harm to the project. It is therefore important for the project leaders to identify gossips and to find ways to avoid them.

And the reason they do not know is that they have not asked and therefore have not been told. … Even people who understand the importance of taking responsibility for relationships often do not communicate sufficiently with their associates. They are afraid of being thought presumptuous or inquisitive or stupid. They are wrong. Whenever someone goes to his or her associates and says, "This is what I am good at. This is how I work. These are my values. This is the contribution I plan to concentrate on and the results I should be expected to deliver," the response is always, "This is most helpful. But why didn't you tell me earlier?" And one gets the same reaction – without exception, in my experience – if one continues by asking, "And what do I need to know about your strengths, how you perform, your values, and your proposed contribution?" In fact, knowledge workers should request this of everyone with whom they work, whether as subordinate, superior, colleague, or team member. And again, whenever this is done, the reaction is always, "Thanks for asking me. But why didn't you ask me earlier?"[20]

An open office environment as described in Chapter 5 encourages the daily use of MBWA. Where MBWA is not possible, for example because teams are geographically too far away, it certainly helps if leaders demonstrate an effort in trying to keep communications flowing as much as possible. Everyone understands the difficulties associated with keeping direct contacts across large distances. A dedicated effort will therefore be even more appreciated.

A leader should demonstrate respect for individuals he or she is communicating with and should take the time to listen carefully to what they have to say. Lack of respect often starts in a humorous way, which nevertheless subsequently may expose a dark side. This may quickly undermine the contributions and commitments of individuals. Leaders should communicate purely on the grounds of the matter under discussion and should deal with individuals as self-leading creative partners rather than as 'human resources' or 'human capital'. This is the best way to encourage people to speak freely and honestly and it adds to raising the level of motivation among them. Creating an open communicative environment will encourage people to voice their opinion and the leader will benefit from this by receiving unfiltered information.

A leader must entirely refrain from any personal offences or hostile aggressiveness. A good captain must not use a whip to bring out the best of his highly-educated team members. It also goes without saying that cultural or ethical backgrounds must not lead to any kind of discrimination of individuals by the leader. Management by 'fear' or discrimination is the worst that can happen in any project, in any company. People should never be afraid to voice their opinion, or be too shy when speaking to their leader. Leaders exhibiting such a behaviour must be removed from their position immediately.

Demonstrate Commitment

Leaders need to demonstrate commitment. According to R.D. Archibald, 'commitment is knowing where you want to go and being persistent in your efforts to get there.'[21] If leaders are committed to something then – according to V. Sathe – they have internalised this something to an extent that it becomes a personal belief or value.[22] Leaders therefore have a clear set of beliefs, values and goals. If leaders want to have committed team members,

---

20 Drucker, P.F. (1999), 'Managing Oneself', *Harvard Business Review* March-April 1999, p. 72.
21 Archibald, R.D. (1992), *Managing High-Technology Programs and Projects*, 2nd Edition, (New York: John Wiley & Sons), p. 112.
22 Sathe, V. (1985), *Culture and Related Corporate Realities* (Homewood: Richard D.Irwin Inc.), p. 12.

it is not enough to just understand what commitment means and how important it is. Leaders need to demonstrate commitment themselves. They visibly need to behave in a way, which is consistent with their personal beliefs and values.

> 'Key organizational values must also be communicated and enforced. This may be the most important area, in which the project leader should be a role model. The pressures of project management can easily lead otherwise ethical individuals to compromise values to get results. ... What is the result on the team member's commitment when the project manager asks them to charge their time incorrectly to another project, in order to minimize a cost overrun on his or her project? ... In this area, the project leader must maintain commitment to the organization of which he or she is a part. Without building this commitment, a project leader cannot expect to receive the support from others necessary to achieve long-term results.'[23]

Note that commitment somewhat contradicts flexibility. For example, a team leader who is completely committed to the original plans – irrespective of changing circumstances – is not effective. Thus, too much commitment on the detailed level can become dysfunctional and is only helpful up to a certain point. Leaders need to find the right balance between commitment and flexibility.

Knowing One's Own Limitations

Leaders should be aware of their personal constrains acting upon them and limiting their capabilities. To this end, good leaders are not only conscious about their own limitations. They also surround themselves with people who are superior to themselves – people who have a better knowledge of strategy or finances or personnel or government relations than themselves.

Having said that, it would also contribute enormously to providing a positive role model if the leaders would be open to constructive feedback from their team members. Remaining open to constructive feedback from others is probably the leader's biggest leadership challenge. But exhibiting the right balance between self-confidence and willingness to listen to feedback and advice from others – instead of showing an attitude of knowing everything better – is a strength well respected by team members.

Demonstrate Discipline

For many people discipline seems to be an old-fashioned behaviour. In fact, it is not. Discipline has many positive aspects such as:

- controlling one's behaviours;
- staying focused by concentrating on the urgent and important matters;
- finding the right balance between work and private life (for example by limiting the workload);
- keeping fit and healthy (through regular sports and diet control);

---

23 Archibald, R.D. (1992), *Managing High-Technology Programs and Projects*, 2nd Edition, (New York: John Wiley & Sons), pp. 120–21.

- being organised; as well as
- sticking to commitments made.

Leaders, who demonstrate good time management as well as respect for the time of others, such as in meetings, will be rewarded by the admiration of their people.

Demonstrate Values

When leaders apply all of the aforementioned behaviours they will realise that this may result in personal conflict situations. Finding the right balance between commitment and flexibility is one example of potential conflict. Conflict between delegation and micro-management is another. In addition, leaders often accept a – sometimes high – level of personal risk. If, for example, required project inputs to a team are not available on time, team leaders might still chose to continue with the project work in an attempt to protect the project rather than to stop and wait for the input. With the inputs missing, they base their decision on as yet unconfirmed assumptions, thus, accepting the risk of later rework.

> 'Real leadership means having the courage to do what you know is right. That's what I did when I refused to drop … 1,000 engineers from my program. I could have obeyed what was arguably a direct order, but doing so would only have ensured my departure down the line, because the 747 program would have failed. Instead, I had stuck to my guns and kept my staffing at existing levels …'[24]

In situations of conflict and risk, values play a crucial role. Leaders need to have values, beliefs and goals in order to intuitively and quickly know how to behave in any given situation in order to resolve a conflict. Values may be company values, to which leaders demonstrate their buy-in, or may originate from a cultural, religious, family or whatever background. A leader's values should be strong and people should be able to find out through the leader's behaviours what those values are. As far as company values are concerned, a team leader must not only be a role model for those values but also their messenger.

SUMMARY

There are certainly many different leadership styles and methods, depending on the one hand on personality, experience, interpersonal skills and technical competence, as well as the characteristics of the project and its environment on the other. Good leaders analyse a situation and modify their behaviours to suit the circumstances. The premise of this 'situational leadership' is that leaders choose or select any of the leadership 'styles' mentioned in this Annex, depending upon the actual situation.

Situational leadership requires a significant amount of knowledge and experience as well as psychological maturity (self-confidence, preparedness to take responsibility, bottom-line orientation) as well as multi-cultural awareness. The latter is particularly

---

24 Sutter, J. (2006), 747: Creating the World's First Jumbo Jet and Other Adventures from a Life in Aviation (New York: Smithsonian Books, HarperCollins Publishers), p. 150.

important for those project management aspects dealing with the increase of productivity, the planning and holding of meetings and discussions, as well as with the increase of pressure to achieve objectives.

> *For example, in cultures with a big Power Distance such as China, a cooperative leadership style will not be understood or may even be misinterpreted as weakness.*

The multi-cultural context should also be taken into account for any of the following:[25]

- rolling out the rules for common ways of working;
- motivation;
- delegation;
- agreement on targets and objectives;
- control;
- feedback: acknowledgment and critics;
- conflict resolution; as well as
- decision process.

In order to be able to lead projects successfully, it is important that project leaders on the one hand clearly have their project objectives in their focus, but on the other constantly observe the reactions, levels of motivation and behaviours of their team members.

However, open, curious, respectful and honest communication, credibility, authenticity, trust, self-reliance, approachability, humour, friendliness, preparedness to learn from other cultures, and the ability to laugh at oneself are features that should always blend together in a leader. Leaders should never be co-notated with any of the following: hostility, aggressiveness, complaints, unresponsiveness, super-agreeable, know-it all expert, being a negativist or indecisive.

Finally, facilitation is a key leadership capability for multi-functional teams. A vital task of a facilitator is to develop a team by regular analysis of 'how' the team is doing its work rather than 'what' it has done. This implies agreeing with the team on team performance measures, monitoring and feeding-back actual team behaviours and performances as well as developing strategies for constant improvement. As H. Eisner points out, 'To be a leader, then, is to know how to build a team. Conversely, and categorically, to not know how to build a team is to not be a leader.'[26] It is this kind of leaders – not bosses – which are needed at the head of multi-functional teams because of the very nature of multi-functionality and because of the imperfections of the complex matrix organisation required to manage multi-functional teams.

---

25 Hoffmann, H.-E., Schoper, Y.-G. and Fitzsimons, C.J. (2004), *Internationales Projektmanagement* (München: Beck-Wirtschaftsberater im Deutschen Taschenbuch Verlag), p. 148, free translation by the author.
26 Eisner, H. (2002), *Essentials of Project and Systems Engineering Management*, 2nd Edition, (New York: John Wiley & Sons), p. 160.

Appendix A3:
Architecture Integration Example

INTRODUCTION

The example in this Appendix describes the type of challenges one encounters when creating integrated, configured Project architectures. For a better understanding of the example, the below convention is introduced:

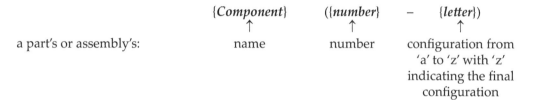

a part's or assembly's:

{Component}	({number}	–	{letter})
↑	↑		↑
name	number		configuration from 'a' to 'z' with 'z' indicating the final configuration

Example: Shell (1-a) represents the 'a' configuration of the component 'Shell' with the number '1'.

PART I

A fuselage Shell (1-z) is to be assembled out of a skin and two stringers, see Figure A3.1.[1] Stringer (1-a) and (2-a) differ in thickness. As can be seen in Figure A3.2, Engineering's initially proposed 'As-Designed[P']' foresees four drawings (D1–D4):

- D1 for Skin (1-a);
- D2 for Stringer (1-a);
- D3 for Stringer (2-a);
- D4 for the complete shell (Shell 1-z).

D1 to D3 are part drawings, whereby D4 is an assembly drawing. Engineering applies a result-oriented view where drawings only describe the result of the various manufacturing processes.[2] In a way, Engineering does not really worry about those processes.

According to Manufacturing, Skin (1-a) is produced in-house from sheet metal as Manufacturing Stage Operation (MSO) [A] at the Manufacturing Stage (MS) [I], see again Figure A3.2. However, things are more complex with the other steps:

---

1 Note that it does not matter that the chosen example deals with aerostructures. It could equally deal with system or equipment units, pipes, harnesses and so on.
2 The *result* of one or several ASO is often called Design As-Planned (DAP). In our example, in the view of Engineering, the shell is a DAP object consisting of a skin and two stringers.

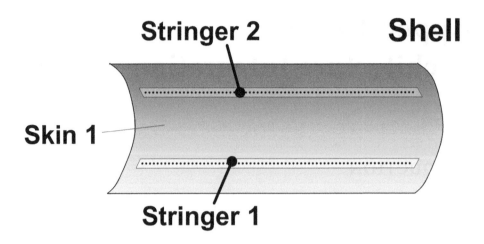

Figure A3.1   Design of aircraft fuselage shell used in the example

- The stringers are procured from an external supplier with the drilling to be done on a special machine in-house. As the stringers differ in thickness (and therefore carry different part numbers) one MSO for drilling each of them is foreseen (MSO [B] and MSO [C], respectively). This is despite being drilled on the same special multi-drilling machine capable of drilling joint stringer/skin combinations (MS [II]). Using MSO [B] Skin (1-a) and Stringer (1-a) are jointly drilled. The skin will receive one set of drilling holes, thus its configuration changes from Skin (1-a) to Skin (1-b), see Figure A3.2. Similarly, Stringer (1-a) changes to Stringer (1-b). The same logic is applied at MSO [C].
- The riveting of both stringers to the skin happens at a common Assembly Stage (AS) [III] representing an automatic riveting machine capable of riveting each stringer individually to the skin. Manufacturing plans nevertheless for two Assembly Stage Operations (ASOs), one for Stringer (1-b) to Skin (1-c) (ASO [D]) resulting in Shell (1-a), and one for Stringer (2-b) to the already partially assembled Shell (1-a) (ASO [E]), yielding the final shell configuration Shell (1-z).

Manufacturing requires from Engineering a drawing for incoming inspections *and* for the outputs of *each* operation. Thus, Manufacturing requires from Engineering nine different drawings (D1-D3, D5-D10), see Figure A3.2. Note that Manufacturing in fact requires more than one drawing for some components to reflect their different successive configuration. Figure A3.3 depicts the configurations required by this Architecture Integration Example for skin and shell. Note that the final *assembly* drawing as requested by Manufacturing (D10) to create Shell (1-z) is different from the drawing D4 proposed by Engineering, although the final Shell (1-z) looks the same. This is because the steps leading to the final shell configuration are different.

However, Engineering might not agree as drawings for the outputs of MSO [B] and [C], respectively, as well as ASOs [D] and [E] are not foreseen in its original 'As-Designed[P]'. At this point, Engineering and Manufacturing would have to start to negotiate what to do. Either Engineering could accept to generate additional drawings to cover the results of all individual MSOs and ASOs, or Manufacturing would

Group	Stage	Stage Operation	Input: Part Drawing	Input: Assembly Drawing	Input: Part	Input: PAT-level	Output: Part Drawing	Output: Assembly Drawing	Output: Part	Output: PAT-level	PAT-solution
Initial Engineering Proposal							D1		Skin (1-a)	N	
		from supplier	D2		Stringer (1-a)	N					
		from supplier	D3		Stringer (2-a)	N					
								D4	Shell (1-z)	N+1	
Manufacturing Request	MS [I]: machine to produce skin from sheet metal	MSO [A]: manufacture of the skin					D1		Skin (1-a)	N-1	
	MS [II]: special multi-drilling machine	MSO [B]: drilling of stringer to skin	D1 D2		Skin (1-a) Stringer (1-a)	N-1	D5 D6		Skin (1-b) Stringer (1-b)	N	
		MSO [C]: drilling of stringer to skin	D5 D3		Skin (1-b) Stringer (2-a)	N	D7 D8		Skin (1-c) Stringer (2-b)	N	
	AS [III]: automatic riveting machine	ASO [D]: riveting of stringer to skin	D7 D6		Skin (1-c) Stringer (1-b)	N		D9	Shell (1-a)	N	
		ASO [E]: riveting of stringer to skin	D8	D9	Shell (1-a) Stringer (2-b)	N		D10	Shell (1-z)	N+1	
Agreed PAT-Solutions	MS [I]: machine to produce skin out of sheet metal	MSO [A]: manufacture of the skin					D1		Skin (1-a)	N-1	N[I]
	AS [IV]: automatic drilling/riveting machine	ASO [F]: drilling and riveting stringer / skin combinations	D1 D2		Skin (1-a) Stringer (1-a)	N-1		D11	Shell (1-b)	N	N[I]
		ASO [G]: drilling and riveting stringer / skin combinations	D3	D11	Shell (1-b) Stringer (2-a)	N		D12	Shell (1-z)	N+1	N+1[I]
Modified PAT-Solutions	MS [I]: machine to produce skin out of sheet metal	MSO [A]: manufacture of the skin					D1		Skin (1-a)	N-1	N[I] N[II]
	AS [V]: stringer manufacture facility	ASO [H]: manufacture of stringers	D13 D14		Coupling (1-a) 2x Stringer Part (1-a)	N-2		D15	Stringer (3-a)	N-1	N-1[I]
	AS [IV]: automatic drilling/riveting machine	ASO [I]: drilling and riveting stringer / skin combinations	D1	D15	Skin (1-a) Stringer (3-a)	N-1		D16	Shell (2-a)	N	N[II]
		ASO [J]: drilling and riveting stringer / skin combinations	D3	D16	Shell (1-d) Stringer (2-a)	N		D17	Shell (2-z)	N+1	N+1[II]

Figure A3.2   Overview on Architecture Integration Example data

for example accept to merge operations [B] to [E] into one. This might require the purchase of a new machine, combining drilling and riveting. In any case, according to Manufacturing every MSO or ASO should correspond to at least one Engineering drawing. This requirement is respected, see again Figure A3.2.[3]

---

3 In the case of MSOs, drawings are part drawings, while for ASOs drawings are installation drawings.

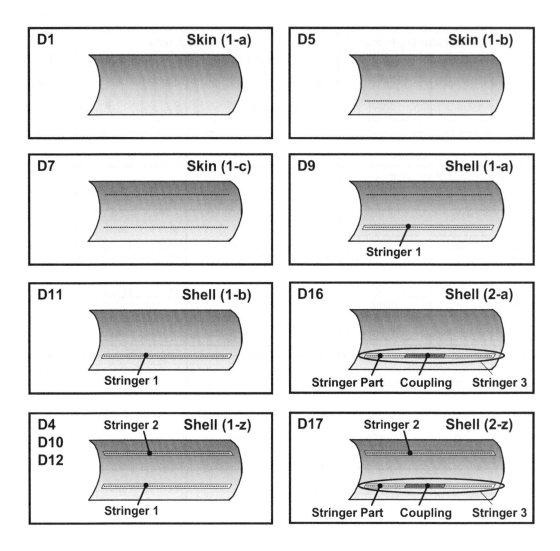

Figure A3.3   Different skin and shell configurations used in the example

PART II

In Figure A3.4 it is depicted in terminology of the Product & Assembly Trees (PATs) what Engineering proposes and Manufacturing requests for the example mentioned above. All drawings but only the components in their configurations as delivered to the next level are shown. Note the difference in number of PAT-levels required, with Manufacturing requiring one more PAT-level compared to Engineering. Note also that, according to Manufacturing, more parts' drawings are requested to produce an assembly PAT-element. Clearly, the Manufacturing approach is more complex but it does reflect its build process in every detail.

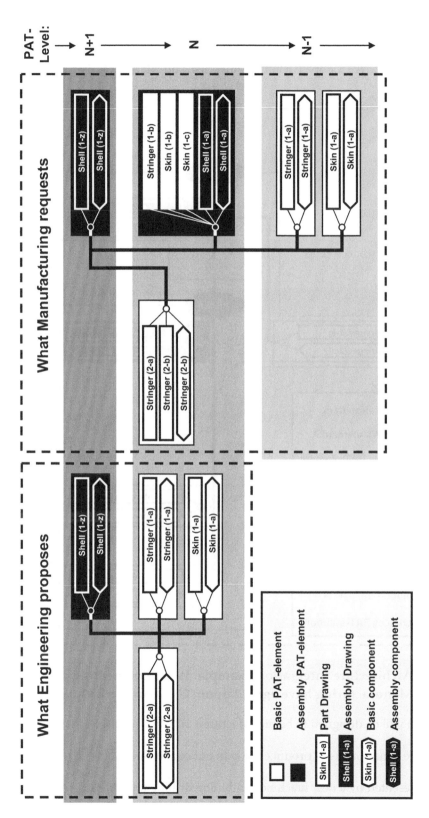

Figure A3.4   Architecture Integration Example: What Engineering proposes and Manufacturing requests (using Product & Assembly Tree terminology)

PART III

Figure A3.2 and Figure A3.5 depict what Engineering and Manufacturing finally have agreed upon.

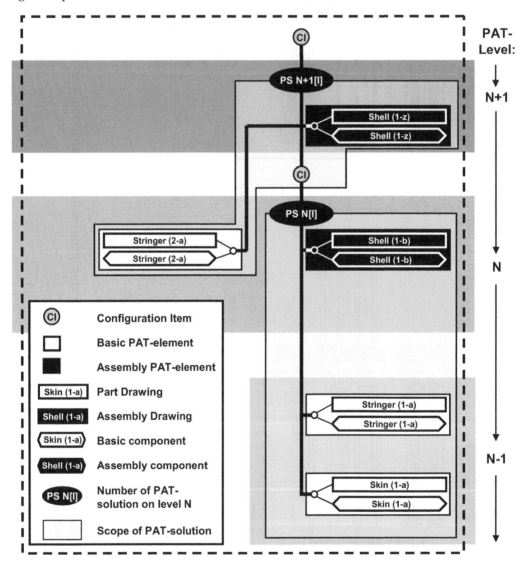

Figure A3.5   Architecture Integration Example: What Engineering and Manufacturing have agreed upon. Compare with Figure A3.4

The Shell (1-z) will be designed to be manufactured at only two stages, that is:

- a MS where a machine is to produce the skin out of sheet metal (MS [I], as before); as well as
- an AS represented by a new machine combining drilling and riveting (AS [IV]).

Engineering and Manufacturing have agreed to structure the PAT using two PAT-solutions (PSs) on two different PAT-levels.[4] As before, the skin is still produced using the MSO [A]. But with one ASO per PS, there are now only two operations for each individual shell:

- ASO [F] to drill and rivet the Stringer (1-a) to Skin (1-a) at AS [IV], resulting in Shell (1-b); as well as
- ASO [G] to drill and rivet the Stringer (2-a) to Shell (1-b) at AS[IV].

Engineering will provide five drawings for the two PSs:

- D1 for Skin (1-a);
- D2 for Stringer (1-a);
- D3 for Stringer (2-a); as well as the new assembly drawings
- D11 for Shell (1-b); and
- D12 for Shell (1-z) replacing the old D4 and D10, respectively, for the complete shell.

Note again that the final assembly drawing now agreed between Engineering and Manufacturing (D11) to create Shell (1-z) is different from the earlier drawings D4 and D10, although the final Shell (1-z) looks the same. This is because the steps leading to the final shell configuration are different.

Note that this agreement between Engineering and Manufacturing features aspects of both Configuration Item-PAT-Solution (CI-PS) structures explained in Chapter 12, Figure 12.3: it has a CI-PS-CI link and is therefore operations-oriented but it is also flatter compared to what was originally required by Manufacturing. There will not be drawings for each of the possible Manufacturing steps.

PART IV

Assuming it to be foreseeable that Stringer (1-a) will no longer be supplied by the supplier after the 200th aircraft has been produced. The contracting aircraft manufacturer may have to prepare for producing this stringer in-house. Unfortunately, the company does not have the right facilities to build the stringer in one piece as the supplier was able to do. However, it decides not to invest in a new facility but rather to change the design so that the stringer can be produced at an existing AS [V]. The stringer will now be produced as an assembly made of three parts: two identical Stringer Parts (1-a) as well as a Coupling (1-a), see Figure A3.6.

---

4 Note that Engineering and Manufacturing could have structured the PAT in many different ways. But for this example, this is what they came up with.

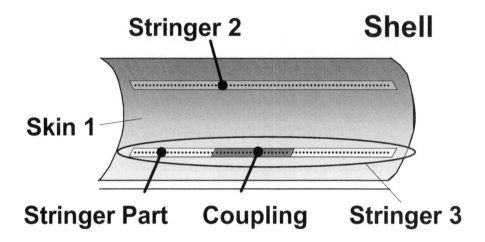

Figure A3.6   Design of modified aircraft fuselage shell used in the example

A new ASO [H] is required to produce Stringer (3-a), see Figure A3.2. Engineering and Manufacturing agree to create a new PS with three new drawings, see Figure A3.7. In addition, while the new Stringer (3-a) can still be fastened to the skin using the AS [IV], another new ASO [I] is nevertheless required as Stringer (3-a) is different from Stringer (1-a), bearing a different part number. As a consequence, the old PS N[I] needs to be replaced by the new PS N[II]. Note that the two PSs are interchangeable. Note also that the drawings for Skin (1-a) as well as Stringer (2-a), respectively, are called-up by the respective old as well as new PSs.

As one can see from this simple example, architecting can indeed get complex very quickly.

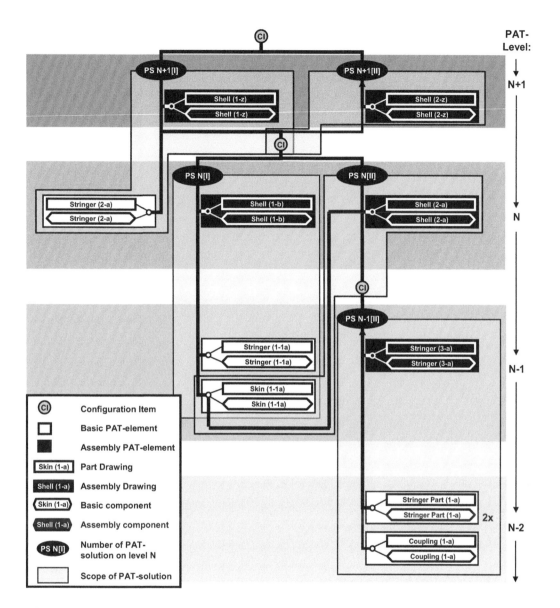

Figure A3.7 Architecture Integration Example: Introducing a modification

Further Reading

SYSTEMS ENGINEERING

Beam, W.R. (1990), *Systems Engineering: Architecture and Design* (New York: McGraw-Hill).

Belev, G.C. (1989), 'Guidelines for specification development', Proceedings of the Annual IEEE Symposium on Reliability and Maintainability.

Blanchard, B.S. and Fabrycky, W.J. (1990), *Systems Engineering and Analysis,* 2nd Edition, (Englewood Cliffs: Prentice-Hall).

Chambers, G.J. and Manos, K. (1992), 'Requirements: Their Origin, Format and Control', Proceedings of the Second Annual International Symposium of the National Council on Systems Engineering, July 1992.

Chorafas, D.N. (1989), *Systems Architecture & Systems Design* (New York: McGraw-Hill).

Eisner, H. (2002), *Essentials of Project and Systems Engineering Management,* 2nd Edition, (New York: John Wiley & Sons).

Hancock, L.R. (1993), 'Enhancing Operability and Reliability Through Configuration Management', Proceedings of the Second ASME/JSME Nuclear Engineering Conference Vol. 2, 21–24 March 1993, pp. 707–9.

Jackson, S. (1997), *Systems Engineering for Commercial Aircraft* (Aldershot: Ashgate Publishing).

Malchow, H.L. and Croopnick, S.R. (1985), 'A methodology for organizing performance requirements for complex dynamical systems', *IEEE Tram. Eng. Manag.* Vol. EM-32 February 1985, pp. 10–15.

Purdy, D.C. (1991), *A Guide for Writing Successful Engineering Specifications* (New York: McGraw-Hill).

Rechtin, E. (1991), *Systems Architecting: Creating and Building Complex Systems* (New York: Prentice-Hall).

Sage, A.P. (1977), *Methodology for Large-Scale Systems* (New York: McGraw-Hill).

Sahal, D. (1976), 'System Complexity: Its Conception and Measurement in the Design of Engineering Systems', *IEEE Trans. Systems, Man, and Cybernetics,* June 1976, pp. 440–45.

Samaras, T.T. and Czerwinski, F.L. (1971), *Fundamentals of Configuration Management* (New York: Wiley-Interscience).

Shishko, R. et al. (1995), *NASA Systems Engineering Handbook* (Pasadena: Jet Propulsion Laboratory), NASA SP-6105.

So, T., Jagannathan, V. and Raman, R.S. (1992), The Role of Configuration Management Systems in a Concurrent Engineering Environment (Anaheim: ASME Winter Annual Meeting), 8–13 November 92-WA/EDB-2.

The International Council on Systems Engineering (INCOSE) http://www.incose.org.

LIFE-CYCLE COSTING

Chow, W.W. (1978), *Cost Reduction in Product Design* (New York: Van Nostrand Reinhold Co.).

Dhillon, B.S. (1989), *Life Cycle Costing* (New York: Gordon and Breach Science Publishers).

Fabrycky, W.J. and Blanchard, B.S. (1991), *Life-Cycle Cost and Economic Analysis* (New Jersey: Prentice Hall).

Gansler, J.S. and Sutherland, G.W. (1974), 'A Design to Cost overview', *Defense Manag.* September 1974, pp. 2–7.

Hall, T.P. (1980), 'Systems Life Cycle Model', *J. Syst. Manag.* April 1980, pp. 29–31.

Taylor, L.L. (1983), 'Systems Engineering and the Total System Life Cost', *Proc. Int. Conf. Systems, Man, and Cybernetics* (New York: IEEE Press), Vol. 2, pp. 1198–201.

RAPID DEVELOPMENT

Anon. (1991), 'A Smarter Way to Manufacture', *Business Week* 30 April 1990, pp. 110–17.

Anon. (1991), 'Concurrent Engineering: Special Report', *IEEE Spectrum* July 1991, pp. 22–37.

Anon. (2001), *People in Projects* (Project Management Institute).

Batson, R.G. (1987), 'Critical Path Acceleration and Simulation in Aircraft Technology Planning', *IEEE Transactions on Engineering Management* Vol. EM-34 No. 4 November 1987, pp. 244–251.

Batson, R.G. and Love, R.M. (1988), 'Risk Assessment Approach to Transport Aircraft Technology Assessment', *AIAA Journal of Aircraft* Vol. 25 No. 2 February 1988, pp. 99–105.

Bell, T.E. (ed.) (1989), 'Special Report: Managing Risk in Large Complex Systems', *IEEE Spectrum* June 1989, pp. 21–52.

Bennis, W. (1989), *On Becoming a Leader* (Addison-Wesley, MA).

Bernstein, J.I. (2000), *Multidisciplinary Design Problem Solving in Product Development Teams* (Cambridge, MA: MIT, Doctoral Thesis).

Beroggi, G.E.G. and Wallace, W.A. (1994), 'Operational Risk Management: A New Paradigm for Decision Making', *IEEE Transactions on Systems, Man, and Cybernetics* Vol. 24 No. 10 October 1994, pp. 1450–57.

Carter, D.E. and Baker, B.S., *Concurrent Engineering: The Product Development Environment for the 1900s* (New York: Addison-Wesley).

Chapman, C.B. (1979), 'Large Engineering Project Risk Analysis', *IEEE Transactions on Engineering Management* Vol. EM-26 No. 3 August 1979, pp. 78–86.

Clausing, D. (1991), *Concurrent Engineering* (Honolulu: Design and Productivity International Conference), 6–8 February 1991.

Condit, P.M. (1996), *Performance, Process, and Value: Commercial Aircraft Design in the 21st Century* (Los Angeles: World Aviation Congress and Exposition), 22 October 1996.

Cooper, D. and Chapman, C. (1987), *Risk Analysis for Large Projects: Models, Methods & Cases* (New York: John Wiley & Sons).

Jassawalla, A.R. and Sashittal, H.C. (2000), 'Strategies of Effective New Product Team Leaders', *California Management Reviews* 42/2 Winter 2000, pp. 34–51.

DePree, M. (1989), *Leadership is an Art* (New York: Bantam Doubleday Dell Publishing Group Inc.).

Fisher, K. (2000), *Leading Self-Directed Work Teams: A Guide to Developing New Team Leadership Skills* (New York: McGraw-Hill).

Fisher, R. and Richardson, J. (1998), *Getting it Done: How to Lead When You're Not in Charge* (New York: HarperCollins Publishers).

Gu, P. and Kusiak, A. (eds.) (1993), *Concurrent Engineering: Methodology and Applications* (Amsterdam: Elsevier).

Harris, E.D., Koshy, T. and Anderson, R.L. (1992), 'Concurrent Engineering and Total Quality Management: The Need for a Seamless Marriage', *Engineering Management Journal* Vol. 4 No. 2 June 1992, pp. 3–8.

Hartley, J.R. (1992), *Concurrent Engineering: Shortening Lead Times, Raising Quality, and Lowering Costs* (Cambridge: Productivity Press).

Lake, J. (1992), 'Implementation of Multi-Disciplinary Teaming', *Engineering Management Journal* Vol. 4 No. 2 June 1992, pp. 9–13.

Lee, D.M.S. (1992), 'Management of Concurrent Engineering: Organizational Concepts and a Framework of Analysis', *Engineering Management Journal* Vol. 4 No. 2 June 1992, pp. 15–25.

March, J.G. and Shapira, Z. (1987), 'Managerial Perspectives on Risk and Risk Taking', *Management Science* Vol. 33 No. 11 November 1987, pp. 1404–418.

Mohrmann, S., Crocker, C. (ed.), and Mohrman, A. (1997), *Designing and Leading Team-Based Organizations, A Workbook for Organizational Self-Design* (San Francisco: Jossey-Bass, Inc.).

Parker, G.M. (1990), *Team Players and Teamwork: The New Competitive Business Strategy* (San Francisco: Jossey-Bass).

Pinto, J.K. and Trailer, J.W. (eds.) (1998), *Leadership Skills for Project Managers* (Project Management Institute).

Pinto, J.K., Thomas, P., Trailer, J., Palmer, T. and Govekar, M. (1998), *Project Leadership: From Theory to Practice* (Project Management Institute).

Rees, F. (1991), *How to Lead Work Teams: Facilitation Skills* (San Diego: Pfeiffer).

Robbins, H. and Finley, M. (1995), *Why Teams Don't Work: What Went Wrong and How to Make It Right* (Princeton: Peterson's/Pacesetter Books).

Rosenau Jr., M.D. and Moran, J.J. (1993), *Managing the Development of New Products: Achieving Speed and Quality Simultaneously Through Multifunctional Teamwork* (New York: John Wiley & Sons).

Scholtes, P. (1988), *The Team Handbook* (Madison, WI: Joiner and Associates).

Schuyler, J. (2001), *Risk and Decision Analysis in Projects*, 2nd Edition, (Project Management Institute).

Smith, P.G. and Reinertsen, D.G. (1998), *Developing Products in Half the Time*, 2nd Edition, (New York: John Wiley & Sons).

Verma, V.K. (1996), *Human Resource Skills for the Project Manager* (Project Management Institute).

Verma, V.K. (1997), *Managing the Project Team* (Project Management Institute).

Wheelwright, S.C. and Clark, K.B. (1992), *Revolutionizing Product Development: Quantum Leaps in Speed, Efficiency, and Quality* (New York: The Free Press).

Wilson, C.C. (1991), 'Potential Pitfalls of Concurrent Engineering: How to Avoid the Risks', *Concurrent Engineering* Vol. 1 No. 1 January/February 1991.

Wideman, R.M. (ed.) (1991), *Project & Program Risk Management: A Guide to Managing Project Risks & Opportunities* (Project Management Institute).

PROJECT MANAGEMENT

Anon. (2002), Practice Standard for Work Breakdown Structures (Project Management Institute).

Adams, J.R. et al. (1997), Principles of Project Management (Project Management Institute).

Archibald, R.D. (1992), Managing High-Technology Programs and Projects, 2nd Edition, (New York: John Wiley & Sons).

Fleming, Q.W. and Koppleman, J.M. (2000), Earned Value Project Management, 2nd Edition, (Project Management Institute).

Flouris, T and Lock, D. (2008), Aviation Project Management (Aldershot: Ashgate).

Görög, M. and Smith, N.J. (1999), Project Management for Managers (Project Management Institute).

Ibbs, W. and Reginato, J. (2002), Quantifying the Value of Project Management (Project Management Institute).

Pinto, J.K. and Trailer, J.W. (ed.) (1999), Essentials of Project Control (Project Management Institute).

Pinto, J.K. (ed.) (1998), Project Management Handbook (Project Management Institute).

Thomas,J., Delisle, C. and Jugdev, K. (2002), Selling Project Management to Senior Executives (Project Management Institute).

Project Management Institute, http://www.pmi.org.

TOTAL QUALITY MANAGEMENT

Anon. (1990), *Total Quality Management – A Practical Approach* (London: Department for Trade and Industry), DTI/PUB260/20K/5/90.

Collard, R. (1993), *Total Quality: Success through People* (London: Institute of Personnel Management).

Dale, B.G. (1994), *Managing Quality*, 2nd Edition, (Englewood Cliffs: Prentice-Hall).

Feduccia, A.J. (1984), 'System Design for Reliability and Maintainability', *Air Force J. Logistics* Spring 1984, pp. 25–9.

Feigenbaum, A.V. (1983), *Total Quality Control*, 3rd Edition, (New York: McGraw-Hill).

Fontenot, G., Behara, R. and Gresham, A. (1994), 'Six Sigma in Customer Satisfaction', *Quality Progress* Vol. 27 No. 12 December 1994, pp. 73–6.

George, S. and Weimerskirch, A. (1994), *Total Quality Management* (New York: J.Wiley & Sons).

Gitlow, H.S. and Gitlow, S.J. (1987), *The Deming Guide to Quality and Competitive Position* (Englewood Cliffs: Prentice-Hall).

Ishikawa, K. (1985), *What is Total Quality Control – the Japanese Way* (Englewood Cliffs: Prentice-Hall).

Juran, J.M. (1979), *Quality Control Handbook*, 3rd Edition, (New York: McGraw-Hill).

Juran, J.M. and Gryna, F.M. (1980), *Quality Planning and Analysis*, 2nd Edition, (New York: McGraw-Hill).

Lock, D. (1994), *Handbook of Quality Management*, 2nd Edition, (Farnham: Gower).

McCloskey, L. and Collett, D. (1993), *TQM: A Basic Text* (Methuen, MA: GOAL/QPC).

Oakland, J.S. (1994), *Total Quality Management*, 2nd Edition, (Oxford: Butterworth-Heineman).

Peters, T.J. and Waterman, Jr., R.H. (1982), *In Search for Excellence* (New York: Harper & Row).

Sethi, R. (2000), 'New Product Quality and Product Development Teams', *Journal of Marketing* Vol 64 N° 2, April 2000, pp. 1–14.

Shuster, H.D. (1999), *Teaming for Quality* (Project Management Institute).

Taguchi, G. and Clausing, D. (1990), 'Robust Quality', *Harvard Business Review* January – February 1990.

Townsend, P.L. and Gebhart, J.E. (1986), *Commit to Quality* (New York: J.Wiley & Sons).

Multi-Cultural Management

Adler, N.J. (1997), *International Dimensions of Organizational Behaviour* (Cincinnatti: South-Western).

Alvesson, M. and Billing, Y.D. (1997), *Understanding Gender and Organisations* (London: Sage).

Ashkanasy, N.M., Wilderom, C.P.M. and Peterson, M.F. (eds.) (2000), *Handbook of Organizational Culture & Climate* (London: Sage).

Chen, G.M. and Starosta, W.J. (1997), *Foundations of Intercultural Communication* (Boston: University Press of America).

Deal, T.E. and Kennedy, A.A. (1982), *Corporate Cultures: The Rites and Rituals of Corporate Life* (Reading, MA: Addison-Wesley).

Dorfman, P.W. and Howell, J.P. (1988), 'Dimensions of National Culture and Effective Leadership Patters: Hofstede Revisited', *Advances in International Comparative Management* Vol. 3, pp. 127–50.

Hall, E.T. (1989), *The Dance of Life* (New York: Anchor Books).

Hoecklin, L. (1995), *Managing Cultural Differences* (Reading, MA: Addison-Wesley Publishing Company).

Hofielen, G. and Broome, J. (2000), 'Leading International Teams – a New Discipline?', *Zeitschrift für Organisationsentwicklung* April 2000, p. 66ff.

Hofstede, G. (1980), *Culture's Consequences: International Differences in World-related Values* (Beverly Hills, CA: Sage).

Hofstede, G. and Hofstede, G.J. (2004), *Cultures and Organizations: Software of the Mind* (New York: McGraw-Hill).

Hofstede, G. and Peterson, M.F. (2000), 'National values and organizational practices' in: Ashkanasy, N.M., Wilderom, C.P.M. and Peterson, M.F. (eds.) (2000), *Handbook of Organizational Culture & Climate* (London: Sage), pp. 401–5.

James, M. (1991), *The Better Boss in Multicultural Organizations* (Walnut Creek, CA: Marshall Publishing).

Lientz, B. and Rea, K. (2003), *International Project Management* (San Diego, CA: Elsevier Science).

Schein, E.H. (1984), 'Coming to a New Awareness of Organizational Culture', *Sloan Management Review* Vol. 25 No. 2, p. 4.

Ting-Toomey, S. (1989), 'Intercultural conflict styles: a face negotiation theory', in: Kim, Y. and Gudykunst, W. (1989), *Theories in Intercultural Communication* (Newbury Park, CA: Sage).

Trompenaars, F. and Hampdan-Turner, C. (1997), *Riding the Waves of Culture. Understanding Diversity in Global Business* (London: McGraw-Hill).

Index

Wherever indexed terms are defined or further explained, page numbers are referenced in **bold**.

Printed and bound by CPI Group (UK) Ltd, Croydon, CR0 4YY

18/10/2024

01776204-0010